Measure Theory and Integration

The most traditional concepts in measure theory and integration receive a rigorous, comprehensive and modern presentation in this work. Building on the classical foundations, the book introduces the theory with full generality and meticulous attention to detail, following the stylistic tradition first introduced by Nicolas Bourbaki. It is designed for graduate students and young researchers seeking a thorough exposition of the theory in an abstract setting. Complete proofs, and the strategies underlying them, foster good mathematical habits in logical reasoning and clarity of deduction.

Beyond standard treatments, *Measure Theory and Integration* features several distinctive elements: some classical results, such as Radon-Nikodým theorem, and Lebesgue and Hahn decompositions, have been presented with original proofs, aimed at clarifying the logic behind the results; some topics that are often overlooked, such as kernels, uniform integrability, the Vitali-Hahn-Saks and Dunford-Pettis theorems are developed in full in dedicated chapters, and a complete account of the disintegration of measures is developed. The book also pays special attention to modern applications, including the construction of product measures for an arbitrary family of measures, by exploiting the properties of kernels, a full account of Daniell's and Carathéodory's methods for constructing and extending measures, and a thorough coverage of the theory of convergence, and shows two paramount applications of the theory to the presentation of the Lebesgue measure and the family of Hausdorff measures.

The book is largely self-contained, with supplementary sections on topology and differential calculus, and an appendix on filters and ultrafilters also included to help the reader to fully understand the notion of convergence with respect to a filter.

Andrea Carpignani graduated *summa cum laude* in Mathematics at the University of Pisa in March 2005. He is a member of the London Mathematical Society and a fellow of the Royal Statistical Society. His academic interests are measure theory and integration, convex and functional analysis, probability theory, mathematical statistics, and data science. Following a few years as a teaching assistant at the University of Pisa, he pursued a career in secondary and further education, teaching Mathematics and Physics in Italy and in the UK, where he is currently KS5 Maths Coordinator at The Radcliffe School, in Milton Keynes. Alongside his teaching activity, Andrea Carpignani continues his studies in mathematics focusing on measure theory, algebraic structures and functional analysis.

Measure Theory and Integration

Andrea Carpignani

CRC Press
Taylor & Francis Group
Boca Raton London New York

CRC Press is an imprint of the
Taylor & Francis Group, an **informa** business
A CHAPMAN & HALL BOOK

Front cover image: Lucy Carpignani

First edition published 2026
by CRC Press
2385 NW Executive Center Drive, Suite 320, Boca Raton FL 33431

and by CRC Press
4 Park Square, Milton Park, Abingdon, Oxon, OX14 4RN

CRC Press is an imprint of Taylor & Francis Group, LLC

ISBN: 978-1-041-20899-0 (hbk)
ISBN: 978-1-041-20900-3 (pbk)
ISBN: 978-1-003-72476-6 (ebk)

DOI: 10.1201/9781003724766

Typeset in Latin Modern font
by KnowledgeWorks Global Ltd.

Publisher's note: This book has been prepared from camera-ready copy provided by the authors.

IN LOVING MEMORY
OF
PAOLA GUERRINI
(1947 – 2011)

Table of Contents

Introduction

Measure theory and integration are tools of fundamental importance both in mathematical analysis and probability theory, and, for this reason, every student of mathematics should be, more or less formally, exposed at least to the basic notions. On the other hand, measure theory and integration form a beautiful subject in its own terms, which, in a Venn's diagram, would be located in the intersection of three main branches of mathematics: Set theory, algebra, and mathematical analysis. With the raise of the axiomatic foundation of Probability Theory, first introduced by the Soviet mathematician A. N. Kolmogorov (Андрей Николаевич Колмогоров, 1903–1987), measure theory has achieved an even more central role in mathematics, acquiring new interesting results from the study of probability and statistics. Thus, being such a central topic, measure theory should be known by any mathematician, regardless of their specific specialisation (analysis, geometry, algebra, applied mathematics), for the techniques and methods used in measure theory, and the tools developed within the theory, are exportable in nearly every other area of mathematics, with the extra 'plus side' of being a very elegant theory, which is, indeed, pleasant to study.

Due to its centrality in mathematics, in the past few decades, many textbooks have been written on measure theory, some of which are extremely beautiful textbooks, and have managed to remain the first and foremost resource for decades, such as Halmos [35], Dudley [21], Cohn [16], and Bogachev's complete monograph [9, 10], not to mention the most comprehensive treatise by Fremlin [30]. Thus, at first glance, one would rightfully wonder the rational for writing yet another textbook on measure theory. The reasons for this are, indeed, several. The first regards the general setting and the style adopted in this book. Being the product of a loyal disciple of Giorgio Letta, this books preserves his formal and abstract style, highly influenced by the work of the Bourbaki group, with the constant search for generality, with the aim to enhance the role of the measure theoretic structures to highlight, with surgical precision, the connection between different cause and effect. Throughout the book, structures are introduced gradually, presenting new concepts and deducing the consequences that they unfold. Uncommon structures, such as most of the topological ones utilised in measure theory, are recalled in dedicated sections, with every result proved in detail, to make the book mostly self-contained to any graduate student of mathematics, but also to the student of physics or engineering which is not put off by the abstraction of mathematics, but conversely appreciates the elegance of the abstract setting.

When presenting a subject that is so deeply entangled with so many areas of mathematics, there is a natural trade-off that needs to be carefully calibrated: The necessity of being complete in the presentation of the concepts, and the overall dimension of the book. The idea behind this book is to maintain the maximum rigour in the development of the theory, presenting all the necessary theorems which find an application in mathematical analysis and probability theory, and which may be thought as the 'theoretical minimum' of measure theory, without indulging in developing topics that, albeit beautiful, are indeed unnecessary, with the intent to maintain the presentation limited in its volume, without renouncing to the overall rigour.

Overview

Chapters I and II are rather classical. The former aims to present the definition of measurable spaces and measures, their properties, and the properties of the maps that preserve this structure. It is in this chapter that Dynkin's λ-π Theorem is discussed, and many equivalent forms of this theorem, such as the 'Monotone Class Theorem', are presented. In Chapter II, instead, the notion of integral is developed, utilising the classic approach of further extending this functional from a simple, special class of functions, to more general and stabler class of functions. Overall, these two chapters, beside introducing the underlying structure, aim to unfold from the start the style of the book and lay down all notations that are going to be subsequently used.

Once the general framework of measure theory has been deployed, it is in Chapter III that we develop the main techniques that will allow the concrete construction of measures. This is done, essentially, using two different but linked methods: Daniell functionals on one side — which sees integrals as monotone linear functionals on a vector space of real functions, which works by extending these linear functionals to bigger spaces, and the use of outer measures and Carathéodory's constructions on the other side — which starts from pre-measures to construct outer measures, to then define measures by taking suitable restrictions thereof.

A few topics have been included in the book, which are generally neglected in classical books, by either not mentioning them at all, or mentioning them only as exercises. One of these is the notion of *kernel*, which is not only developed in detail in the book, but has been dedicated the entire Chapter IV. This chapter starts by introducing the notion of kernel and showing the most common examples, and utilises the properties of kernels to provide a simple and effective proof of the existence of the product of two measures and the famous Fubini's theorem for iterative integration. It shall look rather astonishing the fact that, in the language of kernels, the proof of Fubini's theorem is indeed trivial. The chapter also contains the translation, in the language of pure mathematics, of a pivotal theorem in probability theory: The Ionescu-Tulcea's Theorem. In spite of its importance, this theorem is hardly ever mentioned in a textbook on measure theory, and the few times this is done, it is yet discussed only in the framework of probability theory. With the aid of Ionescu-Tulcea's Theorem, it will also be possible to prove the existence of the product of an arbitrary family of probability measures, which is another example of rarely discussed classical result in measure theory.

It has already been established in Chapter I that monotone Riesz spaces are the protagonist function spaces in measure theory and integration. Nevertheless, from an algebraic point of view, it turns out that the abstract notion of Riesz space is of particular interest. For this reason, Chapter V aims to discuss in detail the properties of Riesz spaces, and then applies this concept to the study of real measures, that is to say measures that may take both positive and negative values. The chapter terminates with a deep dive on Lebesgue's and Hahn's decompositions, which is done, starting from the proof of Lebesgue's decomposition and proving the famous Radon-Nikodým Theorem as a simple corollary of the decomposition. This is perhaps the most original part of this book, for the proof of the decomposition does not rely, as it customarily happens, on external theories, such as the representation of linear forms on Hilbert spaces, nor does it use the Hahn's decomposition, which, conversely, becomes a simple consequence of Lebesgue's decomposition.

Chapter VI is classical and perhaps a little dull, as it is devoted to the study of the structure of $\mathcal{L}^p$ spaces. This short chapter delves into the topology of such spaces, shows a representation theorem for the continuous linear functionals defined on them, and introduces a notion of duality among them. This chapter is therefore exquisitely technical, but a clear attempt has been made to produce elegant proofs and present the topic quickly and effectively.

After having developed the abstract measure theory in Chapters I to VI, we shall start adding some structure, and we shall delve into what could be called the 'topological measure theory', that is to say, measure theory on a topological space. This is done in detail in Chapter VII. Inside this chapter it is possible to recognise two extraneous sections, one on Polish spaces and one on locally compact spaces, which aim to give the reader a full reference of the structures used. These sections are, indeed, a full part of the presentation, and their proofs are also interesting because they contribute in showing techniques that become necessary to prove results that involve both the topological structure and the measure theoretic one. Afterwards, the geometric structure in the underlying space is introduced gradually, to show the link between the more geometrical, topological theory and the more algebraic, abstract measure one. The chapter terminates with the discussion of Riesz-Markov Theorem, as a special case of Daniell's theory developed in the first part of Chapter III, and the theorem of extension of Baire measures, which is also proved using Daniell's theory.

The modes of convergence, for an infinite family of functions, and for a an infinite family of measures, are all collected in Chapter VIII, where convergence is introduced and the different modes of convergence are confronted. This chapter also delves into the abstract notion of uniform integrability in its most general setting, showing necessary and sufficient conditions for uniform integrability with the main, but not unique, aim to prove Vitali's convergence theorem, which extends and generalises Lebesgue's theorem on dominated convergence. The second part of this chapter is devoted to introducing two modes of convergence on the Riesz spaces of measures, and aims to prove three main results, to each of which has been assigned a different section, to highlight their importance: A theorem due to O. Nikodým on the set-wise convergence of measures, Vitali-Hahn-Saks Theorem, which completes the statement of Nikodým's Theorem with an interesting and important extra feature, and Dunford-Pettis Theorem, which offers a characterisation of compactness in the weak topology $\sigma(\mathcal{L}^1, \mathcal{L}^\infty)$. The latter theorem, which is generally proved as a consequence of the famous Eberlein–Shmulyan Theorem on weak compactness in Banach spaces, is instead proved, adapting a proof due to C. Dellacherie and P.-A. Meyer [19], which remains entirely within the realm of measure theory.

While Chapter VIII offers two modes of 'strong' convergence for an infinite sequence of measures — set-wise convergence, and uniform convergence, it is in Chapter IX that the notion of weak convergence, which follows directly from the natural duality between measures and functions given by the linear form $(f, \mu) \mapsto \int f \, d\mu$, is introduced for the space of measures. This chapter is quite short, but it is admittedly rather dense, and offers some particularly interesting results: For example, it is within this chapter that it is proved that, if the underlying space E is compact (resp. Polish), then the space $\mathfrak{P}(E)$ of all probability measures on the Borel σ-field of E retains the same topological structure, that is to say, it is compact (resp. Polish) as well. The chapter terminates with another compactness result, which is referred to as the Prokhorov's Theorem and uses a generalisation of the notion of tightness for a family of functions.

The notion of conditional probability, and the related notion of disintegration of a probability, are of utmost importance in probability theory, but they may also be useful in other contexts like in the study of differential equations, and in optimal transport, as shown in Ambrosio *et al.* [3], Garling [32], and Villani [62, 63]. Chapter X aims to define the general notion of disintegration, showing examples of existence, uniqueness, and non-uniqueness. In particular, an entire section shall also be devoted to the proof of an existence result due to M. Jiřina, which is rarely mentioned even though is very general and particularly useful in the context of probability theory. This chapter wraps up all the concepts studied in the previous sections, for the notion of kernel, and

Ionescu-Tulcea's Theorem come back to define the concept of disintegration, and to prove the existence of a disintegration, respectively; and the topological structure of Polish space shall also reappear, turning out to be a sufficient condition to ensure the existence of a disintegration.

The book ends with two chapters that are, indeed, extremely different from all the previous ones, both in style and aim. One of these deals with the Lebesgue measure in $\mathbb{R}^n$, and the other on the family of Hausdorff measures. Precisely, in Chapter XI, the properties of the Lebesgue measure in $\mathbb{R}$ and $\mathbb{R}^n$ are studied, and the classical theorems connected with integration with respect to this measure are studied. Arriving so late in the book, this chapter may seem extremely easy and perhaps a bit out of place, but to a deeper look, its position is logically chosen to be able to infer, from the abstract work, all properties that Lebesgue measures have and that are, in fact, the result of the wealth of properties that its underlying space have.

The most challenging part of this chapter is perhaps the proof of the change of variables. Indeed, among the many existing proofs of this theorem, none of them is straightforward, and a complete and accurate proof always comes as a burden. The proof that is presented in this book has been selected not only for its elegance, but, most importantly, because it shows that, in the end, the difficulty is not within the realm of measure theory, but in the structure of the maps that make the change of variable. Indeed, the abstract notion of change of variable, $y = f(x)$, from a measure space $(E, \mathcal{E}, \mu)$ to a measurable space $(F, \mathcal{F})$ is a simple concept: If $\nu = f(\mu)$ denotes the image of the measure μ under f, one simply has $\int g \, \mathrm{d}\nu = \int g \circ f \, \mathrm{d}\mu$. The core of the difficulty, here, consists in representing the measure ν in terms of the Lebesgue measure, and that is where the structure of the change of variable, f, comes into effect. For this reason, a section that recaps the basic concepts of diffeomorphisms is included, where the Implicit Function Theorem and the Inverse Function Theorem are stated and proved not just to keep the book self-contained, but to show two beautiful proofs of two beautiful theorems.

Chapter XII completes the book with an overview of the family of Hausdorff measures. While the chapter does not aim to be an introduction, no matter how gentle, to geometric measure theory, the interest for discussing Hausdorff measures is twofold: On one hand, they offer a very important example of measures on a metric space, and as such their definition utilises the finest methods discussed in the previous chapters, but Hausdorff measures are, indeed, a very interesting topic *per se*, introducing the notion of dimension for a set in a metric space, and showing some beautiful examples of the limits of Lebesgue measure in dealing with some Borel measurable sets. In particular, a section of the chapter is devoted to the problem of calculating the dimension of the so-called 'self-similar' sets, such as Sierpiński's triangle, and Koch's curve. The chapter terminates with the area formula, which enables to calculate the measure of the so-called k-surfaces in $\mathbb{R}^n$ that are also going to be accurately defined.

The book was written with two main objectives in mind: To present measure theory in an elegant, Bourbaki-style, format — sharing with the reader the beauty of the theory and the advantages of abstraction, and to empower the reader by giving him or her all the tools that he or she may need when working on topics such as Geometric Measure Theory, the theory of Partial Differential Equation, Optimal Transport, and, of course, Probability Theory. Hopefully, by learning this beautiful mathematical theory, the two objectives merge into the only one of developing an abstract mindset.

Notations and conventions

For the most part, traditional notations have been followed, and, whenever it was possible, well-established customs. The words 'greater than' and 'smaller than' are always meant in the weak sense, i.e. admitting the possibility of the two quantities to be equal. Coherently, a positive number may be zero, and a negative number may be zero as well. In other words, in this book zero is both a positive and a negative number. When the strict inequality is intended, we shall always use the phrases 'strictly positive' and 'strictly negative'. Following this 'weak' format, the set of all natural numbers $\mathbb{N}$ is assumed to contain the number zero.

Sequences in this book start from 1 and, for this reason, the set of indices of a sequence is generally neglected. For example, instead of $(x_n)_{n\geq 1}$, we shall prefer to write (x_n). This is done with the intent to ease off the notation, but we shall maintain the right to sometimes disregard this self-inflicted rule and write the set of indices explicitly, when this will increase readability. For the same reason, bulky symbols such as summations and integrals are generally written in 'text style' rather than in the usual 'display style'.

With the intent to make references throughout the book more readable, all definitions, propositions, theorems, corollary, formulae, etc, are labelled with a single increasing number, preceded by the number corresponding to the section, within the current chapter where the reference belongs, and wrapped into a pair of round brackets. So, the reference $(p.q)$ is the qth label in section p of the current chapter. This label is always on the left-hand side of the page, to make the search even faster. Chapters are labelled with capital Roman numbers, and a reference to a label in a different chapter is made by prepending the Roman number relative to the chapter to the label. So, for example $r(p.q)$ represents the qth label, in section p of Chapter r.

Acknowledgements

Heading towards the end of this work, which has started over 20 years ago, I wish to take a moment to thank all my mentors, colleagues, and lifelong friends, who have contributed, through many years of patience and perseverance, to maintain alive my unconditional love for mathematics. All of them, in their own unique way, have had a pivotal influence in the development of my mathematical taste, my style, my knowledge, and my abilities; they all had, and continue to have, more or less directly, a positive influence in me as a person, as a citizen, and as a mathematician. It is with humble gratitude, immeasurable admiration, and the utmost respect that I wish to thank: Francesca Acquistapace, Paolo Acquistapace, Fabrizio Broglia, Ilaria del Corso, Roberto Dvornicich, Alessandro Iacopetti, Giorgio Letta, Brendan Lowry, Massimo Pappalardo, and Antonio Tarsia. To those who have left us too soon, Luciano Modica and M.K.V. Murthy, I wish to dedicate a memory and a thought, for they were a gem of the finest Italian mathematical tradition, and for I genuinely miss them.

A special thank goes to Lucy Carpignani, who is my first and foremost source of inspiration, without whom I would have never been able to spend so much time 'doing maths', and who has so kindly realised the beautiful image on the cover of this book. My warmest thank and recognition also goes to my parents, Carla and Sandro Carpignani, who have constantly supported me through the joys and sorrows of life.

Finally, my I am deeply indebted to my friend and mentor Paolo Acquistapace, who spent a huge amount of time reading the initial manuscript, finding typos, suggesting improvements, and offering me invaluable suggestions and improvements. Of course, any remaining typos, or any lack of clarity remain entirely my own responsibility.

Bletchley, UK *Andrea Carpignani*

σ-Field or σ-Algebra?

The term *σ-algebra* was first introduced in the study of a special type of algebraic structures, called Boolean algebras. In his book *Lectures on Boolean Algebras*, Paul R. Halmos defines a *Boolean algebra* as a non-empty set, with the two binary operations

$$(a, b) \mapsto a \wedge b, \qquad (a, b) \mapsto a \vee b$$

and a unary operation, $a \mapsto a'$, which satisfy some suitable properties.

He then defines a *field of sets* as a non-empty class of subsets of a given set X, which is stable under the operations of binary intersection, binary union and complement with respect to the whole set X. He observes then, immediately afterwards, that a field of sets is, in particular, a Boolean algebra as soon as one takes as the Boolean operations those of binary intersection, binary union, and complement. However, he hurries to add:

> *"This does not, however, justify the confusion (it is false) that set-theoretic intersection, union, and complement are the only possible operations that convert a class of sets into a Boolean algebra."*

It is indeed this fact, highlighted by Halmos, that makes it necessary to introduce a special keyword (*field of sets*) to denote those special Boolean algebras for which, not only is it true that the elements are subsets of a given set, but it is also true that the Boolean operations $\wedge$ and $\vee$ coincide with the set-theoretic operations of binary intersection and binary union (and, as a consequence, the Boolean operation $'$ coincides with the set-theoretic operation of complementation).

For analogous reasons, after having introduced the (*Boolean*) *σ-algebras* (that is to say those special Boolean algebras in which every countable set of elements admits an infimum and a supremum), Halmos introduces the notion of *σ-field* (*of sets*) which are stable for the operations of countable intersection and countable union.

It is worth noticing that σ-fields are special types of Boolean σ-algebras: Precisely, those Boolean σ-algebras for which, not only is it true that the elements are subsets of a given set, but it is also true that the Boolean operations of countable infimum and supremum coincide with the set-theoretic operations of countable intersection and countable union.

It is thus evident the necessity not to mix up the notion of σ-field with that of (Boolean) σ-algebra. The confusion between *field of sets* and *Boolean algebra* is less dangerous: In fact, every Boolean algebra is isomorphic to a field of sets. However, an analogous statement, where one substitutes the word algebra with σ-algebra, and the word field with σ-field is false.

It is worth mentioning that, within the French tradition, N. Bourbaki had introduced the term *tribu* (tribe) to refer to a σ-field. While this terminology is widely adopted in France and sometimes in Italy, in the English language this terminology never managed to break through, notwithstanding several attempts made by few authors, and it remains customary to continue utilising the term *σ-field*, which is the choice adopted in this book.

Chapter I

The Foundation of Measure Theory

Measure Theory is the branch of Mathematics that studies the general notion of 'measure', which represents the abstract model of the geometric concepts of length, area, and volume, but also of other physical quantities such as the gravitational mass, electrical charge, thermal capacity, or indeed almost everything which one would expect to be 'additive' — which means that the measure of the union of two disjoint sets should be the sum of their measures. For the time being, we shall always assume that these measures are positive numbers, but we shall see later how to further extend this notion to a more general one which is able to describe, for example, the notion of electric charge as well. The actual definition of 'measure' (see §3) is not obvious and depends essentially on certain technical features which are described in detail in a preparatory section (see §1), where the algebraic structure of the family of all those sets that can be measured is introduced, and its properties are discussed. Before giving the notion of measure, it will also be convenient to introduce those maps that 'agree' with the underlying structure necessary to define measures, and that therefore will be able to 'transfer' measures from one set into another. This will be the content of §2. Even though §1 and §2 are rather abstract, it is important that the reader gets sufficiently familiar with the techniques described in these sections because they shall be used constantly in the development of the most interesting tools of measure theory.

1. Measurable spaces

In all the sequel, let us consider a non-empty set E, which may be thought as the ambient space where the objects that we wish to measure live, and let us suppose that we are interested in 'measuring' the 'size' (e.g. length, area, volume) of every subset A of E. Clearly, if the set E is infinite, there will

DOI: 10.1201/9781003724766-I

be a huge amount of such sets and it might be difficult, let alone impossible, to create such a general measure. As a result, it is going to be necessary to choose a non-empty class $\mathcal{E}$ of subsets of E and label as 'measurable' only the elements of such a class. Only for certain, particularly simple choices of E, may this class be chosen as the class $\mathcal{P}(E)$ of all subsets of E, for instance when E is finite or countable. Nevertheless, as soon as the cardinality of the set E exceeds $\aleph_0$, it is necessary to consider a proper subset of $\mathcal{P}(E)$. To the reader concerned on whether this may result in a limitation for the practical usage of the theory, it should be sufficient to mention that, if the family of measurable sets is chosen carefully, then finding a non-measurable set soon becomes a sufficiently challenging task that a mathematician hardly ever encounters such a type of set *by mistake*, that is to say unless, of course, he or she is making a conscious effort to look for one.

Anyway, since, in general, the class $\mathcal{P}(E)$ cannot be used, at the very least it shall be reasonable to require that the class $\mathcal{E}$ be endowed with suitable properties of stability (with respect to the common operations). In other words, the class $\mathcal{E}$ ought to have an algebraic structure that allows the most common operations to take place. Precisely, it would be sensible to require that the complement (with respect to E) of each element of $\mathcal{E}$ be still an element of $\mathcal{E}$, and that the class $\mathcal{E}$ be stable under the operations of union and intersection — which means that the union and the intersection of a finite family of elements of $\mathcal{E}$ be still an element of $\mathcal{E}$. These observations lead to the following definition.

(1.1) Definition. Let E be a non-empty set. A class $\mathcal{E}$ of subsets of E shall be called a **field** on E if the following conditions hold:

(A1) $E \in \mathcal{E}$,
(A2) $A \in \mathcal{E} \Rightarrow A^c \in \mathcal{E}$,
(A3) $A \in \mathcal{E},\ B \in \mathcal{E} \Rightarrow A \cup B \in \mathcal{E}$.

Observe that, in the hypotheses of (1.1), the following also holds:

(A4) $\emptyset \in \mathcal{E}$,
(A5) $A \in \mathcal{E},\ B \in \mathcal{E} \Rightarrow A \cap B \in \mathcal{E}$,
(A6) $A_1, \ldots, A_n \in \mathcal{E} \Rightarrow \bigcup_{i=1}^n A_i \in \mathcal{E}$ and $\bigcap_{i=1}^n A_i \in \mathcal{E}$.

Let us start by showing that (A1) and (A2) imply condition (A4). To see this, it suffices to notice that $E \in \mathcal{E} \Rightarrow \emptyset = E^c \in \mathcal{E}$. Further, let us now show that (A2) and (A3) yield condition (A5). Indeed, if $A, B \in \mathcal{E}$, by means of (A2), one has $A^c, B^c \in \mathcal{E}$, hence, by (A3), $A^c \cup B^c \in \mathcal{E}$. Now, thanks to De Morgan's laws, one has $A \cap B = (A^c \cup B^c)^c$, and this set lies in $\mathcal{E}$ by (A2). Finally, (A6) follows by induction from (A3) and (A5).

Before moving on, let us introduce some useful terminology. Precisely, condition (A2) can be rephrased by saying that the class $\mathcal{E}$ is **stable** under the operation of complementation. Similarly, conditions (A3) and (A5) can be rephrased by saying that $\mathcal{E}$ is also stable under the operation of finite union and intersection, respectively.

The notion of field already presents a certain number of properties that ensure the required stability that was discussed before. Nevertheless, these properties end up not being robust enough: To enable calculations that require an infinite number of sets, it would be sensible to require that the class $\mathcal{E}$ be stable under the union and the intersection of a *countable* family of its elements. All these heuristic ideas lead to adopt the following definition.

(1.2) Definition. Let E be a non-empty set. A class $\mathcal{E}$ of subsets of E shall be called a **σ-field** on E if the following conditions hold:

(T1) $E \in \mathcal{E}$,

(T2) $A \in \mathcal{E} \Rightarrow A^c \in \mathcal{E}$,

(T3) For every family $(A_i)_{i \in I}$, of elements of $\mathcal{E}$, where I is a countable set of indices, the union $\bigcup_{i \in I} A_i$ also lies in $\mathcal{E}$.

Observe that, in the hypotheses of (1.2), the following also holds:

(T4) $\emptyset \in \mathcal{E}$,

(T5) For every family $(A_i)_{i \in I}$, of elements of $\mathcal{E}$, where I is a countable set of indices, the intersection $\bigcap_{i \in I} A_i$ also lies in $\mathcal{E}$.

To prove this, let us start by observing that (T4) follows from the fact that a σ-field is, in particular, a field, and (T4) follows from (A4). To prove (T5), let $(A_i)_{i \in I}$ be a countable family of elements of $\mathcal{E}$. Then observe that, for any index i, one has $A_i \in \mathcal{E}$, and condition (T2) assures that $A_i^c \in \mathcal{E}$ as well. By means of condition (T3), this yields that $\bigcup_{i \in I} A_i^c$ must also be an element of $\mathcal{E}$. Now, by means of De Morgan's laws, the union of the complementaries of the members of a family of sets is the complement of their intersection, so the set $\left(\bigcap_{i \in I} A_i\right)^c = \bigcup_{i \in I} A_i^c$ lies in $\mathcal{E}$ because the right-hand side does, and therefore condition (T2) now ensures that the set $\bigcap_{i \in I} A_i$ must also be an element of the class $\mathcal{E}$.

Besides countable union and intersection, let us observe that a σ-field $\mathcal{E}$ is also stable under limit superior and limit inferior, which are defined, for a sequence (A_n) of elements of the class $\mathcal{E}$, by

$$\limsup_n A_n = \bigcap_{k \geq 0} \bigcup_{n \geq k} A_n \qquad \text{and} \qquad \liminf_n A_n = \bigcup_{k \geq 0} \bigcap_{n \geq k} A_n.$$

In other words, the set $\limsup_n A_n$ is the set of all elements x in E that lie in A_n for an infinite number of indices n, and the set $\liminf_n A_n$ is the set of all elements x in E that lie in A_n, except at most a finite number of indices. This can also be rephrased by saying that $\limsup_n A_n$ is the set of all elements x of E that 'frequently' lie in (A_n), and that $\liminf_n A_n$ is the set of all elements x of E that 'eventually' lie in (A_n).

The following result describes the relationship between these two sets.

(1.3) Proposition. *Let (A_n) be a sequence of subsets of a given set E. Then, the following inclusion holds:*

$$\liminf_n A_n \subseteq \limsup_n A_n.$$

Proof. If x lies in $\liminf A_n$, then there is an index k such that $x \in A_n$ for every index $n \geq k$, whence, x lies in $\bigcup_{n \geq h} A_n$ for every index $h \geq k$, which ultimately ensures that x lies in $\limsup A_n$. □

It is evident that the class $\mathcal{P}(E)$, of all subsets of E, is a σ-field on E, and such is also the family $\{\emptyset, E\}$. It is not so evident, however, how to define explicitly, non-trivial and interesting examples of σ-fields, at least when the set E is infinite — which is, in fact, the very case where this notion is actually interesting. For this reason, the next result is of paramount importance. In order to state it, let us start by recalling that a class $\mathcal{H}$ of subsets of E is included in a class $\mathcal{K}$, if each of the elements of $\mathcal{H}$ is also an element of $\mathcal{K}$, that is to say, if the inclusion holds in the set $\mathcal{P}(E)$. This works, in particular, for σ-fields.

When a σ-field is included in another, we shall say that it is a **sub-σ-field** of the latter. Notice that the σ-field $\mathcal{P}(E)$ contains any other σ-field on E. This is therefore the **biggest** σ-field on E. Conversely, $\{\emptyset, E\}$ is obviously a σ-field, and it is contained in every other σ-field on E. This is therefore the **smallest** σ-field on E. These two extremes are said to be the **trivial** σ-fields on E.

(1.4) Proposition. *The intersection of a family of σ-fields on a non-empty set E is itself a σ-field on E. Furthermore, for each class $\mathcal{I}$ of subsets of E, we can always find a σ-field that contains $\mathcal{I}$ and is contained in every other σ-field including $\mathcal{I}$. This is the minimal (with respect to inclusion) σ-field which contains $\mathcal{I}$.*

Proof. Let $\mathcal{G}$ be a family of σ-fields on E, and denote by $\mathcal{E}$ the intersection of this family, that is to say, the class of all subsets A of E which lie in each of the σ-fields $\mathcal{U} \in \mathcal{G}$. By means of condition (T1), $E \in \mathcal{U}$ for every $\mathcal{U} \in \mathcal{G}$, hence $E \in \mathcal{E}$. Now, if A is an element of $\mathcal{E}$, for every $\mathcal{U} \in \mathcal{G}$, $A \in \mathcal{U}$ and therefore, by means of condition (T2), $A^c \in \mathcal{U}$, whence $A^c \in \mathcal{E}$. Likewise, if $(A_i)_{i \in I}$ is a countable family of elements of $\mathcal{E}$, for every $\mathcal{U} \in \mathcal{G}$ and every index i, $A_i \in \mathcal{U}$ and therefore $\bigcup_{i \in I} A_i \in \mathcal{U}$ by means of condition (T3), thus $\bigcup_{i \in I} A_i \in \mathcal{E}$ as well. This proves that $\mathcal{E}$ is actually a σ-field.

Now, if $\mathcal{I}$ is a class of subsets of E, consider the family of all the σ-fields on E which contain $\mathcal{I}$. This family is not empty because at least $\mathcal{P}(E)$ belongs to it. The intersection of this class is a σ-field, contains $\mathcal{I}$, and is clearly contained in each of the other σ-fields on E containing $\mathcal{I}$. □

The previous result suggests that a quick way of assigning a σ-field is by selecting a suitable 'set of generators'. What do we mean by that? If E is an arbitrary non-empty set, let $\mathcal{I}$ be a non-empty class of subsets of E. Then, at least one σ-field, viz. $\mathcal{P}(E)$, contains $\mathcal{I}$, and therefore the class of all σ-fields containing $\mathcal{I}$ is non-empty. Its intersection, by virtue of Proposition (1.4), is a σ-field itself, and it is the smallest of such σ-fields which contains $\mathcal{I}$. Precisely, we can give the following definition.

(1.5) Definition. Given a non-empty set E, let $\mathcal{I}$ be a non-empty class of subsets of E. The smallest σ-field that contains $\mathcal{I}$ shall be called the σ-field **generated** by $\mathcal{I}$, and it shall be denoted by $\sigma(\mathcal{I})$. Conversely, if $\mathcal{E}$ is a given σ-field on E and if there exists a family $\mathcal{I}$ of subsets of E such that the σ-field generated by $\mathcal{I}$ coincides with $\mathcal{E}$, we shall say that $\mathcal{I}$ **generates** $\mathcal{E}$, or it is a **set of generators** for $\mathcal{E}$.

Let $\mathcal{I}$ and $\mathcal{J}$ be two given collections of subsets of E. If $\mathcal{I}$ is included in the σ-field $\sigma(\mathcal{J})$, then it is also clear that $\sigma(\mathcal{I}) \subseteq \sigma(\mathcal{J})$, because $\sigma(\mathcal{I})$ is the smallest σ-field containing $\mathcal{I}$, and this elementary observation proves the following simple criterion for the coincidence of two σ-fields.

(1.6) Proposition. *Two classes $\mathcal{I}$ and $\mathcal{J}$ of subsets of E generate the same σ-field on E if and only if $\mathcal{I} \subseteq \sigma(\mathcal{J})$ and $\mathcal{J} \subseteq \sigma(\mathcal{I})$.*

Let us see an application of this new notion to define a very important type of σ-fields. To this end, let us start by introducing the following notation.

(1.7) Notation. If E is a non-empty set and $\mathcal{U}$ is a class of subsets of E, we shall denote by $\mathcal{U}_s$ (resp. $\mathcal{U}_d$, $\mathcal{U}_\sigma$, and $\mathcal{U}_\delta$) the class of all finite unions (resp. finite intersections, countable unions, and countable intersections) of elements of $\mathcal{U}$. We shall also set $\mathcal{U}_{sd} = (\mathcal{U}_s)_d$, $\mathcal{U}_{s\delta} = (\mathcal{U}_s)_\delta$, and so on for all the possible combinations. Employing this notation, one can state that a class $\mathcal{U}$ of subsets of E is stable under finite union (resp. finite intersection, etc) by stating that $\mathcal{U}_s = \mathcal{U}$ (resp. $\mathcal{U}_d = \mathcal{U}$, etc).

Let us recall that a class $\mathcal{B}$ of subsets of E is a **base** for a given topology $\mathcal{O}_E$ on E, if the class $\mathcal{B}_\sigma$, of all countable unions of elements of $\mathcal{B}$, coincides with the assigned topology. Similarly, a class $\mathcal{U}$ of subsets of E is called a **sub-base** for $\mathcal{O}_E$ if the class $\mathcal{U}_d$, of all the finite intersections of elements of $\mathcal{U}$, is a base for the topology, that is to say, if the class $\mathcal{U}_{d\sigma}$ coincides with $\mathcal{O}_E$.

Let us consider a topological space $(E, \mathcal{O}_E)$. The σ-field generated by the topology $\mathcal{O}_E$ shall be called the **Borel σ-field** of E and hereinafter shall be denoted by $\mathcal{B}(E)$. In other words, $\sigma(\mathcal{O}_E) = \mathcal{B}(E)$. Also, every element of the σ-field $\mathcal{B}(E)$ shall be called **Borel measurable set** of E. It is quite easy to prove that this σ-field is also generated by the closed sets, as well as by any base $\mathcal{B}$ for the topology $\mathcal{O}_E$.

If the topological space $(E, \mathcal{O}_E)$ has a countable base, it is an easy task to find a set of generators for the Borel σ-field, as shown in the next result.

(1.8) Proposition. *If a topological space $(E, \mathcal{O}_E)$ has a countable base, then each countable base (and more generally each countable sub-base) generates the Borel σ-field on E.*

Proof. Let $\mathcal{U}$ be a countable sub-base of $\mathcal{O}_E$. Then $\mathcal{U} \subseteq \mathcal{B}(E) = \sigma(\mathcal{O}_E)$. Furthermore, since $\mathcal{U}_d$ is a countable base for $\mathcal{O}_E$, this latter set by definition coincides with the set $\mathcal{U}_{d\sigma}$ of all countable unions of elements of $\mathcal{U}_d$. Hence, the thesis follows by noticing that $\mathcal{O}_E = \mathcal{U}_{d\sigma} \subseteq \sigma(\mathcal{U})$. □

Using this result, it is now possible to write a set of generators for the Borel σ-field of several topological spaces. For example, for the Borel σ-field of the Euclidean space $\mathbb{R}^d$, endowed with the usual Euclidean topology, a set of generators is given by the class of all the products of open intervals with rational extremes, that is to say, all sets of the form

$$\prod_{i=1}^{d}(a_i, b_i) \qquad \text{where } a_i, b_i \in \mathbb{Q}.$$

To prove this, it suffices to notice that this class is a countable sub-base for the Euclidean topology of $\mathbb{R}^d$. In particular, when $d = 1$, one can choose the set of open intervals (a, b) with rational extremes, as well as many other sets of generators. For example, let us take a dense subset D of $\mathbb{R}$, which may be but doesn't necessarily need to be the set of rational numbers. Then a set of generators for the Borel σ-field $\mathcal{B}(\mathbb{R})$ is each of the following classes:

$$1)\ \Big\{(-\infty, a) : a \in D\Big\} \qquad 2)\ \Big\{(-\infty, a] : a \in D\Big\}$$

$$3)\ \Big\{(a, +\infty) : a \in D\Big\} \qquad 4)\ \Big\{[a, +\infty) : a \in D\Big\}$$

Bearing in mind the initial goal to study the measure of the 'size' of the subsets of a non-empty set, the first step consists in assigning to it a σ-field and, precisely, the σ-field of the sets that can be measured. This is made precise in the following definition.

(1.9) Definition. The pair $(E, \mathcal{E})$, where E is a non-empty set, and $\mathcal{E}$ is a σ-field on E, shall be called a **measurable space**. If $(E, \mathcal{E})$ is a measurable space, every set A in $\mathcal{E}$ is said to be **measurable** with respect to $\mathcal{E}$ or, simply, **$\mathcal{E}$-measurable**. (In most situations, when it is clear from the context with respect to which σ-field the set is measurable, we are going to simply say that a set is 'measurable' without any further reference to the assigned σ-field.)

If E is a non-empty set, then $(E, \mathcal{P}(E))$ is an example of a measurable space. This construction is useful when E is countable, but it becomes slightly less interesting when the set E is more than countable. To obtain a more meaningful example of measurable space, that is interesting even when E is uncountable, let us consider a topological space $(E, \mathcal{O}_E)$. As we have already mentioned, the topology $\mathcal{O}_E$ generates the Borel σ-field, $\mathcal{B}(E)$. Thus, the pair $(E, \mathcal{B}(E))$ is a measurable space.

Let us now introduce a general method to create a new σ-field from one already given. This is a very similar method to inducing a topology on a subset of a topological space, and entails that every subset of a measurable space may be seen as a measurable space itself. Precisely, if $(E, \mathcal{E})$ is a measurable space, and if D is *any* subset of E, not necessarily a measurable set, one may always equip D with a suitable σ-field $\mathcal{D}$, so that the measurable space $(D, \mathcal{D})$ inherits its structure from $(E, \mathcal{E})$. Such a σ-field is coherently called the 'trace' of the σ-field $\mathcal{E}$ on D, and its formal definition is as follows.

(1.10) Definition. Let $(E, \mathcal{E})$ be a measurable space and let D be any subset of E. The **trace** of the σ-field of $\mathcal{E}$ onto D is denoted by $\mathcal{E}_D$, and it is defined as the class of all the sets of the form $A \cap D$ with $A \in \mathcal{E}$. It is not difficult to show that $\mathcal{E}_D$ is a σ-field on D.

If $\mathcal{I}$ is a set of generators for the σ-field $\mathcal{E}$, then the trace $\mathcal{E}_D$ of $\mathcal{E}$ onto D coincides with the σ-field generated by D and by the class of sets of the form $A \cap D$ with A element of $\mathcal{I}$.

The σ-field $\mathcal{B}([0,1])$ is the trace of $\mathcal{B}(\mathbb{R})$ onto the interval $[0,1]$. More generally, if $(E, \mathcal{O}_E)$ is a topological space and F is a subset of E, the Borel σ-field $\mathcal{B}(F)$ — which is, obviously, the σ-field generated by the topology induced on the set F by E, coincides with the trace onto F of $\mathcal{B}(E)$. To this end, it suffices to recall that the topology induced on F by E is defined as

$$\mathcal{O}_F = \left\{ A \cap F : A \in \mathcal{O}_E \right\},$$

and this is a set of generators for $\mathcal{B}(F)$ as well as the trace onto F of a set of generators of $\mathcal{B}(E)$.

2. Measurable maps

Let us now begin to study the properties of those maps that 'agree' with the measurable structure. We shall see later on that these are also the maps that can be 'integrated' with respect to a given measure. If the definition of σ-field looks somewhat like the definition of topology, the definition of measurable function looks like that of a continuous function. Precisely, let us give the following definition.

(2.1) Definition. Let $(E, \mathcal{E})$ and $(F, \mathcal{F})$ be two measurable spaces, and f a map defined on E, with values in F. We shall say that f is **measurable** from $(E, \mathcal{E})$ to $(F, \mathcal{F})$ if, for every B in $\mathcal{F}$, the set $f^{-1}(B)$, inverse image of B under f, lies in $\mathcal{E}$. Also, we shall call a **measurable isomorphism** from $(E, \mathcal{E})$ to $(F, \mathcal{F})$ any measurable function from $(E, \mathcal{E})$ to $(F, \mathcal{F})$ that is bijective and whose inverse is a measurable function from $(F, \mathcal{F})$ to $(E, \mathcal{E})$.

Let us notice that, if f is a map from E to F, and if $\mathcal{F}$ is a σ-field on F, the set of all the inverse images $f^{-1}(B)$, with $B \in \mathcal{F}$, is a σ-field on E. It is not difficult to show that this is actually the smallest (with respect to the inclusion) of all σ-fields on E that makes f measurable, when F is equipped with the σ-field $\mathcal{F}$. This σ-field is denoted by $\sigma(f)$, and it is called the σ-field **generated** by f. Observe that, if f a map from E to F, it is not generally true that $\sigma(f)$ is included in $\mathcal{E}$. The fact that $\sigma(f)$ is included in $\mathcal{E}$ is actually a necessary condition for f to be measurable.

For example, if $(E, \mathcal{E})$ is a measurable space and D is a subset of E, then the trace $\mathcal{E}_D$ of $\mathcal{E}$ onto D coincides with the σ-field $\sigma(i_D)$ generated by the canonic inclusion i_D of D into E, i.e. the map defined on D, with values in E, by $i_D(x) = x$, for every x of D. As a result, the trace of $\mathcal{E}$ onto D is included in $\mathcal{E}$ if and only if D is measurable.

The next result aims to show that the composition of measurable maps is again measurable.

(2.2) Proposition. *If f is a measurable map from $(E, \mathcal{E})$ to $(F, \mathcal{F})$, and g is a measurable map from $(F, \mathcal{F})$ to $(G, \mathcal{G})$, then the composite map $g \circ f$ is a measurable map from $(E, \mathcal{E})$ to $(G, \mathcal{G})$.*

Proof. It suffices to note that, for every element X in $\mathcal{G}$, one has

$$(g \circ f)^{-1}(X) = f^{-1}\big(g^{-1}(X)\big). \qquad \square$$

Since σ-fields are generally very wide families of sets, it becomes extremely difficult to check that the definition of measurable map holds for every element of the σ-field. It is therefore necessary to establish criteria that simplify the work, such as the following.

(2.3) Theorem (Fundamental criterion of measurability). *Assume that two measurable spaces $(E, \mathcal{E})$ and $(F, \mathcal{F})$ are assigned, and let f be a map from E to F. Also, let $\mathcal{J}$ be a set of generators for $\mathcal{F}$. Then, the following are equivalent:*

(a) *f is a measurable map from $(E, \mathcal{E})$ to $(F, \mathcal{F})$.*

(b) *For every $X \in \mathcal{J}$, one has $f^{-1}(X) \in \mathcal{E}$.*

Proof. The implication (a) ⇒ (b) is obvious, so it suffices to show the implication (b) ⇒ (a). To this end, let us observe that the class

$$\left\{X \in \mathcal{F} : f^{-1}(X) \in \mathcal{E}\right\}$$

is a σ-field. Then, the hypothesis can be rephrased by saying that $\mathcal{J}$ is contained in this class, and therefore it contains the σ-field $\sigma(\mathcal{J}) = \mathcal{F}$ as well. This is sufficient to prove that f is measurable. $\square$

This criterion can be applied, in particular, every time that F is a topological space equipped with its own Borel σ-field. In this special case, it will suffice to check that the inverse image, under f, of each element of a sub-base for the topology of F is measurable.

(2.4) Example. Let $(E, \mathcal{E})$ and $(D, \mathcal{D})$ be two measurable spaces, and assume that D be discrete, $\mathcal{D} = \mathcal{P}(D)$, and f be a map from E to D. In this special case, a set of generators for $\mathcal{D}$ consists of the set of all singletons $\{y\}$ with $y \in D$. Then, the fundamental criterion of measurability entails that if the sets

$$\Big\{x \in E : f(x) = y\Big\}$$

lie in $\mathcal{E}$, for every $y \in D$, then f is a measurable map from $(E, \mathcal{E})$ to $(D, \mathcal{D})$. Such a measurable map shall be called **discrete**.

(2.5) Example. If E and F are two topological spaces, every continuous map from E to F is therefore measurable from $(E, \mathcal{B}(E))$ to $(F, \mathcal{B}(F))$, because, by definition of continuity, the inverse image of every open set of F is an open set of E. A measurable map from $(E, \mathcal{B}(E))$ to $(F, \mathcal{B}(F))$ is often referred to as a **Borel measurable map** from E to F.

(2.6) Example. Given a measurable space $(E, \mathcal{E})$, suppose that f is a real function on E. Since $\mathbb{R}$ is a topological space with its usual Euclidean topology, it is also a measurable space with its Borel σ-field $\mathcal{B}(\mathbb{R})$. Since the topology of $\mathbb{R}$ is generated by the intervals, the fundamental criterion of measurability ensures that a real function f on E is a measurable function if all subsets of the form $\{x \in E : a \leq x \leq b\}$ are measurable, or just if all subsets the even more special form $\{x \in E : f(x) \leq c\}$ are.

(2.7) Example. In measure theory, it is often useful to extend the definition of function, including those taking values in the **extended real line**, i.e. in the set $\overline{\mathbb{R}} = \mathbb{R} \cup \{-\infty, +\infty\}$. In other words, it shall be useful to allow a measurable function to take the values $-\infty$ and $+\infty$ as well. In this case, the definition just given does not change, providing that we also take the intervals of the form $[a, +\infty]$ and $[-\infty, a]$. Such measurable functions shall be referred to as **numeric** functions.

Dealing with numeric functions, we shall always assume that the multiplication extends to the extended real line $\overline{\mathbb{R}}$ by setting $(\pm\infty) \cdot 0 = 0 \cdot (\pm\infty) = 0$, $(\pm\infty) \cdot a = a \cdot (\pm\infty) = \pm\infty$, for every real number $a \geq 0$, and, similarly, for $a \leq 0$, by setting $(\pm\infty) \cdot a = a \cdot (\pm\infty) = \mp\infty$. In general, with the term 'function' we shall always mean, unless otherwise specified, a *numeric* function, rather than a *real* function.

Back to the general case of a measurable space $(E, \mathcal{E})$, let us consider a measurable map f taking values in a measurable space $(F, \mathcal{F})$, and let us introduce a notation borrowed from probability theory. For all sets B in $\mathcal{F}$, the measurable set

$$f^{-1}(B) = \Big\{x \in E : f(x) \in B\Big\}$$

shall be denoted simply by

$$\{f \in B\}$$

In addition if B is a singleton, that is if $B = \{y\}$ with $y \in F$, we shall write $\{f = y\}$ instead of $\{x \in E : f(x) = y\}$. Similarly, if f is a measurable function, or, what is the same, if f is a measurable map taking values in a measurable space in which an order is defined, that is to say where the following expressions make sense, we shall simply write $\{f \leq c\}$ to mean the measurable set $\{x \in E : f(x) \leq c\}$, and $\{a \leq f \leq b\}$ to mean the set $\{x \in E : a \leq f(x) \leq b\}$, and so on.

(2.8) Example. Let $(E, \mathcal{E})$ be a measurable space, and let A be any subset of E. The **indicator function** of A is the function I_A, defined on E, with values in $\{0, 1\}$, by

$$I_A(x) = \begin{cases} 1 & \text{if } x \in A, \\ 0 & \text{if } x \notin A. \end{cases}$$

For the function I_A to be $\mathcal{E}$-measurable, it is necessary and sufficient that the set A be $\mathcal{E}$-measurable, that is to say, that $A \in \mathcal{E}$. Indeed, it is clear that $\{I_A = 1\} = A$, and $\{I_A = 0\} = A^c$.

We now wish to describe the structure of the set $\mathcal{M}(E, \mathcal{E})$ of the measurable functions on $(E, \mathcal{E})$. The following result shows that this set is not only a vector space, but also stable under the operation of supremum and infimum of any two elements, which is called, in algebraic terminology, a vector lattice or, what is the same, a Riesz space.

(2.9) Proposition. *The set $\mathcal{M}(E, \mathcal{E})$ is a vector space stable under the operations $(f, g) \mapsto \sup\{f, g\}$ and $(f, g) \mapsto \inf\{f, g\}$. In particular, if f is a real measurable function on $(E, \mathcal{E})$, such are the functions $f^+ = \sup\{f, 0\}$ and $f^- = \sup\{-f, 0\}$.*

Proof. Let us start by observing that, if f and g are two measurable functions on $(E, \mathcal{E})$, the function $x \mapsto \psi(x) = (f(x), g(x))$, defined on E, with values in $\overline{\mathbb{R}} \times \overline{\mathbb{R}}$, thought as a measurable space with its own Borel σ-field, is measurable, because, for every set of the form $U = (a, b) \times (c, d)$, one has

$$\psi^{-1}(U) = \{a < f < b\} \cap \{c < g < d\},$$

and the sets of this form are a base for the topology in $\overline{\mathbb{R}} \times \overline{\mathbb{R}}$ and therefore a set of generators for the Borel σ-field of $\overline{\mathbb{R}} \times \overline{\mathbb{R}}$. This is therefore sufficient to prove the first statement, for the functions $(u, v) \mapsto u + v$, $(u, v) \mapsto uv$, $(u, v) \mapsto \inf\{u, v\}$ and $(u, v) \mapsto \sup\{u, v\}$ are continuous, hence measurable. In particular, taking g the constant function equal to a, one also has that the function af is measurable as well. □

In the sequel, if f and g are two numeric functions defined on a set E, we shall set $f \vee g = \sup\{f, g\}$ and $f \wedge g = \inf\{f, g\}$. The maps

$$(f, g) \mapsto f \vee g, \qquad (f, g) \mapsto f \wedge g$$

are called **lattice operations**. The following result shows that the vector space $\mathcal{M}(E, \mathcal{E})$ has even stronger properties of stability.

(2.10) Proposition. *Let (f_n) be a sequence of numeric measurable functions defined on a measurable space $(E, \mathcal{E})$. Then, the functions*

$$\sup_n f_n, \qquad \inf_n f_n, \qquad \limsup_n f_n, \qquad \liminf_n f_n$$

are measurable.

In particular, if (f_n) is a pointwise convergent sequence of functions, its limit is measurable.

Proof. Let us start by proving that $\sup_n f_n$ is measurable. To this end, observe that for every number α, one has $\{\sup_n f_n > \alpha\} = \bigcup_n \{f_n > \alpha\}$ and this set is measurable because such are all the sets in the union, as the functions f_n are measurable. In very much the same way, one can also infer that the function $\inf_n f_n$ is measurable. Since $\limsup_n f_n = \inf_n \sup_{k \geq n} f_k$ and $\liminf_n f_n = \sup_n \inf_{k \geq n} f_k$, there is nothing else to prove. □

Let us introduce a class of measurable functions that is particularly important both in measure theory and probability: That of 'simple functions'. Roughly speaking, these are functions that only take a finite number of values. More precisely, let us give the following definition.

(2.11) Definition. On a measurable space $(E, \mathcal{E})$, let f be a function on E, measurable on $(E, \mathcal{E})$. We shall say that f is **simple** if the set $f(E)$ of its values is finite.

The family of all simple functions on $(E, \mathcal{E})$ is the vector space generated by the indicator functions of all measurable sets in $\mathcal{E}$. Furthermore, every simple function f can be written as linear combination of indicator functions of pairwise disjoint elements of $\mathcal{E}$. To see this, it suffices to write

$$f = \sum_{y \in f(E)} y I_{\{f=y\}}.$$

Let us denote by $\mathcal{S}(E, \mathcal{E})$ the set of all simple functions on $(E, \mathcal{E})$. It is not difficult to show that this set is an algebra stable under the lattice operations which contains the constants. In fact, if f and g take only a finite number of values, so do the functions $f + g$, fg, $f \vee g$, and $f \wedge g$. Moreover, taking g the constant function everywhere equal to the real constant c, which is clearly a simple function, then cf is also a simple function.

The importance of simple function is that every positive measurable function can be approximated with them, as shown in the proposition below.

(2.12) Proposition. *Let f be a positive numeric function, measurable on $(E, \mathcal{E})$. Then, f can be written as the supremum of an increasing sequence of positive simple functions on $(E, \mathcal{E})$.*

Proof. Let f be a positive numeric function, measurable on $(E, \mathcal{E})$, and let us note that, for each point x of E, one has

$$f(x) = \sup\Big\{\alpha \in \mathbb{Q}_+ : f(x) \geq \alpha\Big\} = \sup_{\alpha \in \mathbb{Q}_+} \alpha I_{\{f \geq \alpha\}}(x).$$

Thus, the function f can be expressed as the supremum of a countable family of functions, each of which is the product of a constant and the indicator function of a measurable set, hence a simple function. As a result of this, the function f can be written as the supremum of a sequence (f_n) of positive simple functions. As a consequence, f is also the supremum of an 'increasing' sequence $\big((f_1 \vee \cdots \vee f_n)\big)$. Since the terms of this latter sequence are simple functions on $(E, \mathcal{E})$, the theorem is fully proved. □

As a very simple consequence of the previous preposition, we can now prove the following.

(2.13) Proposition. *Every measurable function is the pointwise limit of a sequence of simple functions. In other words, every measurable function can be approximated with arbitrary precision with a simple function.*

Proof. If f is a measurable function on a measurable space $(E, \mathcal{E})$, consider the two functions $f^+ = f \vee 0$ and $f^- = (-f) \vee 0$. Then $f = f^+ - f^-$. Furthermore, by virtue of Proposition (2.12), there are two sequences (g_n) and (h_n) of simple functions on $(E, \mathcal{E})$ such that $g_n \uparrow f^+$ and $h_n \uparrow f^-$ so the sequence $(g_n - f_n)$ converges pointwise to $f^+ - f^- = f$. □

Another consequence of Proposition (2.12) is the following interesting criterion of measurability due to Joseph L. Doob (1910–2004).

(2.14) Theorem (Doob's measurability criterion). *Let $(E, \mathcal{E})$ and $(F, \mathcal{F})$ be two measurable spaces, let f be a map from E to F, and suppose that the σ-field $\mathcal{E}$ coincides with the σ-field generated by f. In other words, suppose that*

$$\mathcal{E} = \Big\{\{f \in A\} : A \in \mathcal{F}\Big\}.$$

For a numeric function g, defined on E, to be measurable on $(E, \mathcal{E})$, it is necessary and sufficient that it exists a function h, measurable on $(F, \mathcal{F})$, such that

$$g = h \circ f.$$

Proof. The condition is clearly sufficient, so let us prove that it is also necessary. To this end, since the extended real line is homeomorphic to $[0, \infty]$, it suffices to prove the statement when g is positive. In this special case, thanks to Proposition (2.12), the function g is the supremum of a countable family of simple functions. With no loss of generality, we can therefore assume that g is a simple function. Actually we can assume that g is the indicator function of a measurable set. In fact, if the statement is true for the indicator function of a measurable set, then it must be also for any simple function, which is a linear combination of indicator functions. Let us assume therefore that g is the indicator function of an element of $\mathcal{E}$, that is to say a set of the form $\{f \in A\}$ with A in $\mathcal{F}$. In this case the thesis is trivial, because $g = I_{\{f \in A\}} = I_A \circ f$. □

Let us complete this section introducing a simple but useful theorem that enables to construct a measurable function by defining it on a suitable class of subsets of E. This is a sort of 'patchworking' construction and for this reason, a function that is constructed in this way shall be said to be 'defined by patchworking'.

(2.15) Theorem (Measurability by patchworking). *Let $(E, \mathcal{E})$ and $(F, \mathcal{F})$ be two measurable space and let f be a function from E to F. Let $\mathcal{H}$ be a countable partition of E, consisting of measurable sets such that, for every element H of $\mathcal{H}$, the restriction of f to the set H is a measurable function φ_H from $(H, \mathcal{E}_H)$ to $(F, \mathcal{F})$. Then, f is a measurable map from $(E, \mathcal{E})$ to $(F, \mathcal{F})$.*

Proof. It suffices to note that, for every element A of $\mathcal{F}$, the set $\{f \in A\}$ coincides with the set

$$\bigcup_{H \in \mathcal{H}} H \cap \{\varphi_H \in A\},$$

which belongs to $\mathcal{E}$, because it is countable union of elements of $\mathcal{E}$. □

3. Measures and measure spaces

In measure theory, the choice of a measurable space, $(E, \mathcal{E})$, is only the very first step. The second and foremost step consists of choosing a 'measure' to weigh all the elements of the σ-field $\mathcal{E}$. What do we mean by that? Given a measurable space $(E, \mathcal{E})$, we wish to associate, to every measurable set A belonging to $\mathcal{E}$, a positive number, $\mu(A)$, which is going to represent the measure of A. Mathematically, this means that we wish to select a numeric function, defined on the σ-field $\mathcal{E}$. A numeric function that is defined on a class of subsets of a non-empty set is usually called a **set function**. If μ is a set function on the σ-field $\mathcal{E}$, we shall say that it is **additive** if

$$\mu(A \cup B) = \mu(A) + \mu(B)$$

whenever A and B are disjoint sets in $\mathcal{E}$.

Likewise, a set function μ defined on $\mathcal{E}$ is said to be **finitely additive** if, for every finite sequence $A_1, \dots, A_n$ of pairwise disjoint sets in $\mathcal{E}$, we have

$$\mu\Big(\bigcup_{i=1}^{n} A_n\Big) = \sum_{i=1}^{n} \mu(A_n).$$

Even though it is obvious, it is worth noting that, if a set function on a σ-field $\mathcal{E}$ is additive, it is also finitely additive.

Thus, it would seem like a natural choice to call a 'measure' on a measurable space $(E, \mathcal{E})$ any finitely additive positive set function on $\mathcal{E}$. However, from a theoretical point of view, in the same way as we had requested that σ-fields be stable for countable unions and intersections, we ought to require that a measure satisfies a stronger property. Precisely, it is going to be a reasonable choice to require that a set function μ defined on a σ-field $\mathcal{E}$ satisfies

$$\mu\Big(\bigcup_{n=1}^{\infty} A_n\Big) = \sum_{n=1}^{\infty} \mu(A_n)$$

whenever (A_n) is a countable family of pairwise disjoint sets in $\mathcal{E}$.

A set function μ, defined on a σ-field $\mathcal{E}$ over a non-empty set E, which satisfies the latter condition is called **countably additive**, or **σ-additive**. Let us observe that, for a countably additive set function μ, if we further assume that $\mu(\emptyset) = 0$, then μ is also finitely additive. In fact, if $(A_1, \dots, A_n)$ is a finite sequence of pairwise disjoint measurable sets, it is sufficient to place $A_i = \emptyset$, for any $i > n$, in order to see that the finite sequence $(A_1, \dots, A_n)$ coincides with the infinite sequence $(A_i)_{i \geq 1}$, and the relation $\mu\left(\bigcup_{i=1}^{\infty} A_i\right) = \sum_{i=1}^{\infty} \mu(A_i)$ reduces to $\mu\left(\bigcup_{i=1}^{n} A_i\right) = \sum_{i=1}^{n} \mu(A_i)$.

All the observations thus far justify the following definition of measure.

(3.1) Definition. A **measure** on an assigned measurable space $(E, \mathcal{E})$ is any map μ, defined on $\mathcal{E}$, with values in $[0, \infty]$, such that

(M1) $\mu(\emptyset) = 0$

(M2) μ is **countably additive**, that is, for any countable family $(A_i)_{i \in I}$ of pairwise disjoint elements of $\mathcal{E}$, one has

$$\mu\Big(\bigcup_{i \in I} A_i\Big) = \sum_{i \in I} \mu(A_i).$$

If μ is a measure on the measurable space $(E, \mathcal{E})$, the triple $(E, \mathcal{E}, \mu)$ is called a **measure space** and, given such a measure space $(E, \mathcal{E}, \mu)$, for each element A of $\mathcal{E}$, the number $\mu(A)$ is called the **measure** of A under μ. If A is an element of $\mathcal{E}$, and $\mu(A) = 0$, we shall say that A is **μ-negligible**, or simply **negligible** under the measure μ.

In the sequel, a positive measure will always be referred to as a 'measure' on a given measurable space. The examples of the electric charge or the magnetic mass clearly show that this definition is not the most general, but we shall see in Chapter V how to generalise this definition to a signed measure, and which models also these physical quantities. For the time being, however, we shall limit ourselves to consider only positive measures, because they lay down the base for any further generalisation.

Let us observe first that a measurable subset E_0 of a measure space $(E, \mathcal{E}, \mu)$ may itself be considered as a measure space $(E_0, \mathcal{E}_0, \mu_0)$, where $\mathcal{E}_0$ is the set of all measurable subsets of E_0 — the 'trace' of the σ-field $\mathcal{E}$ on E_0, and, for every A in $\mathcal{E}_0$, $\mu_0(A) = \mu(A)$. (Note that the trace of a subset E_0 of E does not require E_0 to be a measurable set, but this condition must hold in order to be able to transfer the measure from E to E_0.) Conversely, if a subset E_0 of E is already a measurable space $(E_0, \mathcal{E}_0, \mu_0)$, then E may be made into a measurable space $(E, \mathcal{E}, \mu)$, where $\mathcal{E}$ is the class of all those subsets of E whose intersection with E_0 is in $\mathcal{E}_0$, and for A in $\mathcal{E}$, $\mu(A) = \mu_0(A \cap E_0)$.

A modification of this last construction is frequently useful even if E is already a measure space.

(3.2) Definition. On a measure space $(E, \mathcal{E}, \mu)$, if H is a measurable subset of E, a new measure μ_H may be defined on the class of all measurable subsets A of E by setting $\mu_H(A) = \mu(A \cap H)$; it is easy to verify that $(E, \mathcal{E}, \mu_H)$ is indeed a measure space. This new measure μ_H defined on the measurable space $(E, \mathcal{E})$ is called the **trace** of μ to H.

Before describing the general properties of measures on a given measurable space, let us give the following definitions which are going to help distinguishing different types of measures with similar, but different features. Precisely, we shall distinguish three important types of measures, one of which we already know, and the other two are not much different from it.

(3.3) Definition. On a measure space $(E, \mathcal{E}, \mu)$, the measure $\mu(E)$ of the whole space E under μ is called the **total mass** of μ. If the total mass is finite, then we shall say that the measure μ is **finite**, and the measure space $(E, \mathcal{E}, \mu)$ is **finite**. In particular, when $\mu(E) = 1$, the measure μ is a **probability measure** and the triple $(E, \mathcal{E}, \mu)$ is customarily referred to as a **probability space**, rather than a measure space. We shall also say that μ is **σ-finite** if it is possible to decompose the whole space E as the union of a sequence of measurable sets, each of which has finite measure under μ. It is not difficult to prove that, if needed, such a sequence can always be chosen to be increasing. Furthermore, if μ is a σ-finite measure on $(E, \mathcal{E})$, the triple $(E, \mathcal{E}, \mu)$ is called a **σ-finite** measure space.

While it is fairly evident the reason for the definition of probability measure — the probability of an event should be a number between 0 (almost impossible event) to 1 (almost certain event), the reason for the definition of σ-finite measures lies on the fact that finite (and in particular probability) measures tend to have a large spread of features which makes them particularly easy to use. Thus, it is convenient, whenever it is possible, to take a finite measure instead of a general measure. For this reason, a σ-finite measure turns out to be particularly useful, because it can often be traced back to finite measures, with some standard techniques of reduction.

Let us come back to the general definition of measure, and let us prove the following proposition which contains the basic properties of a measure. These properties are not very difficult to understand and apply, but they are so important that the reader should ensure to master them before moving any further with the theory.

(3.4) Proposition. *Let μ be a measure on a measurable space $(E, \mathcal{E})$. Let A and B be two measurable sets. The following properties hold.*

(a) $\mu(B) = \mu(B \setminus A) + \mu(A \cap B)$.

(b) *(Modularity)* $\mu(A) + \mu(B) = \mu(A \cup B) + \mu(A \cap B)$.

(c) *(Sub-additivity)* $\mu(A \cup B) \leq \mu(A) + \mu(B)$.

(d) *(Monotonicity) If* $A \subseteq B$, *then* $\mu(A) \leq \mu(B)$.

(e) *(Subtractivity) If* $A \subseteq B$ *and* $\mu(A) < \infty$, *then* $\mu(B \setminus A) = \mu(B) - \mu(A)$.

Proof. To prove (a), let us notice that B can always be written as the union of two measurable sets, namely the part of B that lies in A and the part of B that does not lie in A: $(A \cap B)$ and $(A^c \cap B)$. Also notice that the second of these measurable sets coincides with $B \setminus A$. The measurable sets $(A \cap B)$ and $(A^c \cap B)$ are disjoint (the first contains all elements of B that lie in A, the second those which do not), and since their union coincides with B, their measure adds up to the measure of B:

$$\begin{aligned}\mu(B) &= \mu(A^c \cap B) + \mu(A \cap B)\\ &= \mu(B \setminus A) + \mu(A \cap B)\end{aligned}$$

To prove (b), applying the previous identity to both A and B, we get

$$\begin{aligned}\mu(A) &= \mu(A \cap B) + \mu(A \cap B^c),\\ \mu(B) &= \mu(A \cap B) + \mu(A^c \cap B).\end{aligned}$$

Since the measurable sets $(A \cap B^c)$, $(A \cap B)$, and $(A^c \cap B)$ are disjoint, and since their union is the whole set $A \cup B$, adding the two equalities above and using the additivity of μ, we get:

$$\begin{aligned}\mu(A) + \mu(B) &= \left[\mu(A \cap B) + \mu(A \cap B^c)\right] + \left[\mu(A \cap B) + \mu(A^c \cap B)\right]\\ &= \mu(A \cap B) + \left[\mu(A \cap B^c) + \mu(A \cap B) + \mu(A^c \cap B)\right]\\ &= \mu(A \cap B) + \mu(A \cup B).\end{aligned}$$

Property (c) follows straightforwardly from (b). In fact, since $\mu(A \cap B)$ is a positive number,

$$\begin{aligned}\mu(A \cup B) &\leq \mu(A \cup B) + \mu(A \cap B) \\ &= \mu(A) + \mu(B),\end{aligned}$$

In order to prove (d), observe that if $A \subseteq B$, then $A \cap B = A$ and the identity in (a) can be written in the form

$$\mu(B) = \mu(A) + \mu(B \setminus A).$$

Since $\mu(B \setminus A)$ is a positive number, this means that

$$\begin{aligned}\mu(A) &\leq \mu(A) + \mu(B \setminus A) \\ &= \mu(B).\end{aligned}$$

Finally, to prove (e), let us observe that if A is a subset of B, then we have already seen that $\mu(B) = \mu(A) + \mu(B \setminus A)$, and if $\mu(A)$ is finite, then we can subtract this number from each member of the previous identity, obtaining the identity

$$\mu(B \setminus A) = \mu(B) - \mu(A).$$

The theorem is therefore fully proved. □

In light of the set of properties described in Proposition (3.4) which only uses the additivity of the measure μ, it is quite natural to ask at this point why requiring the countable additivity, rather than the mere additivity in the definition of a measure. Is this strongest property really necessary? Is it not sufficient to use additivity to model successfully notions such as length, area, and volume? The answer to these questions is twofold. On one side, the basic idea is that, in order to be able to calculate the area of a complicated region, it is sometimes easier to calculate the 'limit' of a sequence of simpler domains. Now, it is exactly this possibility to take such a limit that requires countable additivity: In addition to that, it is this very property that can be extended to the continuity properties of integrals without which the theory could not be developed nor applied. On the other hand, we shall show later on that the most 'natural' model of the notions of length, area, and volume actually has the property of countable additivity. So, taking everything into account, it seems natural to just accept that a measure must be countably additive.

The first example of this link between continuity and countable additivity is given by the following proposition, which incidentally gives an alternative condition to countable additivity that, sometimes, is easier to prove than the actual one. To this end, let us introduce some useful terminology and notation.

(3.5) Definition. If $(E, \mathcal{E})$ is a measurable space, a sequence (A_n) is said to be **increasing** if, for every index n, one has $A_n \subseteq A_{n+1}$, and, by definition, its **limit** is the union $\bigcup_n A_n$. Similarly, the sequence (A_n) is said to be **decreasing** if, for every index n, one has $A_n \supseteq A_{n+1}$, and, by definition, its **limit** is the intersection $\bigcap_n A_n$. To state that an increasing (resp. decreasing) sequence (A_n) has limit A, instead of the more familiar notation $\lim_n A_n = A$, we shall write shortly $A_n \uparrow A$ (resp. $A_n \downarrow A$).

(3.6) Proposition. *Let $(E, \mathcal{E})$ be a measurable space and suppose that μ is a map from $\mathcal{E}$ to $[0, \infty]$ which satisfies $\mu(\emptyset) = 0$ and which is additive. Then, the following are equivalent:*

(a) *μ is countably additive.*

(b) *If $A_n \in \mathcal{E}$ and $A_n \uparrow A$, then $\mu(A) = \sup_n \mu(A_n) = \lim_n \mu(A_n)$.*

Proof. Let us start by proving (a) $\Rightarrow$ (b). To this end, suppose that μ is countably additive and let (A_n) be an increasing sequence of elements of $\mathcal{E}$. By monotonicity of μ, then $\sup_n \mu(A_n) \leq \mu(A)$. In order to prove that this relationship is, in fact, an identity, we can suppose that the left-hand side is finite. With no loss of generality, we can assume that $A_0 = \emptyset$. Thus, let us set $B_n = A_n \setminus A_{n-1}$ for all $n \geq 1$. Then, (B_n) is a sequence of pairwise disjoint elements of $\mathcal{E}$ and therefore

$$
\begin{aligned}
\mu(A) = \textstyle\sum_{n=1}^{\infty} \mu(B_n) &= \sup_n \textstyle\sum_{i=1}^{n} \mu(B_i) \\
&= \sup_n \textstyle\sum_{i=1}^{n} \big[\mu(A_i) - \mu(A_{i-1})\big] \\
&= \sup_n \big[\mu(A_n) - \mu(A_0)\big] \\
&= \sup_n \mu(A_n).
\end{aligned}
$$

Let us now prove (b) $\Rightarrow$ (a). Assume that condition (b) holds, and let us consider a sequence (A_n) of pairwise disjoint elements of $\mathcal{E}$. Then, the sequence $\big((A_1 \cup \cdots \cup A_n)\big)$ is increasing and

$$
\begin{aligned}
\mu(A) &= \sup_n \mu(A_n \cup \cdots \cup A_n) \\
&= \sup_n \big[\mu(A_1) + \cdots + \mu(A_n)\big] \\
&= \textstyle\sum_{n=1}^{\infty} \mu(A_n).
\end{aligned}
$$

The theorem is thus fully proved. □

Property (b) of Theorem (3.6) may be expressed in words by saying that μ is 'continuous' on every increasing sequence. In symbols, the same property may be summarised as

$$\big[\, A_n \in \mathcal{E} \text{ and } A_n \uparrow A \,\big] \Rightarrow \mu(A_n) \uparrow \mu(A).$$

Another continuity condition holds for decreasing sequences, but only when the measures of each set are finite, as shown in the following proposition.

(3.7) Proposition. *Let μ be a measure on the measurable space $(E, \mathcal{E})$. If $A_n \in \mathcal{E}$, $\mu(A_n) < \infty$ for every index n, and $A_n \downarrow A$, then*

$$\mu(A) = \inf_n \mu(A_n) = \lim_n \mu(A_n).$$

Proof. Since the sequence $(A_1 \setminus A_n)$ is increasing, and since the measure μ is continuous on the increasing sequences (Proposition (3.6)), it follows that

$$\mu(A_1 \setminus A) = \lim_n \mu(A_1 \setminus A_n).$$

Since $\mu(A_n) < \infty$ for every index n (hence $\mu(A) < \infty$), thanks to the property of subtractivity of μ, the previous identity becomes

$$\mu(A_1) - \mu(A) = \mu(A_1) - \lim_n \mu(A_n).$$

This completes the proof. □

The condition stated by this proposition may be expressed in words by saying that μ is continuous on every decreasing sequence of measurable sets with finite measure. In symbols, the same property may be summarised as

$$\big[A_n \in \mathcal{E} \text{ and } \mu(A_n) < \infty \text{ and } A_n \downarrow A\big] \Rightarrow \mu(A_n) \downarrow \mu(A).$$

Another property of measures, which is a consequence of the property of sub-additivity together with the properties of continuity just stated is the following.

(3.8) Proposition (Countable sub-additivity of a measure). *Let $(E, \mathcal{E}, \mu)$ a measure space. For each sequence (A_n) of elements of $\mathcal{E}$, one has*

$$\mu\big(\textstyle\bigcup_{n=1}^{\infty} A_n\big) \le \sum_{n=1}^{\infty} \mu(A_n).$$

Proof. Let us set $B_n = \bigcup_{k=1}^{n} A_k$ for each n. It is clear that (B_n) is an increasing sequence of elements of $\mathcal{E}$, and its union is $\bigcup_{n=1}^{\infty} A_n$. Thanks to the sub-additivity of the measure μ (see Proposition (3.4)(c)) one also has $\mu(B_n) \le \sum_{k=1}^{n} \mu(A_k)$, and therefore, taking the limit as n approaches infinite, we get

$$\mu\big(\textstyle\bigcup_{n=1}^{\infty} A_n\big) = \lim_n \mu(B_n) \le \sum_{n=1}^{\infty} \mu(A_n)$$

which proves the theorem. □

The results about convergence also lead to the following simple proposition, that is an immediate consequence of the definition of limit inferior and limit superior for a sequence of measurable sets. Precisely, the following holds.

(3.9) Proposition. *Let $(E, \mathcal{E}, \mu)$ be a finite measurable space, and (A_n) a sequence of measurable sets. Then,*

$$\mu\big(\liminf_n A_n\big) \le \liminf_n \mu(A_n) \le \limsup_n \mu(A_n) \le \mu\big(\limsup_n A_n\big).$$

Proof. Let (A_n) be a sequence of measurable sets. For every index n, let us set $B_n = \bigcup_{k \geq n} A_n$. It is straightforward to recognise that this is a decreasing sequence and that, by definition, its intersection coincides with the set $\limsup_n A_n$. For every index n and every index k greater than n, one has $A_k \subseteq B_n$, whence $\sup_{k \geq n} \mu(A_k) \leq \mu(B_n)$. Now, taking the infimum on both sides of the latter inequality, and applying Proposition (3.7), yields

$$\limsup_n \mu(A_n) \leq \inf_n \mu(B_n) = \mu\big(\textstyle\bigcap_n B_n\big) = \mu\big(\limsup_n A_n\big).$$

Similarly, let us set $C_n = \bigcap_{k \geq n} A_n$. Likewise, it is straightforward to recognise that this is an increasing sequence and its union coincides with the set $\liminf_n A_n$. For every index n and every index k greater than n, this yields $C_n \leq A_k$, whence $\mu(C_n) \leq \inf_{k \geq n} \mu(A_k)$. Now, taking the supremum on both sides of the latter inequality, and applying Proposition (3.6), yields

$$\mu\big(\liminf_n A_n\big) = \mu\big(\textstyle\bigcup_n C_n\big) = \sup_n \mu(C_n) \leq \liminf_n \mu(A_n),$$

which fully proves the proposition. □

It is now time to examine some examples of measures. Unfortunately, this is not going to be a straightforward task, as σ-fields are generally rather intricate to describe explicitly, making it even more challenging to define a map on a class of sets that cannot be fully described. Ideally, the flow of this discussion would call for the presentation of various physically and geometrically meaningful measures. However, at this stage, we can only introduce them without providing rigorous proofs of their existence. The laborious process of constructing such measures properly cannot be undertaken just yet, as it demands more sophisticated tools from measure theory, which we shall develop in due course. For now, the only examples we can explore in full detail are, admittedly, quite simple — if not downright trivial, though they are absolutely fundamental. The reader ought not to underestimate their significance, as they shall form the foundation of many crucial theoretical and practical problems.

(3.10) Example. Let $(E, \mathcal{E})$ be a measurable space, and x an element of E. For each element A of $\mathcal{E}$ set $\varepsilon_x(A) = 1$ if $x \in A$, and $\varepsilon_x(A) = 0$ if $x \notin A$. In other words, let us set $\varepsilon_x(A) = I_A(x)$, where I_A denotes the indicator function of the set A (see Example (2.8)). The measure whereby constructed is called the **Dirac measure** on x and it models a unit point mass.

In order to generalise the previous example, we shall use the following proposition, which incidentally shows how the set of all measures on the measurable space $(E, \mathcal{E})$ is a cone.

(3.11) Proposition. *Every linear combination with positive coefficients of measures defined on a measurable space is a measure on the same measurable* *space. In other words, if (μ_n) is a sequence of measures on a given measurable space $(E, \mathcal{E})$, and if (α_n) is a sequence of positive real numbers, the set function μ on $\mathcal{E}$ defined by*

$$\mu(A) = \sum_n \alpha_n \mu_n(A) \qquad \text{for every } A \in \mathcal{E},$$

is a measure on $(E, \mathcal{E})$, which we shall be denoted by $\mu = \sum_n \alpha_n \mu_n$.

Proof. Let us start by noting that, if μ is a measure on the measurable space $(E, \mathcal{E})$ and if α is a positive real number, the map ν defined by $\nu(A) = \alpha\mu(A)$ for each $A \in \mathcal{E}$ is a measure. Thus, with no loss of generality, we can assume that each of the coefficients α_n is equal to 1. Hence, let us consider the map $\mu = \sum_n \mu_n$ and let us prove that this is, in fact, a measure. It is obvious that $\mu(\emptyset) = 0$. It suffices to prove that μ is countably additive. To this end, let (A_k) be a sequence of pairwise disjoint measurable sets and let us denote by A their union. Then,

$$\begin{aligned} \mu(A) &= \sum_n \mu_n(A) \\ &= \sum_n \sum_k \mu_n(A_k) \\ &= \sum_k \sum_n \mu_n(A_k) = \sum_k \mu(A_k) \end{aligned}$$

which yields that μ is a measure on $(E, \mathcal{E})$. □

Let us observe that, as a consequence of the previous proposition, whenever a measure is finite, it can be transformed into a probability measure. To see this, if μ is a finite measure on the measurable space $(E, \mathcal{E})$, then the measure $\nu = (1/\mu(E)) \cdot \mu$ is a probability measure on $(E, \mathcal{E})$. This is called the **normalised** measure associated to μ.

(3.12) Example. Generalising Example (3.10), let (x_n) be a sequence of elements of E. Then, for every sequence (α_n) of positive real numbers, the set function $\mu = \sum_n \alpha_n \varepsilon_{x_n}$ is a measure on $(E, \mathcal{E})$. This measure is finite if the series $\sum_n \alpha_n$ is convergent, and it is σ-finite otherwise. In particular, if α is a positive real number, a measure of the form $\mu = \alpha\varepsilon_x$ is called a **point measure** on $(E, \mathcal{E})$.

(3.13) Example. Let E be a non-empty set. On the σ-field $\mathcal{E} = \mathcal{P}(E)$ of all subsets of E, let us denote by $\kappa(A)$ the cardinality of A, if A is finite, and $\kappa(A) = \infty$ otherwise. It is not difficult to show that κ is a measure on the measurable space $(E, \mathcal{P}(E))$. This is called the **counting measure**. The measure κ is generally not σ-finite. If E is countable, however, the counting measure can be thought as a particular type point measure, since it can be written as $\kappa = \sum_{x \in E} \varepsilon_x$, and it is σ-finite.

(3.14) Example (Uniform probability). If E is a finite set, let us denote by n the number of its elements. On the σ-field $\mathcal{P}(E)$ of all subsets of E, the measure $\mu = n^{-1}\kappa = n^{-1}\sum_{x\in E}\varepsilon_x$, which is the normalised measure associated to the counting measure on the measurable space $(E, \mathcal{P}(E))$, is called the **uniform probability** measure on E. Indeed, this is a probability measure, and its value on a set A with k elements is $\mu(A) = k/n$. In other words, the uniform probability is the probability measure that is described by the popular statement 'the ratio between the number of favourable outcomes and the total number of outcomes'.

(3.15) Example (Borel-Lebesgue measure). On the set $\mathbb{R}$ of all real numbers, considered as a topological space with its usual topology, and also considered as a measurable space with the corresponding Borel σ-field, the **Borel-Lebesgue measure** is the measure λ^1 on the σ-field $\mathcal{B}(\mathbb{R})$ that to every interval associates its own length, i.e. the measure defined by

$$\lambda^1([a,b]) = b - a \qquad \text{for every } a, b \in \mathbb{R} \text{ with } a < b.$$

More generally, the Borel-Lebesgue measure in $\mathbb{R}^n$ is the measure, on the Borel σ-field of $\mathbb{R}^n$, which associates every multi-interval $Q = \prod_{i=1}^{n}[a_i, b_i]$ with its volume $\lambda^n(Q) = \prod_{i=1}^{n}(b_i - a_i)$.

We shall not prove yet the existence of the Borel-Lebesgue measure, putting it off until we shall have developed a general criterion to establish the existence and uniqueness of measures. For the time being, however, we shall assume the existence of such a measure and notice how this gives a mathematical model for the notion of the length of a segment, the area of a rectangle and the volume of a cuboid in the Euclidean space. We shall also prove that this measure has other features in common with the measures of length, area, and volume, before proving its existence.

4. Atoms, purely atomic measures, diffuse measures

(4.1) Definition. Let $(E, \mathcal{E}, \mu)$ be a measure space. An element A of $\mathcal{E}$ is said to be an **atom** of μ if:

(a) $\mu(A) > 0$;

(b) For every B in $\mathcal{E}$, if $B \subset A$ and $\mu(B) < \mu(A)$, then $\mu(B) = 0$.

A measure with no atoms is said to be **non-atomic** or **atomless**. This means that for every measurable set A, with $\mu(A) > 0$, there always is a partition made of two non-negligible subsets of A.

Observe immediately that for a measurable set of positive finite measure to be an atom it is sufficient that its only measurable subsets be itself and $\emptyset$. This condition however is evidently not necessary.

To show an important example of atoms of a measure, let us assume that, on the measurable space $(E, \mathcal{E})$, every singleton $\{x\}$, with $x \in E$, belongs to $\mathcal{E}$. If μ is a measure on such a space, every singleton $\{x\}$ with $x \in E$ such that $\mu\{x\} > 0$ is an atom of μ. If, on the other hand, every singleton $\{x\}$ is μ-negligible, the measure μ shall be called **diffuse**. A diffuse measure vanishes on the singletons, but it may have other atoms, nonetheless. For example, if E is an uncountable set, let us denote by $\mathcal{E}$ the σ-field of all subsets A of E such that either A or A^c is countable. On the measurable space $(E, \mathcal{E})$, define the measure μ by setting $\mu(A) = 0$ if A is countable and $\mu(A) = 1$ if A^c is countable. Then every uncountable subset A of E is an atom of μ, but the measure is diffuse, as it vanishes on every countable set, and in particular on all sets with one element only.

For a general measure, the set of atoms may be uninteresting and even cause some technical difficulties, unless there are some additional properties that limit the cardinality of the atoms. An example of this is when a measure is σ-finite, as shown in the following lemma.

(4.2) Lemma. *On a measurable space $(E, \mathcal{E})$, let μ be a σ-finite measure. Then the set $\mathcal{C}(\mu)$ of all pairwise disjoint atoms of μ is countable.*

Proof. Let us start by assuming that μ is finite. Then, for any positive integer n, let us consider the subset $\mathcal{F}_n$ of $\mathcal{E}$ of all atoms A with $\mu(A) > 1/n$. The set $\mathcal{F}_n$ may be finite or infinite. Let us suppose that it has at least k elements, say $A_1, \dots, A_k$. Then, for these elements, $1 \le n\mu(A_i)$. Thus,

$$\begin{aligned} k &\le \textstyle\sum_{i=1}^{k} n\mu(A_i) \\ &= n\mu(A_1 \cup \dots \cup A_n) \\ &\le n\mu(E). \end{aligned}$$

It is thus proved that, for every index n, the collection $\mathcal{F}_n$ is finite. Now, since the union of all $\mathcal{F}_n$, when n runs over all positive integers, coincides with the set $\mathcal{C}(\mu)$ of all atoms of μ, it is proved that the set $\mathcal{C}(\mu)$ is countable. Let us now extend this to the case when μ is σ-finite. To this end, it suffices to consider an increasing sequence (E_n) of measurable subsets of E with finite measure; let us denote by μ_n the trace of μ to E_n, that is to say, the measure defined for every $A \in \mathcal{E}$ by $\mu_n(A) = \mu(A \cap E_n)$. Since each of the μ_n is finite, if $\mathcal{C}(\mu)$ denotes the set of all pairwise disjoint atoms of μ, then each of the sets $\mathcal{C}_n = \mathcal{C}(\mu_n)$ must be countable, hence the set $\mathcal{C}(\mu) = \bigcup_n \mathcal{C}_n$ is also countable. □

Note that, if μ is not σ-finite, then the set $\mathcal{C}(\mu)$ may have any cardinality. For instance, if E is an uncountable set, $\mathcal{E} = \mathcal{P}(E)$, and if μ is the measure defined by $\mu(\emptyset) = 0$ and $\mu(A) = \infty$ for any non-empty subset A of E, it is clear that every singleton $\{x\}$, with $x \in E$, is an atom, and therefore the class $\{\{x\} : x \in E\}$ is included in $\mathcal{C}(\mu)$, which is thereby uncountable.

A measure μ on a measurable space $(E, \mathcal{E})$ is called **purely atomic** if every measurable set of positive measure contains an atom. In light of Lemma (4.2), if μ is a purely atomic σ-finite measure on a measurable space $(E, \mathcal{E})$, then for every measurable set A, the measure $\mu(A)$ can be written as the sum of all measures of the form $\mu(A \cap C)$ with $C \in \mathcal{C}(\mu)$, which is a countable set. In particular, this also shows that, if D is a discrete subset of E, the measure $\mu = \sum_{x \in D} a_x \varepsilon_x$ is a purely atomic measure. This is also arguably the most studied type of purely atomic measure. Furthermore, when a measure μ is σ-finite, there is an interesting decomposition into non-atomic and purely atomic measures. Precisely, the following holds.

(4.3) Theorem. *Let μ be a σ-finite measure on a measurable space $(E, \mathcal{E})$. Then there exists measures λ and ν on $\mathcal{E}$ such that $\mu = \lambda + \nu$, where λ is atomless and ν is purely atomic.*

Proof. Let us denote by $\mathcal{C}$ the class of all atoms of μ. By Lemma (4.2), this is a countable set and therefore the union of all its elements is measurable. Thus, denote by D such union, and for every A in $\mathcal{E}$, set

$$\lambda(A) = \mu(A \cap D^c), \qquad \nu(A) = \mu(A \cap D) = \textstyle\sum_{C \in \mathcal{C}} \mu(A \cap C).$$

It is clear that λ and ν are two measures on $(E, \mathcal{E})$, for they are the trace of μ onto two measurable sets.

Let us prove that λ is atomless and ν is purely atomic. In fact, any atom of λ would also be an atom of μ, but all atoms of μ are included in D. Now, the measure λ vanishes on D, hence λ cannot have any atom. Similarly, if A is a measurable set of E with $\nu(A) > 0$, there must be at least an element C of $\mathcal{C}$, such that $\mu(A \cap C) > 0$. Now, if $\mu(A \cap C) < \mu(C)$, then it would be $\mu(A \cap C) = 0$, because C is an atom of μ. Hence, $\mu(A \cap C) = \mu(C)$, and therefore $A \cap C$ is an atom of ν included in A, and this proves that ν is purely atomic. In order to complete the proof, it now suffices to observe that, by definition, one has $\lambda(A) + \nu(A) = \mu(A \cap D^c) + \mu(A \cap D) = \mu(A)$. □

Let us now focus on atomless measures. This type of measures has quite a few interesting properties. The most important of these properties is stated in the next theorem.

(4.4) Theorem. *Let μ be a finite atomless measure on the measurable space $(E, \mathcal{E})$, and let H be any measurable set. Then, for every real number $a \in [0, \mu(H)]$, there exists a set A in $\mathcal{E}$ with $\mu(A) = a$.*

In order to prove this theorem it is useful to prove the following lemmas.

(4.5) Lemma. *On a measurable space $(E, \mathcal{E})$, let μ be a finite atomless measure. Then, for every set B in $\mathcal{E}$, with $\mu(B) > 0$, and every positive real number ε, there exists a set A in $\mathcal{E}$ with $A \subset B$, and $0 < \mu(A) < \varepsilon$.*

Proof. Given a measurable set B with $\mu(B) > 0$, and given a partition of B with two non-negligible measurable subsets of B, at least one of the two subsets must have a measure which does not exceed a half of the measure of the given measurable set B. Applying this remark, it is easy to see that, for a given non-negligible set B, it is possible to find a decreasing sequence (A_n) of non-negligible measurable sets, with

$$A_0 = B, \qquad \mu(A_n) \leq \tfrac{1}{2}\mu(A_{n-1}) \qquad \text{for } n \geq 1.$$

Then, for every n, $\mu(A_n) \leq \left(\frac{1}{2}\right)^n \mu(B)$, and this proves the lemma. □

(4.6) Lemma. *Let μ be a finite measure on a measurable space $(E, \mathcal{E})$. For any measurable set H and any positive real number a, let us set:*

$$\mathcal{H} = \Big\{A \in \mathcal{E} : A \subseteq H,\, \mu(A) \leq a\Big\}.$$

Then, there exists a maximal element A in $\mathcal{H}$ such that for every element B of $\mathcal{H}$ containing A, one has $\mu(A) = \mu(B)$.

Proof. By induction, it is possible to find an increasing sequence (A_n) of elements of $\mathcal{H}$ such that

$$A_0 = \emptyset, \qquad \sup\Big\{\mu(B) : A_{n-1} \subseteq B \in \mathcal{H}\Big\} \leq \mu(A_n) + \tfrac{1}{n} \qquad \text{for } n \geq 1.$$

Let us set $A = \bigcup_n A_n$. Then A is an element of $\mathcal{H}$. Furthermore, A has the required property of maximality. In fact, if B is an element of $\mathcal{H}$ which contains the set A (hence each of the sets A_n) then thanks to the relation

$$\mu(B) \leq \mu(A) + \tfrac{1}{n} \qquad \text{for } n \geq 1,$$

one has $\mu(B) \leq \mu(A)$ and therefore $\mu(B) = \mu(A)$. □

Using the two lemmas just proved, it is now easy to prove the theorem.

Proof of Theorem (4.4). Suppose that the measure μ is atomless, and given a measurable set H with $\mu(H) < \infty$ and a positive real number a with $a < \mu(E)$, it suffices to prove that there exists a measurable set A, included in H, such that $\mu(A) = a$. To this end, let us define a class $\mathcal{H}$ as in Lemma (4.6) and let us denote by A an element of $\mathcal{H}$ with the maximality property described in the statement of the lemma. It suffices to prove that $\mu(A) = a$. To do so, let us argue by contradiction. Indeed, if it were $\mu(A) < a$, it would be possible, by virtue of Lemma (4.5), to find a measurable set D with $\mu(D) > 0$, included in $H \setminus A$, and with such a small measure to give $\mu(A \cup D) \leq a$. Thus, the set $A \cup D$ is an element of $\mathcal{H}$, is included in A, but whose measure is greater than the measure of A, which is a contradiction. □

(4.7) Corollary. *On the measurable space $(E, \mathcal{E})$, let λ be a finite atomless measure. Then, for every positive real number δ, there exists a finite partition of E, composed of measurable sets whose measures under λ are all strictly smaller than δ.*

Proof. With no loss of generality, we may assume that $\lambda(E) = 1$. Applying Theorem (4.4), it is possible to construct, by induction, a sequence $(\mathcal{P}_n)$ of partitions of E, with a strictly increasing number of elements, and such that the measure of each element of $\mathcal{P}_n$ under λ is equal to 2^{-n}. Thus, taking n sufficiently large, the partition $\mathcal{P}_n$ has the required properties. □

(4.8) Corollary. *Let λ be an arbitrary finite positive measure on the measurable space $(E, \mathcal{E})$. Then, for every strictly positive real number δ, there exists a finite partition of E, of measurable sets, each of which is, under λ, either an atom or a set with measure strictly smaller than δ.*

Proof. With no loss of generality, we may assume that $\lambda(E) = 1$. In addition, thanks to the previous corollary, we may also assume that λ is not atomless. Let us observe that, in these hypotheses, two atoms of λ which are not equivalent are *almost disjoint* (meaning that their intersection is negligible). Hence, for every strictly positive real number n, there cannot be more than n pairwise disjoint atoms whose measures are all greater than $1/n$. As a result, the set of all the equivalence classes modulo λ of the atoms of λ is countable. Thus, there exists a countable family $(A_i)_{i \in I}$ of pairwise disjoint atoms, such that, denoting by E_0 the complement of their union, and by $\mathcal{E}_0$ the trace of $\mathcal{E}$ on E_0, the trace of λ to $\mathcal{E}_0$ is an atomless measure. Thanks to Corollary (4.7), if the set E_0 is not empty, there exists a finite partition $\mathcal{P}_0$ of E_0 composed of measurable sets with measure, under λ, strictly smaller than δ. If the set I is finite, then a partition of E that satisfies the required properties is obtained by taking the finite family $(A_i)_{i \in I}$ of atoms of E and (if E_0 is not empty) the elements of $\mathcal{P}_0$. If, on the other hand, the set I is infinite, then there exists a finite subset J of I, with $\sum_{i \notin J} \lambda(A_i) < \delta$, and a partition with the required properties is composed of the finite family $(A_i)_{i \in J}$, the set $\bigcup_{i \in I \setminus J} A_i$ and (if the set E_0 is not empty) the elements of $\mathcal{P}_0$. □

5. Dynkin's π-λ theorem

In this section, we shall develop some abstract tools which are going to be particularly useful to construct and handle σ-fields, and that shall be very helpful in providing effective conditions of measurability. The main goal of this section consists in proving the so-called Dynkin's π-λ Theorem, whose applications will lead to some very interesting results about σ-fields.

To this end, let us start by introducing the notions of π- and λ-systems, the latter sometimes also called a 'Dynkin system', after the Soviet/American probabilist Eugene Dynkin (Евгений Борисович Дынкин, 1924-2014), who first introduced them and used them to prove the π-λ Theorem. Roughly speaking, the notions of π- and λ-systems are somewhat similar to the definition of σ-field, but, in the end, they turn out to be much more manageable.

(5.1) Definition. Let E be a non-empty set. A non-empty class $\mathcal{P}$ of parts of E is called a **π-system** if it is closed under finite intersection. In other words, $\mathcal{P}$ is a π-system if, from $A, B \in \mathcal{P}$ follows $A \cap B \in \mathcal{P}$.

(5.2) Definition. Let E be a non-empty set. A class $\mathcal{L}$ of subsets of E is called a **λ-system** if the following conditions hold:

(λ_1) E is an element of $\mathcal{L}$,
(λ_2) For every $A, B \in \mathcal{L}$, with $A \subseteq B$, also $B \setminus A \in \mathcal{L}$,
(λ_3) The union of an increasing sequence of elements of $\mathcal{L}$ lies in $\mathcal{L}$.

Note immediately that from the hypotheses of (5.2), the following also hold.

(λ_4) $A \in \mathcal{L} \Rightarrow A^c \in \mathcal{L}$
(λ_5) The intersection of a decreasing sequence of elements of $\mathcal{L}$ lies in $\mathcal{L}$.

To prove this, let us start by observing that properties (λ_1) and (λ_2) obviously entail (λ_4), whence a λ-system is also stable under complementation; likewise, from (λ_4) and (λ_3) it follows that the intersection of a decreasing family of elements of $\mathcal{L}$ is an element of $\mathcal{L}$, whence (λ_5) also holds. We shall say, briefly, that *a λ-system is closed under monotone union and intersection.*

A π-system, instead, is a much simpler structure, which only requires to be stable under finite intersection. In other words, $\mathcal{P}$ is a π-system if and only if $\mathcal{P} = \mathcal{P}_d$ (see Notation (1.7)). Moreover, if $\mathcal{I}$ is a non-empty class of sets, the set $\mathcal{I}_d$ of all finite intersections of elements of $\mathcal{I}$ is manifestly a π-system, and it is not difficult to check that it is included in every π-system which contains $\mathcal{I}$. For this reason, it shall be called the π-system **generated** by $\mathcal{I}$.

A similar argument can be used for λ-systems. Precisely, if $\mathcal{J}$ is a non-empty class of subsets of E, there always exists the smallest (with respect to inclusion) λ-system which contains $\mathcal{J}$, that is to say a λ-system that contains $\mathcal{J}$ and that is included in every other λ-system that contains $\mathcal{J}$. It suffices to take the intersection of all λ-systems that contain $\mathcal{J}$, and observe that this is a λ-system itself. Consistently, this shall be called it the λ-system **generated** by $\mathcal{J}$, and shall be denoted by $\lambda(\mathcal{J})$.

Observe as well that $\lambda(\mathcal{J}) \subseteq \sigma(\mathcal{J})$, for a σ-field is, in particular, a λ-system. Furthermore, it is easy to check that, if $\mathcal{I} \subseteq \mathcal{J}$, then $\lambda(\mathcal{I}) \subseteq \lambda(\mathcal{J})$.

As we have already mentioned, every σ-field is in particular a λ-system; the reciprocal, however, is generally false. Nevertheless, if a λ-system is also a π-system, i.e. if the λ-system is also stable under finite intersection, then it is indeed a σ-field. This is the content of the next proposition.

(5.3) Proposition. *A class $\mathcal{L}$ of parts of a non-empty set E that is as the same time a π-system and a λ-system is a σ-field.*

Proof. The class $\mathcal{L}$ contains E by (λ_1), and, by (λ_4), it is closed under complementation. Since, by hypothesis, $\mathcal{L}$ is stable under finite intersection, it is also stable under finite union, whence a field (see Definition (1.1)). It remains to prove that $\mathcal{L}$ is also stable under countable union. To this end, let (A_n) be a sequence of elements of $\mathcal{L}$, and let, for every index n, let us set $B_n = A_1 \cup \cdots \cup A_n$. Then, (B_n) is an increasing sequence of elements of $\mathcal{L}$ whose union $\bigcup_n B_n = \bigcup_n A_n$ lies in $\mathcal{L}$ by (λ_3). Thus, $\mathcal{L}$ is a σ-field. □

Finally, we can now state and prove Dynkin's π-λ Theorem.

(5.4) Theorem (Dynkin's π-λ Theorem). *If $\mathcal{L}$ is a λ-system, and $\mathcal{P}$ is a π-system contained in $\mathcal{L}$, then the σ-field generated by $\mathcal{P}$ is contained in $\mathcal{L}$. In other words, if $\mathcal{P} \subseteq \mathcal{L}$, then $\sigma(\mathcal{P}) \subseteq \mathcal{L}$. In particular, $\lambda(\mathcal{P}) = \sigma(\mathcal{P})$.*

Proof. Let us denote by $\mathcal{S}$ the λ-system generated by $\mathcal{P}$, that is to say the intersection of all λ-systems containing $\mathcal{P}$. Since $\mathcal{P} \subseteq \mathcal{L}$, then $\mathcal{P} \subseteq \mathcal{S} \subseteq \mathcal{L}$. Let us show that $\mathcal{S}$ is also a π-system. To this end, let us start by observing that, if $A \in \mathcal{P}$, then the class $\{X \in \mathcal{L} : A \cap X \in \mathcal{S}\}$ is a λ-system which contains $\mathcal{P}$, whence $\mathcal{S}$. Thus, if $B \in \mathcal{S}$, the class $\{X \in \mathcal{L} : X \cap B \in \mathcal{S}\}$ is a λ-system which contains $\mathcal{P}$, whence $\mathcal{S}$; and this ultimately proves that $\mathcal{S}$ is stable under binary intersection, hence a π-system.

So, $\mathcal{S}$ is both a π- and a λ-system, whence a σ-field, by Proposition (5.3). Since $\mathcal{S}$ is a σ-field and since $\mathcal{P} \subseteq \mathcal{S}$, it follows that $\sigma(\mathcal{P}) \subseteq \mathcal{S} \subseteq \mathcal{L}$, and the theorem is fully proved. □

Let us immediately see an important application of Dynkin's π-λ Theorem: A criterion that helps to establish when two measures coincide, checking if they agree on a particular set of generators. To this end, let us give the following definition.

(5.5) Definition. Let $(E, \mathcal{E})$ be a measurable space, and let $\mathcal{I}$ be a set of generators for the σ-field $\mathcal{E}$. We shall say that $\mathcal{I}$ is a **base** if the following conditions hold: (a) $\mathcal{I}$ is a π-system, that is to say, is stable under finite intersection, and (b) the whole space E is the union of a countable family of elements of $\mathcal{I}$.

We are now ready to prove the following important theorem, which, adopting the perhaps slightly pompous, but indeed well-earned, terminology introduced by G. Letta, shall hereinafter be referred to as the 'Fundamental criterion for the coincidence of two measures'.

(5.6) Theorem (Fundamental criterion for the coincidence of two meas's). *On a measurable space $(E, \mathcal{E})$, let μ and ν be two measures. Suppose also that $\mathcal{I}$ is a base for the σ-field $\mathcal{E}$, such that $\mu(A) = \nu(A) < \infty$ for every element A of $\mathcal{I}$. Then, the two measures μ and ν are identical.*

Proof. With no loss of generality, we can suppose that the two measures are finite. In fact, if the theorem is proved in this special case, we can always take a sequence (E_n) of elements of $\mathcal{I}$ whose union is E and, with no loss of generality, we can suppose that this sequence is made of pairwise disjoint elements. For every index n, we shall also denote by μ_n the trace of μ on E_n and with ν_n the trace of ν on E_n. By hypothesis, the two measures μ_n and ν_n are identical for each n and therefore $\mu = \sum_n \mu_n = \sum_n \nu_n = \nu$.

It remains to prove the theorem in the special case when the two measures μ and ν are finite. To this end, let us observe that, by virtue of Proposition (3.4)(e) and Proposition (3.6), the collection

$$\Big\{A \in \mathcal{E} : \mu(A) = \nu(A)\Big\}$$

is manifestly a λ-system, and it contains $\mathcal{I}$, which is a π-system. Since this class is a λ-system, it contains the λ-system generated by $\mathcal{I}$ that, by virtue of Dynkin's π-λ Theorem, coincides with the σ-field generated by $\mathcal{I}$, which is $\mathcal{E}$. This completes the proof. □

(5.7) Example. Let us use this criterion to show that the Lebesgue measure is invariant under translations. To this end, let us start by observing that the class $\mathcal{Q}$ of all the products of intervals of the form $[a, b]$, with $a < b$, is a set of generators for the Borel σ-field of $\mathbb{R}^d$, it is stable under intersection, and the space $\mathbb{R}^d$ is union of a sequence of its elements, hence it is a base for the Borel σ-field of $\mathbb{R}^d$. Now, for any subset A of $\mathbb{R}^d$, the translation of A by a vector x is

$$A + x = \Big\{y + x : y \in A\Big\}.$$

As a consequence, with the expression 'λ is invariant under translation' one means that $\lambda(A) = \lambda(A + x)$ for every Borel set A in $\mathbb{R}^d$. To show this, let us consider the two measures

$$A \mapsto \lambda(A) \quad \text{and} \quad A \mapsto \lambda(A + x).$$

These two measures clearly coincide on $\mathcal{Q}$ and therefore, by means of the fundamental criterion for the coincidence of two measures, these two measures coincide for any Borel set A.

6. The monotone class theorem

In measure theory λ- and π-systems are sometimes avoided, and similar theorems, involving functions rather than sets, are sometimes preferred. For this reason, we shall now introduce the most common of these alternative theorems, which is referred to as the 'Monotone Class Theorem'. In order to state this theorem, however, it becomes necessary to introduce some new terminology. Let us start with the following definition.

(6.1) Definition. A vector space $\mathcal{R}$ of real functions defined on a non-empty set E is called a **Riesz space** (or a **vector lattice**) if it is stable for the 'lattice operations' $(f,g) \mapsto f \wedge g$ and $(f,g) \mapsto f \vee g$.

Let us also set

$$f^+ = f \vee 0, \qquad f^- = -f \wedge 0.$$

Then, the following equalities hold:

$$f = f^+ - f^-, \qquad |f| = f^+ + f^-,$$
$$f \vee g = f + (g-f)^+ = \tfrac{1}{2}\big(f + g + |g - f|\big).$$
$$f \wedge g = -\big[(-f) \vee (-g)\big], \qquad f^- = (-f)^+.$$

Using these relations, it turns out that a vector subspace $\mathcal{R}$ of $\mathbb{R}^E$ is a Riesz space if it is stable under any of the following operations:

$$f \mapsto |f|, \qquad f \mapsto f^+, \qquad f \mapsto f^-.$$

Let us observe that the vector space $\mathcal{M}(E,\mathcal{E})$ of all measurable functions on a measurable space $(E,\mathcal{E})$ is a Riesz space (Proposition (2.9)) and such is also the space $\mathcal{S}(E,\mathcal{E})$ of all simple functions on E.

(6.2) Definition. A subset $\mathcal{M}$ of $\mathbb{R}^E$ is a **monotone class** if it is stable under 'dominated monotone convergence of sequences', which means that the two following properties hold:

(a) For every increasing sequence (f_n) of elements of $\mathcal{M}$, all bounded from above by the same element of $\mathcal{M}$, one has $\lim_n f_n = \sup_n f_n \in \mathcal{M}$.

(b) For every decreasing sequence (f_n) of elements of $\mathcal{M}$, all bounded from below by the same element of $\mathcal{M}$, one has $\lim_n f_n = \inf_n f_n \in \mathcal{M}$.

If $\mathcal{M}$ is a Riesz space and a monotone class, $\mathcal{M}$ is called a **monotone Riesz space**. Further, given a class $\mathcal{H}$ of real functions defined on E, the intersection of all monotone classes containing $\mathcal{H}$ shall be called the monotone class **generated** by $\mathcal{H}$ and shall be denoted by $\mathcal{M}(\mathcal{H})$. This is the 'smallest' (with respect to inclusion) among all monotone classes containing $\mathcal{H}$.

If $\mathcal{M}$ is the monotone class generated by a subset $\mathcal{H}$ of $\mathbb{R}^E$, every element of $\mathcal{M}$ is bounded between two elements of $\mathcal{H}$. In fact, the class of all functions satisfying this property is monotone and contains $\mathcal{H}$, hence $\mathcal{M}$. More generally, the following theorem shows that if $\mathcal{H}$ is a Riesz space, such is $\mathcal{M}$.

(6.3) Theorem. *If a subset $\mathcal{H}$ of $\mathbb{R}^E$ is a Riesz space, then such is the monotone class $\mathcal{M}$ generated by $\mathcal{H}$.*

Proof. Let $\mathcal{H}$ be a Riesz space and let us prove that $\mathcal{M}$ is a Riesz space. In order to do this, take a real number α, and let us prove that $\mathcal{M}$ is stable for the operation $f \mapsto \alpha f$. To this end, it suffices to observe that the class

$$\{f \in \mathbb{R}^E : \alpha f \in \mathcal{M}\}$$

is a monotone class which contains $\mathcal{H}$, hence $\mathcal{M}$.

Now, for each element f of $\mathcal{H}$, the monotone class $\mathcal{M}$ is stable for the operation $g \mapsto f + g$. To prove this, it suffices to note that the class

$$\{g \in \mathbb{R}^E : f + g \in \mathcal{M}\}$$

is a monotone class which contains $\mathcal{H}$, hence $\mathcal{M}$. Finally, using this result and repeating the same argument, it is easy to prove that, for every element g of $\mathcal{M}$, the class $\mathcal{M}$ is stable for the operation $f \mapsto f + g$, which proves that $\mathcal{M}$ is a vector space. To see that $\mathcal{M}$ is a Riesz space, i.e. that it is stable for the lattice operations, it suffices to show, with the same argument used twice before, that $\mathcal{M}$ is stable for the operation $f \mapsto f^+$. □

It is now possible to prove the famous 'Monotone Class Theorem'.

(6.4) Theorem (The Monotone Class Theorem). *Let $(E, \mathcal{E})$ be an assigned measurable space, and let $\mathcal{H}$ be a collection of real functions on E containing the constant 1, stable under the operation of multiplication and containing the indicator functions of all the elements of a set of generators for $\mathcal{E}$. Then every monotone Riesz space $\mathcal{L}$ containing $\mathcal{H}$ also contains all bounded measurable functions on $(E, \mathcal{E})$.*

Proof. Let us denote by $\mathcal{I}$ a set of generators for the σ-field $\mathcal{E}$, and let us suppose that $\mathcal{H}$ contains the indicator functions of every element of $\mathcal{I}$. Since $\mathcal{H}$ is stable under the operation of multiplication, this latter class contains also the indicator functions of all the elements of the class $\mathcal{I}_d$ of all finite intersections of elements of $\mathcal{I}$. Thus, if $\mathcal{L}$ is a monotone Riesz space, let us set:

$$\mathcal{M} = \Big\{A \in \mathcal{E} : I_A \in \mathcal{L}\Big\}.$$

Since $1 \in \mathcal{H} \subseteq \mathcal{L}$, we have $E \in \mathcal{M}$. In addition, since $\mathcal{L}$ is a vector space, if A and B are any two elements of $\mathcal{M}$, with $A \subseteq B$, then $B \setminus A \in \mathcal{M}$. Further, since $\mathcal{L}$ is stable under dominated monotone convergence of increasing sequences, the union of increasing sequence of elements of $\mathcal{M}$ is an element of $\mathcal{M}$. As a result $\mathcal{M}$ is a λ-system which contains the class $\mathcal{I}_d$ of all finite intersections of elements of $\mathcal{I}$, which on its hand is a π-system. Thus, by Dynkin's π-λ Theorem, the λ-system generated by $\mathcal{I}_d$ coincides with the σ-field generated by $\mathcal{I}$, which is $\mathcal{E}$.

Now, since $\mathcal{L}$ is a vector space which contains the indicator functions of all elements of the σ-field $\mathcal{E}$, it also contains all simple functions, and since $\mathcal{L}$ is stable under dominated monotone convergence it also contains all positive measurable functions on $(E, \mathcal{E})$, by means of Proposition (2.12), and again, since $\mathcal{L}$ is a vector space, from the decomposition $f = f^+ - f^-$, it turns out that it contains all measurable functions on $(E, \mathcal{E})$. □

A very simple consequence of the monotone class theorem is the following.

(6.5) Proposition. *On the measurable space $(E, \mathcal{E})$, let $\mathcal{I}$ be a π-system which contains the whole space E and generates the σ-field $\mathcal{E}$. Let $\mathcal{L}$ be a monotone vector space of $\mathbb{R}^E$, containing the class $\{I_A : A \in \mathcal{I}\}$. Then $\mathcal{L}$ contains every bounded real measurable function on $(E, \mathcal{E})$.*

Proof. It suffices to note that $\{I_A : A \in \mathcal{I}\}$ is a class of bounded measurable functions on E which generates the σ-field $\mathcal{E}$, is stable under multiplication and contains the constant 1. The conclusion follows therefore by virtue of the monotone class theorem. □

In spite of being a simple statement (and a simple proof), the previous theorem is not particularly useful because hardly ever does a π-system contain the whole space. At least, this hardly ever happens with useful π-systems. In order to show a similar result that eliminates this hypothesis, let us begin by proving a lemma. To this end, let us first give a definition.

(6.6) Definition. If $\mathcal{R}$ is a Riesz space of real functions on a non-empty set E, we shall say that it is endowed with **Stone's property** if, for every element f of $\mathcal{R}$, we have $f \wedge 1 \in \mathcal{R}$.

Thanks to the relation $f \wedge 1 = (f^+ \wedge 1) - f^-$, in order for the Riesz space $\mathcal{R}$ to be endowed with Stone's property it suffices that the relation $f \wedge 1 \in \mathcal{R}$ holds for every element f of $\mathcal{R}_+$, the cone of all positive elements of $\mathcal{R}$. It is also easy to check that, if $\mathcal{R}$ is endowed with Stone's property, so does the monotone class generated by $\mathcal{R}$.

(6.7) Lemma. *If $\mathcal{I}$ is a π-system of subsets of E, and if $\mathcal{R}$ denotes the vector space generated by the class $\{I_A : A \in \mathcal{I}\}$, then $\mathcal{R}$ is a Riesz space endowed with Stone's property.*

Proof. Let us start by observing that, since the class $\{I_A : A \in \mathcal{I}\}$ is stable under multiplication, the vector space $\mathcal{R}$ is in fact an algebra. Furthermore, the class $\mathcal{I}$ is contained into the class $\mathcal{J}$ defined by

$$\mathcal{J} = \{A \in \mathcal{P}(E) : I_A \in \mathcal{R}\}.$$

To complete, we must prove that, for every element f of $\mathcal{R}$, the functions f^+ and $f \wedge 1$ also belong to $\mathcal{R}$. To show this, it suffices to write f in the form

$$f = \textstyle\sum_{i \in J} a_i I_{A_i},$$

where $(A_i)_{i \in J}$ is a finite family of *pairwise disjoint* elements of $\mathcal{J}$. In fact, if f can be written in this form, one has

$$f^+ = \textstyle\sum_{i \in J} a_i^+ I_{A_i} \in \mathcal{R}, \qquad f \wedge 1 = \textstyle\sum_{i \in J} (a_i \wedge 1) I_{A_i} \in \mathcal{R}.$$

In order to write the function f in this form, let us start by writing it in the form $f = \sum_{i \in J} a_i I_{A_i}$, where $(A_i)_{i \in J}$ is a finite family of not necessarily disjoint elements of $\mathcal{I}$. Let us denote by $\mathcal{H}$ the class of all non-empty subsets of I, and, for any element H of $\mathcal{H}$, let us set

$$C_H = \bigcap_{i\in H} A_i, \qquad B_H = C_H \cap \bigcap_{i\in J\setminus H} A_j^c.$$

Then, the sets B_H are pairwise disjoint and one has

$$f = \sum_{H\in\mathcal{H}} b_H I_{B_H}, \qquad \text{with} \quad b_H = \sum_{i\in H} a_i.$$

Furthermore, for every index H, since $C_H \in \mathcal{I}$, one also has $B_H \in \mathcal{J}$, because

$$I_{B_H} = I_{C_H} \prod_{i\in J\setminus H} (1 - I_{A_i}) \in \mathcal{R}.$$

The theorem is therefore proved. □

We are now ready to prove the following.

(6.8) Theorem. *On the measurable space $(E, \mathcal{E})$, let $\mathcal{I}$ be a π-system which generates the σ-field $\mathcal{E}$. Let $\mathcal{L}$ be a monotone vector subspace of $\mathbb{R}^E$, containing the indicator functions of all sets in $\mathcal{I}$. Then $\mathcal{L}$ also contains all functions of the form*

$$fI_{A_1\cup\cdots\cup A_n},$$

where f is a bounded real measurable function on $(E, \mathcal{E})$ and $(A_1, \ldots, A_n)$ is a finite sequence of elements of $\mathcal{I}$. In other words, $\mathcal{L}$ contains every every real bounded measurable function on $(E, \mathcal{E})$ which is zero outside of an element of $\mathcal{I}_s$.

Proof. Let $A_1, \ldots, A_n$ be a finite sequence of elements of $\mathcal{I}$, and set

$$\mathcal{L}' = \left\{ f \in \mathbb{R}^E : fI_{A_1\cup\cdots\cup A_n} \in \mathcal{L} \right\}.$$

In order to prove the theorem, it suffices to show that $\mathcal{L}'$ contains every bounded real measurable function on $(E, \mathcal{E})$. To this end, since $\mathcal{L}'$ is a monotone vector space, it suffices to show that $\mathcal{L}'$ contains the indicator functions of every element of the class $\mathcal{I}' = \mathcal{I} \cup \{E\}$ (obtained adding the whole space to the π-system $\mathcal{I}$). In other words, it is sufficient to prove that, for every element A of $\mathcal{I}'$,

$$I_A I_{A_1\cup\cdots\cup A_n} \in \mathcal{L}.$$

To this end, it suffices to observe that

$$I_A I_{A_1\cup\cdots\cup A_n} = (I_{A\cap A_1}) \vee \cdots \vee (I_{A\cap A_n}) \in \mathcal{R} \subseteq \mathcal{L},$$

where $\mathcal{R}$ denotes the vector space generated by the indicator functions of the elements of $\mathcal{I}$, which is, in particular, a Riesz space, by Lemma (6.7). □

Let us finally prove a further variation of the monotone class theorem.

(6.9) Theorem. *Let $\mathcal{L}$ a monotone Riesz space of real functions defined on E endowed with Stone's property. Let us denote by $\mathcal{E}$ the smallest σ-field which makes all the functions in $\mathcal{L}$ measurable. In other words, $\mathcal{E}$ is the intersection of all σ-fields $\sigma(f)$, with $f \in \mathcal{L}$. Then, every real measurable function on $(E, \mathcal{E})$, bounded between two elements of $\mathcal{L}$, belongs to $\mathcal{L}$.*

Proof. Let us set

$$\mathcal{I} = \Big\{ \{f > c\} : f \in \mathcal{L}_+, c > 0 \Big\}.$$

Since the elements of the form $\{f > c\}$ are a set of generators for $\sigma(f)$, it turns out that the class $\mathcal{I}$ is a set of generators for $\mathcal{E}$. Also, since $\mathcal{L}$ is a vector space, every element of this class can be written in the form $\{f > 1\}$ with $f \in \mathcal{L}_+$. Furthermore, since $\mathcal{L}$ is stable under the lattice operation, the class $\mathcal{I}$ is, in particular, stable under binary intersection, and therefore it is actually a base for $\mathcal{E}$. On the other hand, since $\mathcal{L}$ is a monotone class, thanks to the two obvious relations

$$I_{\{f>1\}} \le f, \qquad I_{\{f>1\}} = \sup_n \Big\{ 1 \wedge \big[n(f - f \wedge 1) \big] \Big\},$$

the indicator function of every element of $\mathcal{I}$ belongs to $\mathcal{L}$. Now, since $\mathcal{L}$ is a monotone Riesz space, and by means of Theorem (2.12), every positive measurable function on $(E, \mathcal{E})$ is the supremum of an increasing sequence of simple functions, it will suffice to prove that every positive simple function on $(E, \mathcal{E})$ belongs to $\mathcal{E}$, as long as it is bounded from above by an element of $\mathcal{L}$. To this end, since $\mathcal{L}$ is a vector space, it will only suffice to prove that the indicator function of an element A of $\mathcal{E}$ belongs to $\mathcal{L}$, as long as it is bounded from above by an element of $\mathcal{L}$. To prove this, let us take an element A_0 in $\mathcal{I}$, and let us observe that the class

$$\Big\{ A \in \mathcal{E} : I_{A \cap A_0} \in \mathcal{L} \Big\}$$

is a σ-field on E which contains $\mathcal{I}$, and therefore it coincides with $\mathcal{E}$. This means that, for every element A of $\mathcal{E}$, one has $I_{A \cap A_0} \in \mathcal{L}$ and therefore, in particular, $I_A \in \mathcal{L}$ if A is included in A_0. Since the choice of A_0 is arbitrary, the conclusion follows. □

7. Product of a family of σ-fields

Given a non-empty set E, and a family $(\mathcal{E}_i)_{i \in I}$ of σ-fields on E, we have already mentioned (Proposition (1.4)) that the intersection of this family, $\bigcap_{i \in I} \mathcal{E}_i$, is a σ-field on E. Unfortunately, the union of an arbitrary family of σ-fields is generally not a σ-field. As a result of this, it may be convenient to introduce a special notation to identify the σ-field generated by the union of a family of σ-fields. Precisely, if $(\mathcal{E}_i)_{i\in I}$ is a family of σ-fields on E, the σ-field generated by the union $\bigcup_{i\in I}\mathcal{E}_i$ is denoted by $\bigvee_{i\in I}\mathcal{E}_i$. In symbols, let us define

$$\bigvee_{i\in I}\mathcal{E}_i = \sigma\left(\bigcup_{i\in I}\mathcal{E}_i\right).$$

Furthermore, given a non-empty set E, a family $((F_i,\mathcal{F}_i))_{i\in I}$ of measurable spaces, and, for every index i in I, a map f_i defined on E with values in F_i, the σ-field $\bigvee_{i\in I}\sigma(f_i)$ shall also be denoted by $\sigma(f_i : i\in I)$, and it shall be called the σ-field **generated** by the family $(f_i)_{i\in I}$. This is, by definition, the smallest σ-field on E which makes all the maps f_i simultaneously measurable, when each of the sets F_i is equipped with the σ-field $\mathcal{F}_i$.

Let us observe immediately that, from the fundamental criterion of measurability, it follows that, if $\mathcal{J}_i$ is a set of generators for $\mathcal{F}_i$, then the class

$$\Big\{\{f_i\in A\} : i\in I,\, A\in\mathcal{J}_i\Big\}$$

is a set of generators for the σ-field $\sigma(f_i : i\in I)$.

The σ-field generated by a family of maps is useful, for example, when one seeks the smallest σ-field which makes them all measurable.

However, one of the most interesting applications of this new notion is that it allows to define a natural σ-field to the product of a family of sets, each of which is already equipped with a σ-field, in order to maintain a link between the measure structure of each set and the global measurable structure. In order to do this, let us give the following definition.

(7.1) Definition. Let $((F_i,\mathcal{F}_i))_{i\in I}$ be a family of measurable spaces, let us denote by F the product $\prod_{i\in I}F_i$ and, for every index i, let us also denote by ξ_i the projection of F onto F_i. The σ-field $\sigma(\xi_i : i\in I)$ is called the **product** of the family $(\mathcal{F}_i)_{i\in I}$, and it is denoted by $\bigotimes_{i\in I}\mathcal{F}_i$. In symbols:

$$\bigotimes_{i\in I}\mathcal{F}_i = \sigma\big(\xi_i : i\in I\big).$$

In other words, the product σ-field is the smallest σ-field on the product space that makes all the projections measurable.

When the family $(\mathcal{F}_i)_{i\in I}$ is finite, for instance taking $(\mathcal{F}_1,\ldots,\mathcal{F}_n)$, instead of $\bigotimes_{i=1}^n\mathcal{F}_i$ we shall preferably write $\mathcal{F}_1\otimes\cdots\otimes\mathcal{F}_n$. In particular, if all the given σ-fields are identical, and if $\mathcal{F}$ is one (hence each) of these σ-fields, we shall write $\mathcal{F}^{\otimes n}$ instead of $\mathcal{F}\otimes\cdots\otimes\mathcal{F}$.

(7.2) Proposition. *In the same hypotheses as in Definition (7.1), let us suppose that a measurable space $(E,\mathcal{E})$, and, for every index i, a map f_i from E to F_i, are given. Let us denote by f the map from E to $\prod_{i\in I}F_i$ defined by*

$$f(x) = \big(f_i(x)\big)_{i\in I} \quad \text{for every element } x \text{ in } E.$$

Then, the following conditions are equivalent:

(a) *f is a measurable map from $(E, \mathcal{E})$ to $\left(\prod_{i\in I} F_i, \bigotimes_{i\in I} \mathcal{F}_i\right)$.*

(b) *For every index i, f_i is a measurable map from $(E, \mathcal{E})$ to $(F_i, \mathcal{F}_i)$.*

Proof. A set of generators for the σ-field $\bigotimes_{i\in I} \mathcal{F}_i$ is the class

$$\mathcal{H} = \Big\{\{\xi_i \in A\} : i \in I, A \in \mathcal{F}_i\Big\}.$$

Thus, the fundamental criterion of measurability yields that condition (a) holds if and only if, whenever a set lies in the class $\mathcal{H}$, i.e. for every set of the form $\{\xi_i \in A\}$, with $i \in I$ and $A \in \mathcal{F}_i$, its inverse image under f lies in $\mathcal{E}$. Now, this condition is equivalent to condition (b), because for every set of the form $\{\xi_i \in A\}$, its inverse image under f can be written in the form

$$\begin{aligned}\Big\{f \in \{\xi_i \in A\}\Big\} &= \{x \in E : f(x) \in \{\xi_i \in A\}\} \\ &= \big\{x \in E : \xi_i(f(x)) \in A\big\} \\ &= \big\{x \in E : f_i(x) \in A\big\} \\ &= \big\{f_i \in A\big\}. \qquad \square\end{aligned}$$

In particular, as a simple consequence of the previous proposition, it is not difficult to recognise that, if $(E, \mathcal{E})$ and $(F, \mathcal{F})$ are two measurable spaces, the product $\mathcal{E} \otimes \mathcal{F}$ coincides with the σ-field generated by the sets of the form $A \times B$, with $A \in \mathcal{E}$ and $B \in \mathcal{F}$.

The function f defined in the proposition is sometimes called the **block** of the family $(f_i)_{i\in I}$, and it may be denoted by $[f_i]_{i\in I}$. Further, for any index i, the map f_i is called the **component** of f with index i, or simply the ith component of f. For instance, if $(\xi_i)_{i\in I}$ is the family of the projections of the product set $\prod_{i\in I} F_i$ to F_i, then the block $[\xi_i]_{i\in I}$ is the identity on $\prod_{i\in I} F_i$.

Using this notation, the previous remark can be written in a simpler form by saying that

$$\sigma\big(f_i : i \in I\big) = \sigma\big([f_i]_{i\in I}\big).$$

This latter remark shows in particular that the σ-field generated by a family of maps can always be thought as the σ-field generated by a single map.

Let us consider a fundamental example. The Euclidean space $\mathbb{R}^n$ is the product of n copies of the real line, which is a measurable set with its Borel σ-field. For this reason, the product space $\mathbb{R}^n$ may be equipped with the product σ-field $\mathcal{B}(\mathbb{R})^{\otimes n}$. On the other hand, $\mathbb{R}^n$ is also a topological space and therefore it has its own Borel σ-field, $\mathcal{B}(\mathbb{R}^n)$. The next theorem will show that these two σ-fields are identical. Actually, in the more general setting of a family of topological spaces, the inclusion of the product σ-field into the Borel σ-field of the product space always hold, but the equality only holds in a special case. Precisely, let us prove the following theorem.

(7.3) Theorem. *Let $(E_i)_{i\in I}$ be a family of topological spaces, and consider the topological space $E = \prod_{i\in I} E_i$ equipped with its natural product topology.*

Then,

$$\mathcal{B}(E) \supseteq \bigotimes_{i\in I} \mathcal{B}(E_i).$$

Furthermore, this inclusion becomes an identity when the set of indices I is countable and each of the topological spaces E_i has a countable base.

Proof. Let us denote by ξ_i the projection of E onto E_i. By definition of product topology, all the maps ξ_i, with i in I, are continuous, hence Borel measurable. As a result of this, the Borel σ-field $\mathcal{B}(E)$ contains the σ-field $\sigma(\xi_i : i \in I) = \bigotimes_{i\in I} \mathcal{B}(E_i)$.

Let us now suppose that I is countable, and that, for every index i, the space E_i has a countable base $\mathcal{U}_i$. Let us consider the class $\mathcal{U}$ of all the sets of the form $\{\xi_i \in A\}$ with i element of I and A element of $\mathcal{U}_i$. On one side, $\mathcal{U}$ is a set of generators for the σ-field $\bigotimes_{i\in I} \mathcal{B}(E_i)$; on the other side, $\mathcal{U}$ is a countable sub-base for the topology of E.

As a result $\mathcal{B}(E) = \bigotimes_{i\in I} \mathcal{B}(E_i)$. □

Let us go back to Proposition (7.2), and note that the class $\mathcal{H}$ that is defined in the proof of the proposition is a set of generators for $\bigotimes_{i\in I} \mathcal{F}_i$, but this class is not (in general) stable for the operation of finite intersection. In order to get a system of generators which has this further property, i.e. a base for the product σ-field, it suffices to take the class $\mathcal{H}_d$ of all finite intersections of elements of $\mathcal{H}$. It is straightforward to show that the class $\mathcal{H}_d$ contains exactly the products of the form $\prod_{i\in I} A_i$, where $(A_i)_{i\in I}$ is any family of sets, such that, for each index i, A_i lies in $\mathcal{F}_i$, and with the further condition that the set $\{i \in I : A_i \neq E_i\}$ is finite. This type of sets is sometimes called a 'measurable rectangle'.

Since the product of two spaces is often particularly important, let us conclude this section by showing a useful result.

(7.4) Theorem. *Let E and F be two topological space. Then the equality*

$$\mathcal{B}(E \times F) = \mathcal{B}(E) \otimes \mathcal{B}(F)$$

holds as soon as one of the two spaces has a countable base.

Proof. With no loss of generality, let us suppose that E has a countable base $\{U_n : n \geq 1\}$. Given an open set W of $E \times F$, for any whole number n, let us denote by V_n the largest of all open sets V in F such that $U_n \times V \subseteq W$.

For each pair (x, y) in W, there exist an integer n and an open set V in F such that

$$(x, y) \in U_n \times V \subseteq W,$$

and therefore also

$$(x, y) \in U_n \times V_n \subseteq W.$$

As a consequence, one has also $W = \bigcup_{n \geq 1}(U_n \times V_n) \in \mathcal{B}(E) \otimes \mathcal{B}(F)$.

Since the set W is an arbitrary open set of $E \times F$, this proves the inclusion

$$\mathcal{B}(E \times F) \subseteq \mathcal{B}(E) \otimes \mathcal{B}(F),$$

and therefore the equality $\mathcal{B}(E \times F) = \mathcal{B}(E) \otimes \mathcal{B}(F)$. □

Chapter II

Integration

This chapter is devoted to the introduction and the study of arguably the most fundamental tool in Measure Theory, that is to say that of integral with respect to a measure. In particular, in §1 and §2, we shall develop the construction of the integral of a real valued function with respect to a measure. The idea of this construction, which shall be developed mainly in §1, shall be to see the measure of a set as a functional of the indicator function of that set, then to extend by linearity the notion of measure to linear combinations of indicator functions, and finally to extend it to all positive measurable functions which are, by means of Proposition I(2.12), the supremum of an increasing sequence of simple functions. In §2, instead, we shall complete the construction extending, again by linearity, the notion of integral to all real valued functions. Once the integral is completely defined, we shall delve, throughout §3 to §6, into a full exploration of the main properties to underpin the full potential of this newborn integral. In particular, the chapter concludes with the discussion of the so-called 'convergence theorems' that are amongst the most important features of the integral with respect to a measure.

1. Integral of a positive measurable function

In the sequel, we shall suppose that a measure space $(E, \mathcal{E}, \mu)$ is assigned and, whenever we say that a real valued function is 'measurable', unless otherwise specified, we shall always mean that the function is defined on E, and it is measurable on $(E, \mathcal{E})$. Observe that a positive simple function f on $(E, \mathcal{E})$ can always be put in the form

$$f = \sum_x x I_{\{f=x\}}$$

where the sum on the right-hand side ranges over all real numbers x, even though it naturally reduces to a finite sum, because only for a finite number of values x will the set $\{f = x\}$ be non-empty. Also recall that, by definition,

DOI: 10.1201/9781003724766-II

the product $0 \cdot \infty$ shall be deemed to be equal to 0. In these hypotheses, we may give the following definition.

(1.1) Definition. Let f be a positive simple function on $(E, \mathcal{E}, \mu)$. The **integral** of f with respect to μ is the number $\int f \, \mathrm{d}\mu$ defined by

$$\int f \, \mathrm{d}\mu = \sum_x x\mu\{f = x\}.$$

Observe immediately that, if f is a positive simple function, written as

$$f = \sum_{i \in I} a_i I_{A_i},$$

where $(A_i)_{i \in I}$ is a finite partition of E, such that, for every index i, the set A_i is measurable, one also has

$$\int f \, \mathrm{d}\mu = \sum_{i \in I} a_i \mu(A_i).$$

In fact, thanks to the fact that μ is finitely additive, the value of this sum does not change when passing from a given partition to a finer one. In other words, given two such partitions, the corresponding sums coincide with the sum of the coarser, and the partition $(\{f = x\})$ is the coarsest of all these possible partitions.

The first consequence of this definition is the following property.

(1.2) Proposition. *If A is a measurable set, then*

$$\int I_A \, \mathrm{d}\mu = \mu(A).$$

Let us also observe that, if f is a positive simple function which is not the constant 0, and if a denotes the minimum of its non-zero values, then, by setting $A = \{f \geq a\} = \{f > 0\}$, it turns out that the relation $\int f \, \mathrm{d}\mu < \infty$ is equivalent to $\mu(A) < \infty$.

Let us now state, in the next few propositions, the fundamental properties of the integral of a positive simple function with respect to a measure.

(1.3) Proposition (Positive homogeneity and additivity). *For every positive real number α and every pair f, g of positive simple functions, we have*

$$\int (\alpha f) \, \mathrm{d}\mu = \alpha \int f \, \mathrm{d}\mu, \qquad \int (f + g) \, \mathrm{d}\mu = \int f \, \mathrm{d}\mu + \int g \, \mathrm{d}\mu.$$

Proof. Let us start by proving the first statement. To this end, let us observe that, if α is a positive real number and if $f = \sum_x x I_{\{f=x\}}$ is a positive simple function, then $\alpha f = \sum_x \alpha x I_{\{f=x\}}$ and therefore

$$\begin{aligned}\int (\alpha f) \, \mathrm{d}\mu &= \sum_x \alpha x \mu\{f = x\} \\ &= \alpha \Big(\sum_x x\mu\{f = x\} \Big) \\ &= \alpha \int f \, \mathrm{d}\mu.\end{aligned}$$

Let us now prove the second statement: If f and g are any two positive simple functions, we have to prove that $\int(f+g)\,\mathrm{d}\mu = \int f\,\mathrm{d}\mu + \int g\,\mathrm{d}\mu$. To this end, let us start by noting that, thanks to the additivity of the measure μ, the integral $\int f\,\mathrm{d}\mu$ can be written in the form

$$\begin{aligned}\int f\,\mathrm{d}\mu &= \textstyle\sum_x x\mu\{f=x\}\\ &= \textstyle\sum_x x\sum_y \mu\{f=x,\, g=y\}\\ &= \textstyle\sum_{x,y} x\mu\{f=x,\, g=y\}.\end{aligned}$$

Likewise, the integral $\int g\,\mathrm{d}\mu$ can be written in the form

$$\int g\,\mathrm{d}\mu = \textstyle\sum_{x,y} y\mu\{f=x,\, g=y\}.$$

Adding up these two expressions gives therefore

$$\begin{aligned}\int f\,\mathrm{d}\mu + \int g\,\mathrm{d}\mu &= \textstyle\sum_{x,y}(x+y)\mu\{f=x,\, g=y\}\\ &= \textstyle\sum_z \sum_{x+y=z}(x+y)\mu\{f=x,\, g=y\}\\ &= \textstyle\sum_z z\sum_{x+y=z}\mu\{f=x,\, g=y\}\\ &= \textstyle\sum_z z\mu\{f+g=z\}\\ &= \int(f+g)\,\mathrm{d}\mu,\end{aligned}$$

and the theorem is fully proved. □

(1.4) Proposition (Monotonicity and subtractivity). *Let f and g be two simple functions with $0 \le f \le g$. Then,*

$$\int g\,\mathrm{d}\mu = \int f\,\mathrm{d}\mu + \int(g-f)\,\mathrm{d}\mu \ge \int f\,\mathrm{d}\mu.$$

As a result, if $\int f\,\mathrm{d}\mu$ is finite, and also

$$\int(g-f)\,\mathrm{d}\mu = \int g\,\mathrm{d}\mu - \int f\,\mathrm{d}\mu.$$

Proof. The proprieties of monotonicity and subtractivity follow from the property of additivity, observing that the equality $g=(g-f)+f$ holds, and then taking the integral of this expression. □

(1.5) Proposition (Continuity on decreasing sequences). *If (f_n) is a decreasing sequence of positive simple functions, such that $\inf_n f_n = 0$ and if, for every index n, $\int f_n\,\mathrm{d}\mu < \infty$, then $\inf_n \int f_n\,\mathrm{d}\mu = \lim_n \int f_n\,\mathrm{d}\mu = 0$.*

Proof. Let us set $A=\{f_0>0\}$. Then $\mu(A)<\infty$. Moreover, the functions f_n and f are zero outside A. Taken a strictly positive number ε, and denoted by c the highest value of f_0, one has

$$f_n \le cI_{\{f_n>\varepsilon\}} + \varepsilon I_A$$

and therefore $\int f_n\,\mathrm{d}\mu \le c\mu\{f_n>\varepsilon\} + \varepsilon\mu(A)$.

On the other hand, $(\{f_n > \varepsilon\})$ is a decreasing sequence of measurable sets, with finite measure, with empty intersection. Applying Theorem (3.7) of Chapter I, it turns out that $\lim_n \int f_n \, d\mu \leq \varepsilon \int f \, d\mu$. Since ε is arbitrary, it follows that $\lim_n \int f_n \, d\mu = 0$. □

(1.6) Proposition (Continuity on increasing sequences). *Let (f_n) be an increasing sequence of positive simple functions, and let us suppose that the function $f = \sup_n f_n$ is simple (and obviously positive).*

Then, $\int f \, d\mu = \sup_n \int f_n \, d\mu = \lim_n \int f_n \, d\mu$.

Proof. Let us start by observing that, by monotonicity of the integral, one has $\sup_n \int f_n \, d\mu \leq \int f \, d\mu$. Hence, if $\sup_n \int f_n \, d\mu = \infty$, then $\int f \, d\mu = \infty$. Thus, with no loss of generality, one may suppose that $\sup_n \int f_n \, d\mu < \infty$ and that f is not the constant 0. Then, if a denotes the minimum non-zero value of f, if we set $A = \{f \geq a\}$ and α is a real number with $0 < \alpha < a$, then, from the inequality $\alpha I_{\{f_n > \alpha\}} \leq f_n$, it follows that $\alpha \mu\{f_n > \alpha\} < \int f_n \, d\mu$, hence,

$$\begin{aligned} \mu(A) &= \mu\{f > \alpha\} \\ &= \sup_n \mu\{f_n > \alpha\} \\ &\leq \alpha^{-1} \sup_n \textstyle\int f_n \, d\mu < \infty. \end{aligned}$$

As a result $\int f \, d\mu < \infty$, hence $\int (f - f_n) \, d\mu < \infty$ for every n. So, using the properties of subtractivity and continuity under decreasing sequences, we finally get $\int f \, d\mu - \int f_n \, d\mu = \int (f - f_n) \, d\mu \downarrow 0$ and this completes the proof. □

We may now extend the definition of integral to any positive measurable function. To this end, let us start by observing that, with the properties proved thus far, the function $f \mapsto \int f \, d\mu$ is a positive linear functional defined on the class $\mathcal{S}_+(E, \mathcal{E})$ of all positive simple functions on the measurable space $(E, \mathcal{E})$ which is continuous on monotone (increasing or decreasing) sequences. This functional can now be easily extended to the class $\mathcal{M}_+(E, \mathcal{E})$, constituted of all positive measurable functions on $(E, \mathcal{E})$, using the fact that, as shown in Theorem I(2.12), every positive measurable function is the supremum of an increasing sequence of positive simple functions.

Precisely, let us give the following definition.

(1.7) Definition. Let $(E, \mathcal{E}, \mu)$ be a measure space, and f a positive measurable function defined on $(E, \mathcal{E})$. The **integral** of f with respect to μ is denoted by $\int f \, d\mu$, and it is defined as the supremum of all the numbers of the form $\int g \, d\mu$, with g positive simple function such that $g \leq f$. In other words, the integral of f with respect to measure μ is defined as

$$\int f \, d\mu = \sup\Big\{\int g \, d\mu : g \in \mathcal{S}_+(E, \mathcal{E}),\, g \leq f\Big\}$$

Let us observe immediately that, in the particular case where f is simple, this supremum turns out to be a maximum, and it is precisely the positive number $\int f\,\mathrm{d}\mu$. Thus, the integral just defined in the class of all positive measurable functions is an extension, to the class of all positive measurable functions, of the integral of a positive simple functions. This is the reason why, for this extended integral, it is allowed to use the same symbol.

Also, beside the already introduced notation, we shall sometimes use some others. Amid the most common, let us mention the following:

$$\langle \mu, f \rangle, \qquad \int \mu(\mathrm{d}x) f(x), \qquad \int f(x)\,\mu(\mathrm{d}x), \qquad \int f(x)\,\mathrm{d}\mu(x), \qquad \text{etc.}$$

Regardless of the dimension of the symbol, which in this presentation we shall always endeavour to keep in *text style*, whenever this is possible, each of these notations has its specific interest and corresponds, from different points of view, to different interpretations of the notion of integral. For example, the notation $\langle \mu, f \rangle$ reminds a scalar product and aims to highlight the 'duality' that exists between measurable functions and measures, that is to say that the function $(f, \mu) \mapsto \int f\,\mathrm{d}\mu$ is bilinear and induces a duality between functions and measures, as we shall see later. Precisely, we shall show that the integral with respect to a measure has the three following fundamental properties:

(a) *Positivity*: $\langle \mu, f \rangle \geq 0$ and $\langle \mu, f \rangle = 0$ if $f = 0$.

(b) *Linearity*: $\langle \mu, af + bg \rangle = a\langle \mu, f \rangle + b\langle \mu, g \rangle$.

(c) *Monotone convergence*: If $f_n \uparrow f$, then $\langle \mu, f_n \rangle \uparrow \langle \mu, f \rangle$.

All the other notations, instead, aim to display the dummy variable x and, for this reason, are quite useful in practical calculations, especially when there are changes of variables involved, as well as when dealing with measures in product spaces, where it can be useful to distinguish between the components of the space.

Let us also introduce a useful notation and terminology.

(1.8) Definition. Let f be a positive function defined on a part A of E. Then f may though as defined on the whole set E by setting $f(x) = 0$ when x in not in A. Let us assume that the function $I_A f$ is measurable, so that the integral $\int (I_A f)\,\mathrm{d}\mu$ is defined. This integral shall be denoted by $\int_A f\,\mathrm{d}\mu$, and it is called the integral of f **on** A (or **over** A).

We can now prove the properties of this extended integral, which are natural extension of the properties of the integral for a positive simple function, starting with a pretty straightforward one.

(1.9) Proposition (Monotonicity). *Let f and g be two measurable functions, with $0 \leq f \leq g$. Then*

$$\int f\,\mathrm{d}\mu \leq \int g\,\mathrm{d}\mu.$$

Let us now prove the property of monotone convergence, which is also known as 'Beppo Levi's theorem on monotone convergence', in honour of the Italian mathematician Beppo Levi (1875–1961) who first proved it.

(1.10) Theorem (Beppo Levi). *If (f_n) is a sequence of positive measurable functions, and $f_n \uparrow f$, then $\int f_n \, \mathrm{d}\mu \uparrow \int f \, \mathrm{d}\mu$.*

Proof. Let us start by observing that, by means of the monotonicity of the integral, $\sup_n \int f_n \, \mathrm{d}\mu \leq \int f \, \mathrm{d}\mu$. To conclude, it suffices to prove the converse inequality, that is to say

$$\int f \, \mathrm{d}\mu \leq \sup_n \int f_n \, \mathrm{d}\mu,$$

which means that, for every simple function g, with $0 \leq g \leq f$,

$$\int g \, \mathrm{d}\mu \leq \sup_n \int f_n \, \mathrm{d}\mu.$$

Given such a simple function g, let us observe that $g \wedge f_n \uparrow g \wedge f = g$. In other words, g is the supremum of the increasing sequence $(g \wedge f_n)$. Now, each term of this sequence is, in turn, the supremum of a sequence of positive simple functions, each of which is bounded from above by f_n. As a result, g can be written as the supremum of an *increasing sequence* of simple function, each of which is bounded from above by some of the f_n. Let us denote by (g_k) such a sequence of simple functions. Then, thanks to the continuity on increasing sequences of the integral for simple functions,

$$\int g \, \mathrm{d}\mu = \sup_k \int g_k \, \mathrm{d}\mu \leq \sup_n \int f_n \, \mathrm{d}\mu.$$

The theorem is thus proved. □

The integral of positive measurable functions has the further property of additivity and homogeneity.

(1.11) Proposition (Homogeneity and additivity). *For every positive real number α and every pair f, g of positive measurable functions, we have*

$$\int (\alpha f) \, \mathrm{d}\mu = \alpha \int f \, \mathrm{d}\mu, \qquad \int (f+g) \, \mathrm{d}\mu = \int f \, \mathrm{d}\mu + \int g \, \mathrm{d}\mu.$$

Proof. As we already mentioned, there are two increasing sequences (f_n) and (g_n) such that $\sup_n f_n = f$ and $\sup_n g_n = g$. Hence, thanks to the corresponding properties of the integral of positive simple functions, we have $\int (\alpha f_n) \, \mathrm{d}\mu = \alpha \int f_n \, \mathrm{d}\mu$ and $\int (f_n + g_n) \, \mathrm{d}\mu = \int f_n \, \mathrm{d}\mu + \int g_n \, \mathrm{d}\mu$, and the properties required follow immediately from Beppo Levi's theorem. □

(1.12) Proposition (Series integration). *If (f_n) is a sequence of positive measurable functions, let us set $f = \sum_{n \geq 1} f_n$. Then,*

$$\int f \, \mathrm{d}\mu = \sum_{n \geq 1} \int f_n \, \mathrm{d}\mu.$$

Proof. From the additive property, by induction, we get

$$\int (f_1 + \cdots + f_n) \, \mathrm{d}\mu = \sum_{i=1}^{n} \int f_i \, \mathrm{d}\mu.$$

The thesis follows applying Beppo Levi's theorem. □

2. Integrable functions

In the sequel, we shall complete the definition of integral with respect to a measure, extending it by linearity to the measurable functions of any sign. In this section, we shall suppose again that a measure space $(E, \mathcal{E}, \mu)$ is given and whenever we say that a function is measurable, we shall mean that the function is measurable over the measurable space $(E, \mathcal{E})$.

(2.1) Definition. Let f be a numeric function defined on E. We shall say that f is **lower semi-integrable** with respect to μ if it is measurable and $\int f^- \, \mathrm{d}\mu < \infty$. Likewise, we shall say that f is **upper semi-integrable** with respect to μ if it is measurable and $\int f^+ \, \mathrm{d}\mu < \infty$. If f is either lower semi-integrable or upper semi-integrable, then f is said to be **semi-integrable**, and its **integral** is defined as

$$\int f \, \mathrm{d}\mu = \int f^+ \, \mathrm{d}\mu - \int f^- \, \mathrm{d}\mu.$$

Finally, we shall say that the function f is **integrable** with respect to μ, or simply **μ-integrable**, if f is at the same time lower and upper semi-integrable, or, what is the same, if f is measurable and the integral of its absolute value $|f| = f^+ + f^-$ is finite, i.e. if the number

$$\int |f| \, \mathrm{d}\mu = \int f^+ \, \mathrm{d}\mu + \int f^- \, \mathrm{d}\mu$$

is finite.

Before diving into any further remarks about the notion of integral, let us prove the basic important properties of the integral. Let us start with the following inequality.

(2.2) Theorem. *If f is an integrable function, then*

$$\left|\int f \, \mathrm{d}\mu\right| \leq \int |f| \, \mathrm{d}\mu.$$

If the measure μ is finite, every real valued function f which is measurable and bounded on the measurable space $(E, \mathcal{E})$ is integrable with respect to μ.

Proof. This inequality follows from the triangle inequality for the absolute value. In fact, since f^+ and f^- are positive functions, whence the numbers $\int f^+ \, \mathrm{d}\mu$ and $\int f^- \, \mathrm{d}\mu$ are also positive, and since $f = f^+ - f^-$, it follows that

$$\begin{aligned}\left|\int f \, \mathrm{d}\mu\right| &= \left|\int f^+ \, \mathrm{d}\mu - \int f^- \, \mathrm{d}\mu\right| \\ &\leq \int f^+ \, \mathrm{d}\mu + \int f^- \, \mathrm{d}\mu \\ &= \int |f| \, \mathrm{d}\mu.\end{aligned}$$

In order to prove the second statement, suppose that μ is finite. Then, if c is any strictly positive real constant such that $|f| < c$, by means of the monotonicity of the integral for a positive function, then

$$\int |f| \, \mathrm{d}\mu \leq c\mu(E) < \infty. \qquad \square$$

We are now able to prove the remaining basic properties of the integral, which incidentally extend the same properties already proved for the integral of a positive function. These properties are summarised in the following theorem.

(2.3) Theorem. *On the measure space $(E, \mathcal{E}, \mu)$, the following properties of the integral hold.*

(a) *In order for the function f to be upper semi-integrable, it is necessary and sufficient that $-f$ is lower semi-integrable. Furthermore, if this happens, then*

$$\int(-f)\,\mathrm{d}\mu = -\int f\,\mathrm{d}\mu.$$

(b) *In order for a function to be upper (resp. lower) semi-integrable, it is necessary and sufficient that it is bounded from above (resp. from below) by an integrable function.*

(c) *If f is a lower (resp. upper) semi-integrable function, such is αf for any positive real number α, and $\int(\alpha f)\,\mathrm{d}\mu = \alpha \int f\,\mathrm{d}\mu$.*

(d) *If f and g are two lower (resp. upper) semi-integrable functions, such is their sum $f+g$, and $\int(f+g)\,\mathrm{d}\mu = \int f\,\mathrm{d}\mu + \int g\,\mathrm{d}\mu$.*

In particular, from (c) and (d), it turns out that the functional $f \mapsto \int f\,\mathrm{d}\mu$, defined in the space of all measurable function on $(E, \mathcal{E})$, integrable with respect to μ, is linear.

Proof. The first three properties are obvious. Let us instead prove (d). With no loss of generality, we shall suppose that f and g are lower semi-integrable. Indeed, the corresponding proof for upper semi-integrable functions follows from (a), taking $-f$ and $-g$ instead of f and g. To start off, let us observe that from the inequality $(f+g)^- \leq f^- + g^-$, integrating with respect to the measure μ and using the linearity of the integral of positive functions, we get $\int(f+g)^-\,\mathrm{d}\mu \leq \int f^-\,\mathrm{d}\mu + \int g^-\,\mathrm{d}\mu$, hence the sum $f+g$ is lower semi-integrable, because such are f and g. Moreover,

$$f + g = (f+g)^+ - (f+g)^-$$
$$f + g = (f^+ - f^-) + (g^+ - g^-)$$

equating the right-hand side of the previous two equations gives

$$(f+g)^+ - (f+g)^- = f^+ - f^- + g^+ - g^-,$$

and rearranging the equation thereby obtained gives

$$(f+g)^+ + f^- + g^- = (f+g)^- + f^+ + g^+.$$

Thus, from the additivity of the integral for positive functions, we get

$$\int(f+g)^+\,\mathrm{d}\mu + \int f^-\,\mathrm{d}\mu + \int g^-\,\mathrm{d}\mu = \int(f+g)^-\,\mathrm{d}\mu + \int f^+\,\mathrm{d}\mu + \int g^+\,\mathrm{d}\mu$$

and since the three numbers $\int f^- \, d\mu$, $\int g^- \, d\mu$ and $\int (f+g)^- \, d\mu$ are finite, the latter equation can be again rearranged to get

$$\int (f+g)^+ \, d\mu - \int (f+g)^- \, d\mu = \int f^+ \, d\mu - \int f^- \, d\mu + \int g^+ \, d\mu - \int g^- \, d\mu,$$

which is equivalent to the equation to be proved. □

The next proposition, named after the Soviet mathematician and probabilist Andrey Andreyevich Markov (Андрей Андреевич Марков, 1856–1922), is an incredibly useful result, whose real beauty lies in its extreme simplicity.

(2.4) Proposition (Markov's inequality). *For every integrable function f and every positive real number α, we have*

$$\mu\{|f| \geq \alpha\} \leq \alpha^{-1} \int |f| \, d\mu.$$

Proof. This is immediate consequence of the relation $\alpha I_{\{|f| \geq \alpha\}} \leq |f|$. □

3. Negligible functions and sets

Let us continue assuming that a measure space $(E, \mathcal{E}, \mu)$ is given and that every function is defined on E and whenever we say that a function is measurable, we shall mean that it is measurable on the measurable space $(E, \mathcal{E})$.

(3.1) Definition. A numeric function f is **negligible** with respect to the measure μ, or simply **μ-negligible**, if f is measurable and if $\int |f| \, d\mu = 0$. Similarly, we shall also say that a subset A of E is **negligible** with respect to the measure μ, or simply **μ-negligible**, if it is a measurable set, i.e. if it belongs to $\mathcal{E}$, and its indicator function I_A is μ-negligible, i.e. if $\mu(A) = 0$.

It is clear that a measurable subset of a negligible subset is also negligible. Furthermore, from the sub-additivity of the measure μ it follows that the countable union of negligible sets is still a negligible set. Moreover, it is crystal clear that, for a function f to be negligible, it is necessary and sufficient that, for any strictly positive real number ε, the set $\{|f| > \varepsilon\}$ is negligible, i.e. that the set $\{f \neq 0\}$, is negligible.

Finally, if f is a positive measurable function, with $\int f \, d\mu < \infty$, then the set $\{f = \infty\}$ is negligible. When this happens, we shall often say that the function f is **finite almost everywhere** with respect to μ. To see this, it suffices to observe that, for any real number α, we have $\alpha I_{\{f=\infty\}} \leq f$, whence, integrating with respect to μ,

$$\alpha\mu\{f = \infty\} \leq \int f \, d\mu < \infty.$$

Since α is arbitrary, it follows that $\mu\{f = \infty\} = 0$, which means that the function f is finite almost everywhere.

The notion of negligible set induces an equivalence relation in the set of all measurable maps from a measure set $(E, \mathcal{E}, \mu)$ to a measurable set $(F, \mathcal{F})$.

(3.2) Definition. Given two maps f and g, defined on E, with values in F, we shall say that f and g are **equivalent modulo** μ, if there exists a negligible set A such that f and g coincide on the complement of A.

It is not difficult to show that this is an equivalence relation on the set of all applications from E to F. As a result, if f and g are two maps from E to F, if they are equivalent modulo μ, we shall write

$$f \equiv g \pmod{\mu}$$

and we shall say that f and g coincide **almost everywhere** (with respect to μ) or, what is the same, that the coincide μ**-almost everywhere**. In fact, the phrase 'almost everywhere' may be used in an even more general sense, by saying that a property holds *μ-almost everywhere* if there exists a set, which is negligible with respect to μ, such that the property holds on its complement.

(3.3) Theorem. *Let f and g be two positive measurable functions which are equivalent modulo μ. Then, we have $\int f \, d\mu = \int g \, d\mu$.*

Proof. Let A be a negligible set such that f and g coincide on $B = A^c$. Set $h = f I_B = g I_B$, so that $f = h + f I_A$ and $g = h + g I_A$. The thesis follows by observing that, since each of the functions $f I_A$ and $g I_A$ is negligible with respect to μ, because they both vanish outside the μ-negligible set A, then $\int f \, d\mu$ and $\int g \, d\mu$ both coincide with $\int h \, d\mu$. □

Let us denote by

$$\mathcal{L}^1(\mu)$$

the set of all real valued integrable functions with respect to the measure μ.

It is not difficult to show that $\mathcal{L}^1(\mu)$ is a vector space, which is stable under $f \mapsto |f|$, hence a Riesz space. Furthermore, from the properties of the integral with respect to a measure, it is not difficult to check that the function $f \mapsto \int |f| \, d\mu$ is a semi-norm on $\mathcal{L}^1(\mu)$. The convergence under the topology defined by this semi-norm shall be called the **convergence in mean** or, equivalently, **convergence in $\mathcal{L}^1$**. More precisely, a sequence (f_n) converges in $\mathcal{L}^1$ towards a function f in $\mathcal{L}^1(\mu)$, if the sequence $\int |f_n - f| \, d\mu$ of real numbers converges towards zero.

Observe that, thanks to the inequality

$$\left|\int f \, d\mu\right| \leq \int |f| \, d\mu \qquad \text{for all } f \in \mathcal{L}^1(\mu),$$

the linear functional $f \mapsto \int f \, d\mu$ is continuous over the space $\mathcal{L}^1(\mu)$.

The quotient of $\mathcal{L}^1(\mu)$ under the equivalence relation of 'equivalence modulo the measure μ' shall be denoted by $L^1(\mu)$, and the canonical injection of the space $\mathcal{L}^1(\mu)$ into $L^1(\mu)$ shall be denoted by $f \mapsto f^\bullet$.

By definition, $L^1(\mu)$ is the set of all equivalence classes modulo μ of all numeric functions on $(E, \mathcal{E})$ which are integrable with respect to μ. More precisely, if $f^\bullet$ is such an equivalence class, and if f and g are two of its representatives, that is two say two numeric functions such that $f \equiv g$ modulo μ, Theorem (3.3) ensures that $\int f \, \mathrm{d}\mu = \int g \, \mathrm{d}\mu$, which means that the value of the integral with respect to a measure ultimately depends only on the equivalence class of the function and not on the specific representative. As a result, it makes perfect sense to define, for every element $f^\bullet$ of $L^1(\mu)$, the integral $\int f^\bullet \, \mathrm{d}\mu$ as the common value of the integral with respect to μ on any representative of the class $f^\bullet$.

Note that, since $\mathcal{L}^1(\mu)$ is a semi-normed space with $f \mapsto \int |f| \, \mathrm{d}\mu$, then $L^1(\mu)$ is a normed space. We shall see in Chapter VI, that $L^1(\mu)$ is actually a complete normed space with the norm $f^\bullet \mapsto \int |f^\bullet| \, \mathrm{d}\mu$, that is to say, a Banach space.

4. Integration with respect to the image measure

A measurable map can 'transfer' the measure from a measurable space onto another. To see this, let f be a measurable map defined on measurable space $(E, \mathcal{E})$ with values in another measurable space $(F, \mathcal{F})$. Suppose that a measure μ is assigned on the measurable space $(E, \mathcal{E})$. Then, observe that, for every measurable set B in $\mathcal{F}$, we already know that the set $\{f \in B\}$ of all x in E such that $f(x) \in B$, is an element of $\mathcal{E}$; in turn, this set may be measured using the measure μ. Now, it is easy to check that the map

$$B \mapsto \mu\{f \in B\}$$

is a measure on $(F, \mathcal{F})$. This is called the 'image' of μ under f.

(4.1) Definition. Let f be a measurable map defined on a measurable space $(E, \mathcal{E})$ with values in the measurable space $(F, \mathcal{F})$, and let μ be a measure on the measurable space $(E, \mathcal{E})$. The set function defined on $\mathcal{F}$ by $B \mapsto \mu\{f \in B\}$ is a measure on $(F, \mathcal{F})$, and it is called the **image** of μ under f, and it is denoted by $f(\mu)$.

(4.2) Example. Let $(E, \mathcal{E}, \mu)$ be a measure space and H is a subset of E. We have already mentioned that H can be equipped with a structure of measurable space with the trace $\mathcal{E}_H$ of $\mathcal{E}$ onto H, viz. with the σ-field $\sigma(i_H)$ generated by the canonic injection i_H of H into E. If H is a measurable set, then the canonic injection i_H is a measurable map from $(E, \mathcal{E})$ to $(H, \mathcal{E}_H)$ and the image of the measure μ under i_H is the restriction to $\mathcal{E}_H$ of the trace μ_H of μ onto H.

It is readily shown that the image of a measure is transitive. More precisely, if $(E, \mathcal{E})$, $(F, \mathcal{F})$, and $(G, \mathcal{G})$ are three measurable spaces, if f a measurable map from the first to the second, g is a measurable map from the second to the third, and if μ is a measure on $\mathcal{E}$, then

$$g(f(\mu)) = (g \circ f)(\mu).$$

The following theorem shows how the integral with respect to the image of μ can be traced back to an integral with respect to the measure μ. Let us notice that this theorem is a sort of generalised version of a change of variable, which tells how to adapt the measure when one changes the variable inside the integral.

(4.3) Theorem. *Given two measurable spaces $(E, \mathcal{E})$ and $(F, \mathcal{F})$, let f be a measurable map from the first to the second, and let μ be a measure on $\mathcal{E}$. Let us set $\nu = f(\mu)$. Then,*

$$\int g \, \mathrm{d}\nu = \int (g \circ f) \, \mathrm{d}\mu$$

for every positive measurable function g on $(F, \mathcal{F})$.

Proof. Let us start by observing that, when g is the indicator function of a measurable set A, the equation $\int g \, \mathrm{d}\nu = \int (g \circ f) \, \mathrm{d}\mu$ simply reduces to the form $\nu(A) = \mu\{f \in A\}$ which is true by definition of ν. This relation remains true also when g is a positive simple function: In fact, such a function is a finite linear combination of indicator functions of measurable sets (for which the equation holds). Finally, when g is a positive measurable function, there exists a sequence (g_n) of positive simple functions such that $g_n \uparrow g$ and since the equation $\int g_n \, \mathrm{d}\nu = \int (g_n \circ f) \, \mathrm{d}\mu$ holds for each n, it suffices to take the limit of this equation and to apply Beppo Levi's theorem on both sides in order to completely prove the theorem. □

As a result of the previous theorem, the following also holds.

(4.4) Theorem. *Given two measurable spaces $(E, \mathcal{E})$ and $(F, \mathcal{F})$, let f be a measurable map from the first to the second, and let μ be a measure on $\mathcal{E}$. Let us set $\nu = f(\mu)$. Assume that g is a real function on $(F, \mathcal{F})$, not necessarily positive. Then, the following conditions are equivalent:*

(a) *g is lower (resp. upper) semi-integrable with respect to $\nu = f(\mu)$.*

(b) *$g \circ f$ is lower (resp. upper) semi-integrable with respect to μ.*

Furthermore, when these two conditions hold, we have

$$\int g \, \mathrm{d}\nu = \int (g \circ f) \, \mathrm{d}\mu.$$

5. Measure defined by a density

In the sequel, assume we are given a measure space $(E, \mathcal{E}, \mu)$, a real measurable function f defined on $(E, \mathcal{E})$, and a measurable set A. Let us recall that the function f is said to be integrable over A (with respect to the measure μ) if the function $I_A f$ is integrable. Similarly, f is said to be semi-integrable over A if the function $I_A f$ is semi-integrable. In this case, we shall write

$$\int_A f \, \mathrm{d}\mu = \int (I_A f) \, \mathrm{d}\mu,$$

and the number defined in this way shall be called the integral of f 'over' A. Let us prove the following theorem.

(5.1) Theorem. *For every positive measurable function f, the numeric set function ν, defined on $\mathcal{E}$ by setting $\nu(A) = \int_A f \, \mathrm{d}\mu$ is a measure.*

Proof. Observe that the equality $\nu(\emptyset) = 0$ immediately follows because the indicator function of the empty set is the constant 0. The property of series integration (Proposition (1.12)) shows that the function is countably additive, hence a measure. □

As a result of this, we can see that integrals provide an incredibly powerful tool to create new measures from a given measure. Precisely, let us give the following definition.

(5.2) Definition. Let f be a positive measurable function. The measure ν defined by

$$\nu(A) = \int_A f \, \mathrm{d}\mu \qquad \text{for all } A \in \mathcal{E}$$

shall be denoted by $\nu = f \cdot \mu$ and shall be called a measure **with base** μ, or, equivalently, **defined by the density** f with respect to μ, and the function f shall be called (a version of) the **density** of ν with respect to μ.

(5.3) Example. If μ is a measure on the measurable space $(E, \mathcal{E})$, and H is a measurable subset of E, the trace of μ to H, that is to say the measure μ_H defined for every measurable set A by $\mu_H(A) = \mu(A \cap H)$ can be written as $I_H \cdot \mu$, hence it is, evidently, a measure defined by the density I_H with respect to μ.

Observe that for the measure $f \cdot \mu$ to be finite it is necessary and sufficient that f is integrable with respect to μ. Furthermore, if μ is σ-finite, then in order for $f \cdot \mu$ to be σ-finite, it is necessary and sufficient that the function f is finite μ-almost everywhere. It is also clear that, if f and g are two positive measurable functions on $(E, \mathcal{E})$, such that $f \leq g$ holds μ-almost everywhere, then $f \cdot \mu \leq g \cdot \mu$, that is $\int_A f \, \mathrm{d}\mu \leq \int_A g \, \mathrm{d}\mu$ for every measurable set A. Conversely, the following theorem holds.

(5.4) Theorem. *Let f and g be two positive real measurable functions on the measurable space $(E, \mathcal{E})$, such that $f \cdot \mu \leq g \cdot \mu$. Then, this condition is true 'locally almost everywhere' with respect to μ. Equivalently, for any measurable set A, with finite measure with respect to μ, the set $\{f > g\} \cap A$ is μ-negligible.*

Proof. The set $\{f > g\} \cap A$ can be written as the countable union of sets of the form

$$\{f \geq r > s \geq g\} \cap A,$$

where r and s are rational numbers. It suffices to prove that each of these sets is negligible with respect to μ. Set $B = \{f \geq r > s \geq g\} \cap A$, and suppose, by contradiction, that it were $\mu(B) > 0$. Then, from

$$I_B f \geq r I_B > s I_B \geq I_B g,$$

and from the fact that $0 < \mu(B) < \infty$, we get

$$\int_B f \, \mathrm{d}\mu \geq r\mu(B) > s\mu(B) \geq \int_B g \, \mathrm{d}\mu,$$

and this contradicts the hypothesis that $f \cdot \mu \leq g \cdot \mu$. □

As a consequence of the previous theorem, if μ is a σ-finite measure and if f and g are two positive real valued measurable functions on $(E, \mathcal{E})$, then, in order to have $f \cdot \mu \leq g \cdot \mu$, it is necessary and sufficient that the relation $f \leq g$ holds almost everywhere with respect to μ. In particular, in order to have $f \cdot \mu = g \cdot \mu$, it is necessary and sufficient that the function f and g are equivalent modulo μ.

The following theorem shows how it is possible to calculate the integral with respect of a measure with base μ by tracing it back to an integral with respect to the measure μ itself.

(5.5) Theorem. *Given, on the measure space $(E, \mathcal{E}, \mu)$, a positive real measurable function f on $(E, \mathcal{E})$, let us set $\nu = f \cdot \mu$. Then, one has*

$$\int g \, \mathrm{d}\nu = \int g f \, \mathrm{d}\mu$$

for any positive real valued measurable function g on $(E, \mathcal{E})$.

Proof. Let us start by observing that, when g is the indicator function of a measurable set A, the equation $\int g \, \mathrm{d}\nu = \int g f \, \mathrm{d}\mu$ simply reduces to the form $\nu(A) = \int (I_A f) \, \mathrm{d}\mu$ which is true by definition of ν. This relation remains true also when g is a positive simple function: In fact, such a function is a finite linear combination of indicator functions of measurable sets (for which the equation holds). Finally, when g is a positive measurable function, there exists a sequence (g_n) of positive simple functions such that $g_n \uparrow g$ and since the equation $\int g_n \, \mathrm{d}\nu = \int g_n f \, \mathrm{d}\mu$ holds for each n, it suffices to take the limit of this equation and to apply Beppo Levi's theorem on both sides in order to completely prove the theorem. □

As a result of the previous theorem, the following result also holds.

(5.6) Theorem. *Given, on the measure space $(E, \mathcal{E}, \mu)$, a positive real measurable function on $(E, \mathcal{E})$, let us set $\nu = f \cdot \mu$. Assume that g is a real function on $(F, \mathcal{F})$, not necessarily positive. Then, the following conditions are equivalent:*

(a) *g is lower (resp. upper) semi-integrable with respect to $\nu = f(\mu)$.*

(b) *gf is lower (resp. upper) semi-integrable with respect to μ.*

Furthermore, when these two conditions hold, we have

$$\int g \, \mathrm{d}\nu = \int gf \, \mathrm{d}\mu.$$

6. Convergence theorems

Let us now prove the convergence properties of the integral, which are inherited by the convergence properties of a measure. To this end, let us assume that a measure space $(E, \mathcal{E}, \mu)$ is given. In the sequel, all measurable functions shall be assumed to be defined on the measurable space $(E, \mathcal{E})$.

(6.1) Theorem (Fatou's lemma). *Let (f_n) be a sequence of measurable functions, all bounded from below by the same integrable function g. Let us set $f = \liminf_n f_n$. Then, f and f_n are lower semi-integrable, and*

$$\int f \, \mathrm{d}\mu \leq \liminf_n \int f_n \, \mathrm{d}\mu.$$

Proof. With no loss of generality, we shall suppose that g is finite. If not, since g is integrable, then $\{g = \infty\}$ is negligible, and all the f_n are bounded from below by g, thus it suffices to change f_n and g on a negligible set. Also, taking the difference $f_n - g$, we can also suppose that all the functions f_n are positive. Suppose therefore that $f_n \geq 0$ for all n, and set

$$g_n = f_n \wedge f_{n+1} \wedge \cdots$$

Thus, $0 \leq g_n \leq f_n$, and $g_n \uparrow f$, whence, thanks to Beppo Levi's theorem,

$$\int f \, \mathrm{d}\mu = \lim_n \int g_n \, \mathrm{d}\mu \leq \liminf_n \int f_n \, \mathrm{d}\mu. \qquad \square$$

Beside the result just proved there is another (with the opposite sign and the limit superior instead of the limit inferior) which holds for measurable functions which are all bounded from above by the same integrable function. Naturally, these two statements are completely equivalent: The passage from one to the other coincides with the passage from the sequence (f_n) to the sequence $(-f_n)$. Putting these two results together, it is now easy to prove Lebesgue's theorem on dominated convergence.

(6.2) Theorem (Lebesgue's theorem on dominated convergence). *On a measure space $(E, \mathcal{E}, \mu)$, let (f_n) be a sequence of integrable functions convergent almost everywhere towards a function f, and let us suppose that there exists a positive integrable function g such that $|f_n| \leq g$ for all n. Then, the function f is also integrable, and*

$$\int f \, \mathrm{d}\mu = \lim_n \int f_n \, \mathrm{d}\mu.$$

Proof. Observe that, by means of Fatou's lemma, if (f_n) is a sequence of integrable functions such that $|f_n| \leq g$ for every n, then the function $\liminf_n f_n$ and $\limsup_n f_n$ are integrable and

$$\int (\liminf_n f_n) \, \mathrm{d}\mu \leq \liminf_n \int f_n \, \mathrm{d}\mu \leq \limsup_n \int f_n \, \mathrm{d}\mu \leq \int (\limsup_n f_n) \, \mathrm{d}\mu$$

Since the sequence (f_n) pointwise converges to f, the functions $\liminf_n f_n$ and $\limsup_n f_n$ coincide with f, and this completes the proof. □

As a consequence of this theorem, it turns out that the set $\mathcal{L}^1(\mu)$ is actually a monotone Riesz space. This theorem is often used in Mathematical Analysis and Probability Theory. It can be also stated for more general families of functions. For example, if $(f_t)_{t \in T}$ is a filtered family of measurable functions which converges towards a measurable function f, and if all the f_t are bounded from above, in absolute value, by an integrable function g, then Lebesgue's theorem enables to state that

$$\lim_{t \in T} \int f_t \, \mathrm{d}\mu = \int f \, \mathrm{d}\mu.$$

In fact, if, by contradiction, that were not the case, it would be possible to extract from (f_t) a sequence of functions that, in spite of satisfying all the hypotheses of Lebesgue's theorem, would contradict its thesis.

Now, several important functions in Mathematical Analysis (such as the Gamma function, or the Fourier transform), are defined as the integral of a function which depends on a parameter; Lebesgue's theorem is a fundamental tool to study this kind of functions.

Let us complete this section by showing two results that help to study this kind of functions.

(6.3) Theorem. *Let T be a metric space and let f be a real function defined on the product $E \times T$ such that, for each element t of T, the function $f(\cdot, t)$ is measurable, and, for each element x of E, the function $f(x, \cdot)$ is continuous. In addition, assume that all the functions $f(\cdot, t)$ are bounded from above, in absolute value, by the same element g of $\mathcal{L}^1(\mu)$.*

Then the function

$$t \mapsto \int f(x, t) \, \mu(\mathrm{d}x)$$

is continuous on T.

Proof. Let (t_n) be a sequence of elements of T which converges towards t. Then, the sequence (f_n) of measurable functions defined by $f_n = f(\cdot, t_n)$ for each n, satisfies the hypotheses of Lebesgue's theorem on dominated convergence, and pointwise converges towards the function $f(\cdot, t)$. Applying Lebesgue's theorem, the thesis follows immediately. □

Similarly, the following theorem also holds.

(6.4) Theorem. *Let T be an open subset of $\mathbb{R}$ and let f be a numeric function on $E \times T$, such that all functions $f(\cdot, t)$, with $t \in T$, are measurable, all function $t \mapsto f(t, x)$, with $x \in E$, are differentiable, and all the partial derivatives $\partial f / \partial t$ are all bounded from above, in absolute value, by the same element g of $\mathcal{L}^1(\mu)$.*

Then the function F defined on $t \in T$ by

$$F(t) = \int f(t, x)\, \mu(\mathrm{d}x)$$

is differentiable, and its derivative is given by

$$\frac{\mathrm{d}F}{\mathrm{d}t} = \int \left(\frac{\partial f}{\partial t} \right) \mathrm{d}\mu.$$

Proof. By linearity of the integral, it turns out that

$$\frac{F(t+h) - F(t)}{h} = \int \left[\frac{f(t+h, x) - f(t, x)}{h} \right] \mu(\mathrm{d}x).$$

When h approaches 0, the integrand tends towards $(\partial f / \partial t)(x, t)$ and all these functions, thanks to Lagrange's mean value theorem, are bounded from above, in absolute value, by the integrable function g. The thesis follows therefore from Lebesgue's theorem. □

Chapter III

Construction and Extension of Measures

This chapter is devoted to the study of different methods to construct measures. Up to now, in fact, we have only studied the general properties of measures without giving any concrete method to construct a measure, if not in some very elementary cases. In particular, even though the basic properties of the Borel-Lebesgue measure have already been mentioned, the existence of such a measure has yet to be ascertained. This is going to be one of the challenges of this chapter.

In particular, we shall now introduce not one, but two different methods to construct measures, which correspond to two completely different ways to look at measures. Precisely, in §1 and §2, we shall introduce an abstract notion that generalises that of integral, as a monotone functional on a suitable function space. After that, in §3 and §4, we shall focus on two important theorems, both due to Percy John Daniell (1889–1946), which shall be pivotal in constructing measures. In particular, these two theorems will facilitate the proof in §5 of the so-called Carathéodory-Hahn's extension theorem, which in turn allows to construct a measure by defining it on a particular set of generators for the σ-field. The main application of this extension theorem will give the Lebesgue measure as we shall see in §6.

Once Daniell's theory has been fully developed, in §7, we shall radically change our point of view and discuss the notion of outer measure, which was introduced by Constantin Carathéodory (Κωνσταντίνος Καραθεοδωρή, 1873–1950) in 1914, and which may be thought as a slightly weaker version of a measure, but that has the advantage of being defined for every set. Using the notion of outer measure as a starting point, we shall then discuss, in §8 and §9, two methods to construct measures, the first that starts from the notion of pre-measure, and does not require any structure, and the second that does require the ambient space to be a metric space. Finally, we shall use Carathéodory's theory to discuss, in §10, the problem of completion of a measure.

DOI: 10.1201/9781003724766-III

1. Introduction

As we saw in Chapter II, starting from a measure space $(E, \mathcal{E}, \mu)$, we are able to develop the theory of integration, ultimately leading to the construction of the monotone Riesz space $\mathcal{L}^1(\mu)$ of all integrable functions on E with respect to the measure μ. Moreover, if we define

(1.1) $$J(f) = \int f \, d\mu \qquad \text{for every element } f \text{ of } \mathcal{L}^1(\mu),$$

we find that J is a monotone linear form on $\mathcal{L}^1(\mu)$ satisfying the condition

(1.2) If $f_n \in \mathcal{L}^1(\mu)$, such that $f_n \downarrow 0$, then $J(f_n) \downarrow 0$.

Sometimes, however, it may be convenient to introduce the notion of integral directly, without relying on the underlying concept of measure. This idea was originally developed by P.J. Daniell who studied, in two papers which were published between 1918 and 1920, an abstract notion of functional that generalises the one of integral with respect to a measure — namely, the notion of a monotone linear form on a given Riesz space of real functions, equipped with property (1.2). Such a mathematical object is known as a 'Daniell functional' or 'Daniell integral'.

Once this notion has been introduced, it is natural to seek answers to two fundamental questions.

(a) Given a Daniell functional on a Riesz space $\mathcal{R}$ of real functions on a non-empty set, is it possible to extend it to a Daniell functional defined on a suitable monotone Riesz space containing $\mathcal{R}$?

(b) Given a Daniell functional on a Riesz space $\mathcal{R}$ of real functions on a non-empty set, is it possible to 'represent' it as an integral with respect to a suitable measure, defined on a suitable σ-field, or more precisely, as the restriction to $\mathcal{R}$ of a functional of the form (1.1)?

We shall first provide an affirmative answer to the first of these questions in §3. Following the approach introduced by G. Letta, this will be achieved by extending the functional from $\mathcal{R}$ to the class of all $\mathcal{R}$-*bounded* functions. Once this is accomplished, in §4, we shall see that, under suitable mild restrictions on the Riesz space $\mathcal{R}$, this extended integral is in fact an integral with respect to a suitable measure, thereby providing an affirmative answer to the second question as well.

2. Daniell functional

In the sequel, E shall be a given non-empty set. If $\mathcal{R}$ is a vector space of real functions defined on E (i.e. a vector subspace of $\mathbb{R}^E$), a linear form on $\mathcal{R}$ is a map L, defined on $\mathcal{R}$, with values in $\mathbb{R}$, such that

$$L(f+g) = L(f) + L(g), \qquad L(\alpha f) = \alpha L(f),$$

for any pair of functions f, g in $\mathcal{R}$, and any real number α. The linear form L is said to be **monotone** if $L(f) \leq L(g)$ for any pair f, g of functions in $\mathcal{R}$, with $f \leq g$. In order for a linear form to be monotone it is necessary and sufficient that it be **positive**, that is to say that $L(f) \geq 0$ for any positive function f in $\mathcal{R}$ — that is to say, for any $f \in \mathcal{R}_+$, where $\mathcal{R}_+$ denotes the cone of all positive elements of $\mathcal{R}$.

As we have already mentioned in §1, to any monotone linear form L, defined on a vector space $\mathcal{R}$ of real functions on a non-empty set E, which satisfies the property that, if $f_n \in \mathcal{R}$, such that $f_n \downarrow 0$, then $L(f_n) \downarrow 0$, we shall give the name of *Daniell functional*. This is not, however, the only possible condition which can be chosen, but there are other alternative ones that may be handier, depending on the situation. Precisely, let us prove the following result that gives four equivalent conditions.

(2.1) Theorem. *Let L be a monotone linear form on the vector space $\mathcal{R}$. Then, the following conditions are equivalent:*

(a) *If $f_n \in \mathcal{R}$, such that $f_n \downarrow 0$, then $L(f_n) \downarrow 0$.*

(b) *If $f, f_n \in \mathcal{R}$, such that $f_n \downarrow f$, then $L(f_n) \downarrow L(f)$.*

(c) *If $f, f_n \in \mathcal{R}$, such that $f_n \uparrow f$, then $L(f_n) \uparrow L(f)$.*

(d) *If $f, f_n \in \mathcal{R}_+$, such that $f = \sum_n f_n$, then $L(f) = \sum_n L(f_n)$.*

Proof. (a) $\Rightarrow$ (b): Take $f, f_n \in \mathcal{R}$, such that $f_n \downarrow f$, and set $g_n = f_n - f$. Then $g_n \downarrow 0$, thus, by hypothesis, $L(g_n) = L(f_n - f) = L(f_n) - L(f) \downarrow 0$, whence $L(f_n) - L(f) \downarrow 0$, which is equivalent to $L(f_n) \downarrow L(f)$.

(b) $\Rightarrow$ (c): This is trivial, and the two conditions (b) and (c) are manifestly equivalent, because these two statements follow one from the other by simply swapping the functions f, f_n with the functions $-f, -f_n$ and using the property of linearity of L.

(c) $\Rightarrow$ (d): Take $f, f_n \in \mathcal{R}$, such that $f = \sum_n f_n$ and set $g_k = \sum_{n \leq k} f_n$. Then, $g_k \uparrow f$, and observe that, by virtue of condition (c),

$$L(g_k) = \textstyle\sum_{n \leq k} L(f_n) \uparrow L(f).$$

On the other hand, $\sum_{n \leq k} L(f_n) \uparrow \sum_n L(f_n)$, hence $L(f) = \sum_n L(f_n)$, by uniqueness of limits.

(d) $\Rightarrow$ (a): Take $f_n \in \mathcal{R}$, such that $f_n \downarrow 0$. Since L in monotone, then $(L(f_n))$ is a decreasing sequence of positive real numbers, and hence converges to its infimum, say $c = \inf_n L(f_n)$. It suffices to prove that $c = 0$. To this end, set $g_n = f_n - f_{n+1}$. Then $\sum_n g_n = \sum_n (f_n - f_{n+1}) = f_0$, so

$$\begin{aligned} L(f_0) &= L(\textstyle\sum_n g_n) \\ &= \textstyle\sum_n L(f_n - f_{n+1}) \\ &= \textstyle\sum_n \big[L(f_n) - L(f_{n+1})\big] \\ &= L(f_0) - c. \end{aligned}$$

Hence, $c = 0$, and therefore $L(f_n) \downarrow 0$. □

We can now properly state the definition of Daniell functional.

(2.2) Definition. A functional L defined on a vector space $\mathcal{R}$ of real functions defined on E satisfies **Daniell's condition** if one (whence each) of the conditions stated in Theorem (2.1) holds. A real functional L, defined on $\mathcal{R}$, is called a **Daniell functional** if:

(D1) If $f, g \in \mathcal{R}$ and $\alpha \in \mathbb{R}$, then $f + g \in \mathcal{R}$, and $\alpha f \in \mathcal{R}$

(D2) If $f, g \in \mathcal{R}$, such that $f \leq g$, then $L(f) \leq L(g)$

(D3) L satisfies Daniell's condition.

In general, the limits of functions in $\mathcal{R}$ are not elements of $\mathcal{R}$, and therefore a Daniell functional is not stable under limiting operations. For this reason, the theorem that we are going to state and prove in the next section, consists in extending a Daniell functional L to a suitable class of functions on E, containing $\mathcal{R}$, and closed under suitable limiting operations (viz. a monotone Riesz space).

3. Daniell's extension theorem

Let us continue to assume that E is a non-empty set, and let us now assume that the vector space $\mathcal{R}$ of real functions on the set E is a Riesz space.

(3.1) Definition. A real function f defined on E is **$\mathcal{R}$-bounded** if there are two real functions g and h in $\mathcal{R}$ such that $h \leq f \leq g$, that is to say, if f is bounded by two elements of $\mathcal{R}$. We shall also denote by $\mathcal{R}^\mathfrak{b}$ the set of all $\mathcal{R}$-bounded functions on E.

It is easy to check that, from the fact that $\mathcal{R}$ is a Riesz space, it follows that the set $\mathcal{R}^\mathfrak{b}$ is a monotone Riesz space which contains $\mathcal{R}$, and therefore also the monotone class $\mathcal{M}(\mathcal{R})$, generated by $\mathcal{R}$.

Now, if L is a Daniell functional on the Riesz space $\mathcal{R}$, for any real function f in $\mathcal{R}^\mathfrak{b}$, let us set

$$L^*(f) = \inf\Big\{ \sup_n L(g_n) : g_n \in \mathcal{R},\ g_n \uparrow \sup_n g_n \geq f \Big\},$$

and

$$L_*(f) = \sup\Big\{ \inf_n L(h_n) : h_n \in \mathcal{R},\ h_n \downarrow \inf_n h_n \leq f \Big\}.$$

Observe immediately that these two numbers are always finite, because the real function f is assumed to be $\mathcal{R}$-bounded. In fact, since f is $\mathcal{R}$-bounded, there exist two functions, g and h in $\mathcal{R}$ such that $h \leq f \leq g$. Now, since among all increasing sequences of elements of $\mathcal{R}$ whose supremum is greater than f there is the constant sequence whose terms all coincide with g, one has $L^*(f) \leq L(g)$. Likewise, since among all decreasing sequences of elements of $\mathcal{R}$ whose infimum is lower than f there is the constant sequence whose terms all coincide with h, one has $L_*(f) \geq L(h)$, and this guarantees that both $L_*(f)$ and $L^*(f)$ are finite.

Thus, the maps $f \mapsto L^*(f)$ and $f \mapsto L_*(f)$ define two real valued functions defined on the monotone Riesz space $\mathcal{R}^{\mathfrak{b}}$. Furthermore, from the fact that

$$L_*(f) = -L^*(-f), \tag{3.2}$$

any statement involving one of these functionals has a corresponding dual statement involving the other. In particular, once the statement involving one is proved, the dual statement is automatically proved as well.

In addition, observe that, if f is $\mathcal{R}$-bounded, and if (h_m) and (g_n) are two sequences of functions in $\mathcal{R}$, such that $g_n \uparrow \sup_n g_n \geq f$ and $h_m \downarrow \inf_n h_n \leq f$, then there must exist two indices $\bar{m}$ and $\bar{n}$, such that $h_m \leq f \leq g_n$ as soon as $m \geq \bar{m}$ and $n \geq \bar{n}$. Now, since L is monotone, then, for $m \geq \bar{m}$ and $n \geq \bar{n}$, we have $L(h_m) \leq L(g_n)$, and therefore $\inf_m L(h_m) \leq \sup_n L(g_n)$. Hence

$$L_*(f) \leq L^*(f).$$

Finally, observe that, if f is a function in $\mathcal{R}$, then, among all increasing sequences of elements of $\mathcal{R}$ whose supremum is greater than f, there is the constant sequence whose terms all coincide with f, and this shows that $L^*(f) \leq L(f)$. On the other hand, if (g_n) is any increasing sequence of elements of $\mathcal{R}$ with $f \leq \sup_n g_n$, then $f_n = g_n \wedge f \uparrow f$, and therefore by property (c) of Theorem (2.1), one also has

$$L(f) = \sup_n L(f_n) \leq \sup_n L(g_n).$$

Now, since (g_n) is arbitrary, this entails $L(f) \leq L^*(f)$. This proves that the restriction of L^* to $\mathcal{R}$ coincides with L, and the same happens to L_*, thanks to the relation (3.2). For these reasons, the two functions L^* and L_* may and shall be called the **upper extension** and the **lower extension** of L, respectively. Note that these two extensions are generally not linear on the whole monotone Riesz space $\mathcal{R}^{\mathfrak{b}}$, but their restriction to $\mathcal{R}$ is.

The main algebraic properties of the upper extensions are summarised in the following proposition.

(3.3) Proposition. *Let $\mathcal{R}$ be a Riesz space of real functions on a non-empty set, and let L be a Daniell functional on $\mathcal{R}$. Then:*

(a) *L^* is a monotone extension of L.*

(b) *$L^*(\alpha f) = \alpha L^*(f)$, for each positive real number α.*

(c) *$L^*(f+g) \le L^*(f) + L^*(g)$.*

(d) *$L^*(f \wedge g) + L^*(f \vee g) \le L^*(f) + L^*(g)$.*

Proof. We have already shown that L^* is an extension of L and the fact that it is a monotone function is also an obvious consequence of its definition. The identity $L^*(\alpha f) = \alpha L^*(f)$, in the special case when $\alpha = 0$, becomes $L^*(0) = 0$, which is automatically true, for L^* is an extension of L and L is linear. In order to prove the same identity when α is strictly positive, observe that, for any increasing sequence (g_n) of elements of $\mathcal{R}$, $\sup_n L(\alpha g_n) = \alpha \sup_n L(g_n)$, and that, as (g_n) ranges over the collection of all increasing sequences of elements of $\mathcal{R}$ whose supremum is greater than f, the sequence (αg_n) describes the collection of all increasing sequences of elements of $\mathcal{R}$ whose supremum is greater than αf.

Finally, in order to prove properties (c) and (d), take a strictly positive real number ε and choose two increasing sequences (f_n) and (g_n) of functions in $\mathcal{R}$, such that

$$\begin{aligned} f &\le \sup\nolimits_n f_n, & g &\le \sup\nolimits_n g_n, \\ \sup\nolimits_n L(f_n) &< L^*(f) + \tfrac{1}{2}\varepsilon, & \sup\nolimits_n L(g_n) &< L^*(g) + \tfrac{1}{2}\varepsilon. \end{aligned}$$

Then, $(f_n + g_n)$, $(f_n \wedge g_n)$, and $(f_n \vee g_n)$ are three increasing sequences of functions in $\mathcal{R}$ whose suprema are greater than $f+g$, $f \wedge g$, and $f \vee g$, respectively. Thus,

$$\begin{aligned} L^*(f+g) &\le \sup\nolimits_n L(f_n + g_n) \\ &= \sup\nolimits_n L(f_n) + \sup\nolimits_n L(g_n) \\ &< L^*(f) + \tfrac{1}{2}\varepsilon + L^*(g) + \tfrac{1}{2}\varepsilon \\ &= L^*(f) + L^*(g) + \varepsilon \end{aligned}$$

and

$$\begin{aligned} L^*(f \wedge g) + L^*(f \vee g) &= \sup\nolimits_n L(f_n \wedge g_n) + \sup\nolimits_n L(f_n \vee g_n) \\ &= \sup\nolimits_n L(f_n \wedge g_n + f_n \vee g_n) \\ &= \sup\nolimits_n L(f_n + g_n) \\ &< L^*(f) + L^*(g) + \varepsilon. \end{aligned}$$

Since ε is arbitrary, this is sufficient to prove properties (c) and (d). □

The properties of the lower extensions are summarised in the following proposition, whose proof follows by duality from (3.3), by applying (3.2).

(3.4) Proposition. *Let $\mathcal{R}$ be a Riesz space of real functions on a non-empty set, and let L be a Daniell functional on $\mathcal{R}$. Then:*

(a) *L_* is a monotone extension of L.*

(b) *$L_*(\alpha f) = \alpha L_*(f)$, for each positive real number α.*

(c) *$L_*(f+g) \geq L_*(f) + L_*(g)$.*

(d) *$L_*(f \wedge g) + L_*(f \vee g) \geq L_*(f) + L_*(g)$.*

The sets of properties of the upper and lower extensions of a Daniell functional L suggest that, if the class of all $\mathcal{R}$-bounded functions on E such that $L^*(f) = L_*(f)$ had enough properties of stability (i.e. be a monotone Riesz space), then the restrictions of the upper and lower extensions of L to this class not only coincide, but this common restriction is also necessarily linear. This idea leads to Daniell's extension theorem.

(3.5) Theorem (Daniell's extension theorem). *On a Riesz space $\mathcal{R}$, let L be a Daniell functional, and let us denote by $\mathcal{L}$ the class of all $\mathcal{R}$-bounded functions on E for which the upper extension and the lower extension of L coincide. In other words, let us set*

$$\mathcal{L} = \left\{ f \in \mathcal{R}^{\mathfrak{b}} : L^*(f) = L_*(f) \right\}.$$

Let us also denote by J the common restriction of these two extensions to this class. Then, $\mathcal{L}$ is a monotone Riesz space, which contains $\mathcal{R}$, and J is a Daniell functional, which extends L.

Proof. Using the properties (a), (b), (c) for the upper and lower extensions, L^* and L_*, it is clear that the class $\mathcal{L}$ is a vector space and that J is a monotone linear form defined on $\mathcal{L}$. In order to prove that $\mathcal{L}$ is also a Riesz space, let us observe that, thanks to property (d) of the upper and lower extensions, if f and g are any two elements of $\mathcal{L}$, then

$$L^*(f \wedge g) + L^*(f \vee g) \leq J(f) + J(g) \leq L_*(f \wedge g) + L_*(f \vee g).$$

This chain of inequalities, since the left-hand side is greater than the right-hand side, must be in fact a chain of identities. Hence, rearranging this expression gives

$$L^*(f \wedge g) - L_*(f \wedge g) = L_*(f \vee g) - L^*(f \vee g).$$

Since in this latter identity the left-hand side is positive, while the right-hand side is negative, it follows that these two terms must both be zero. This proves that $f \wedge g$ and $f \vee g$ both belong to $\mathcal{L}$, and this proves that $\mathcal{L}$ is a Riesz space.

It remains to prove that the space $\mathcal{L}$ is monotone, and that the monotone linear form J on $\mathcal{L}$ is a Daniell functional. Thanks to condition (b) of Theorem (2.1), in order to prove these two statements simultaneously, it suffices to show that, if f is $\mathcal{R}$-bounded, and if (f_n) is an increasing sequence of elements of $\mathcal{L}$, such that $f = \sup_n f_n$, then f belongs to the class $\mathcal{L}$ and $J(f) = \sup_n J(f_n)$. To this end, choose a strictly positive real number ε and take an increasing sequence $(f_{n,k})$ of elements of $\mathcal{R}$, such that

$$f_n \le \sup_k f_{n,k}, \qquad \sup_k L(f_{n,k}) \le J(f_n) + \varepsilon.$$

Then $f = \sup_n f_n \le \sup_n \sup_k f_{n,k}$. Now, since L is a Daniell functional, using property (D2),

$$\begin{aligned} L^*(f) &\le \sup_n L(\sup_k f_{n,k}) \\ &= \sup_n \sup_k L(f_{n,k}) \\ &\le \sup_n J(f_n) + \varepsilon. \end{aligned}$$

Since ε is arbitrary, this means that

$$L^*(f) \le \sup_n J(f_n) \le L_*(f),$$

which proves that $L^*(f) = L_*(f) = \sup_n J(f_n)$. □

Thus, the previous theorem shows that every Daniell functional L defined on a Riesz space $\mathcal{R}$ can be extended to a Daniell functional J on a monotone Riesz space $\mathcal{L}$, containing $\mathcal{R}$. This new Daniell functional J is shall be called the **Daniell's extension** of L.

4. Daniell's representation theorem

The first application of the previous extension theorem consists in finding a theorem that responds to the question asked in §1, on the possibility of 'representing' a given Daniell functional as an integral with respect to a suitable measure. In order to tackle this problem, it may be useful to introduce the following terminology and notation.

If $\mathcal{R}$ is a class of numeric functions defined on a non-empty set E, let us denote by $\mathcal{R}_\sigma$ (resp. $\mathcal{R}_\delta$) the class of all numeric functions that are the supremum (resp. infimum) of a sequence of functions in $\mathcal{R}$. Note that, when $\mathcal{R}$ is a Riesz subspace of $\mathbb{R}^E$, then the two classes $\mathcal{R}_\sigma$ and $\mathcal{R}_\delta$ are stable under addition (and the elements of one are the opposite of the elements of the other).

(4.1) Theorem (Daniell's representation theorem). *Let $\mathcal{R}$ be a Riesz space of real functions on E, such that the constant 1 belongs to $\mathcal{R}_\sigma$. If L is a Daniell functional on $\mathcal{R}$, then on the σ-field $\mathcal{E}$ generated by $\mathcal{R}$, there exists one and only one measure μ such that*

(4.2) $\qquad \mathcal{R} \subseteq \mathcal{L}^1(\mu), \qquad L(f) = \int f \, \mathrm{d}\mu \quad$ *for every element f of $\mathcal{R}$.*

Proof. Let us consider an increasing sequence (g_n) of elements of $\mathcal{R}_+$, such that $\sup_n g_n = 1$. Let us also denote by $\mathcal{M}$ the monotone class generated by the Riesz space $\mathcal{R}$. Then, $\mathcal{M}$ (which is a monotone Riesz space) is endowed with Stone's property. In fact, for every element f of $\mathcal{M}$, the function $f \wedge 1$, being the supremum of the increasing sequence $(f \wedge g_n)$, whose terms belong to $\mathcal{M}$ and are bounded from above by f, belongs to $\mathcal{M}$. As a result, by means of Theorem I(6.9), the space $\mathcal{M}$ consists of all measurable functions on $(E, \mathcal{E})$ that are $\mathcal{M}$-bounded, hence $\mathcal{R}$-bounded.

Let us also denote by J the Daniell extension of L. Since $\mathcal{M}$ is the smallest monotone Riesz space containing $\mathcal{R}$, and since J is defined on a monotone Riesz space containing $\mathcal{R}$, by minimality, its domain also contains $\mathcal{M}$. For any positive integer n, the function μ_n defined on $\mathcal{E}$ by

$$\mu_n(A) = J(g_n I_A) \tag{4.3}$$

is a measure. In fact, if (A_k) is a sequence of pairwise disjoint elements of $\mathcal{E}$, whose union is A, from the identity $g_n I_A = \sum_k g_n I_{A_k}$, it follows, by means of condition (d) of Theorem (2.1) that

$$\begin{aligned}\mu_n(A) &= J(g_n I_A) \\ &= J(\textstyle\sum_k g_n I_{A_k}) \\ &= \textstyle\sum_k J(g_n I_{A_k}) \\ &= \textstyle\sum_k \mu_n(A_k),\end{aligned}$$

thanks to the fact that J is a Daniell integral.

Moreover, observe that (μ_n) is an increasing sequence of finite measures on $\mathcal{E}$, thus, by setting $\mu(A) = \sup_n \mu_n(A) = \lim_n \mu_n(A)$, it is defined a set function μ on the σ-field $\mathcal{E}$ which is not only additive, but also continuous on increasing sequences, and therefore a measure. For every element A of $\mathcal{E}$, whose indicator function is bounded from above by an element of $\mathcal{R}$ (and therefore belonging to $\mathcal{M}$), taking the limit of identity (4.3), gives $\mu(A) = J(I_A)$. By linearity, then the identity $\int f \, d\mu = J(f)$ holds for every positive simple function f on $(E, \mathcal{E})$ which is bounded from above by an element of $\mathcal{R}_+$. On the other hand, every element of $\mathcal{R}_+$ is the supremum of an increasing sequence of positive simple functions on $(E, \mathcal{E})$, so the identity $\int f \, d\mu = J(f)$ keeps on holding for every element f of $\mathcal{R}_+$. This proves that μ is a measure on $(E, \mathcal{E})$ such that $\mathcal{R} \subseteq \mathcal{L}^1(\mu)$ and $J(f) = \int f \, d\mu$ for every f in $\mathcal{R}$, as requested. To complete the proof, it suffices to show that if ν is a measure with the required properties, then ν and μ are identical. To this end, let us consider the class

$$\Big\{ f \in \mathcal{L}^1(\mu) \cap \mathcal{L}^1(\nu) : \int f \, d\mu = \int f \, d\nu \Big\}.$$

This is a monotone class which contains $\mathcal{R}$, hence $\mathcal{M}$. In particular, for every element A of $\mathcal{E}$, and every integer n, since $g_n I_A$ belongs to $\mathcal{M}$, we have

$$\int g_n I_A \, d\mu = \int g_n I_A \, d\nu$$

and taking the limit as n tends towards infinity, we get $\mu(A) = \nu(A)$. $\square$

In the hypotheses and with the notation of the previous theorem, the only measure μ on $\mathcal{E}$ such that $L(f) = \int f \, d\mu$ for every $f \in \mathcal{R}$, shall be called the **representation** of Daniell functional L, and it is sometimes denoted by μ_L to emphasise the relationship with L.

In order to immediately show some applications of Daniell's Theory, we shall now prove an interesting result on the approximation of integrable functions through a special class of integrable functions.

(4.4) Theorem. *Let $(E, \mathcal{E}, \mu)$ be a measure space and let $\mathcal{R}$ be a Riesz subspace of $\mathcal{L}^1(\mu)$ endowed with Stone's property and such that the σ-field $\mathcal{E}$ is generated by $\mathcal{R}$. Then, for every element f of $\mathcal{L}^1(\mu)$ which is $\mathcal{R}$-bounded,*

$$\sup\left\{\int g \, d\mu : g \in \mathcal{R}_\delta, \, g \leq f\right\} = \int f \, d\mu = \inf\left\{\int h \, d\mu : h \in \mathcal{R}_\sigma, \, f \leq h\right\}.$$

If, in addition, the constant 1 belongs to $\mathcal{R}_\sigma$, then the previous identity holds for any element f of $\mathcal{L}^1(\mu)$ and the Riesz space $\mathcal{R}$ is dense in $\mathcal{L}^1(\mu)$.

Proof. Let us begin by observing that the inclusion $\mathcal{R} \subseteq \mathcal{L}^1(\mu)$ implies that every element of $\mathcal{R}_\sigma$ is lower semi-integrable and every element of $\mathcal{R}_\delta$ is upper semi-integrable. Let us now prove the first statement. To this end, observe that, since the space $\mathcal{L}^1(\mu)$ contains $\mathcal{R}$, it also contains the monotone class $\mathcal{M}$, generated by $\mathcal{R}$. However, this latter class, being a monotone Riesz space endowed with Stone's property, coincides with the class of all measurable functions on $(E, \mathcal{E})$ which are $\mathcal{M}$-bounded (hence $\mathcal{R}$-bounded), by means in Theorem I(6.9).

Consider the monotone linear form J defined on $\mathcal{L}^1(\mu)$ by

$$J(f) = \int f \, d\mu \qquad \text{for every } f \text{ in } \mathcal{L}^1(\mu),$$

and denote by L its restriction to $\mathcal{R}$. Then J coincides, on $\mathcal{M}$, with the Daniell extension of L. In fact, the subset of $\mathcal{M}$ where the required coincidence holds is monotone and contains $\mathcal{R}$, hence $\mathcal{M}$. Thus, for every function f in $\mathcal{M}$, the number $\int f \, d\mu$ is equal to $L^*(f)$, which in turn is the infimum of all numbers of the form $\sup_n L(f_n)$, where (f_n) is an increasing sequence of elements of $\mathcal{R}$ whose supremum is greater than f. Now, these numbers are exactly those of the form $\int h \, d\mu$, where h is an element of $\mathcal{R}_\sigma$ greater than f. The second of the required equalities is therefore proved. To prove the first, it suffices to apply the second to the function $-f$.

Let us now prove the second statement. To this end, thanks to the decomposition $f = f^+ - f^-$, it suffices to prove that the desired identity holds only for every positive function f in $\mathcal{L}^1(\mu)$. Now, the hypothesis $1 \in \mathcal{R}_\sigma$ implies that there exists an increasing sequence (u_n) of functions in $\mathcal{R}_+$ such that $u_n \uparrow \infty$. As a result, given a positive function f in $\mathcal{L}^1(\mu)$, it is possible to write this function as the sum of a sequence (f_n) of measurable functions, which all are both positive and $\mathcal{R}$-bounded. Indeed, it suffices to take

$$f_1 = f \wedge u_1, \qquad f_n = f \wedge u_n - f \wedge u_{n-1} \qquad \text{for every } n > 1.$$

Let ε be a strictly positive real number. Then, for every index n, thanks to the fact that f_n are all $\mathcal{R}$-bounded, it is possible, applying to f_n the first part of the theorem, to find a pair g_n, h_n of positive functions, such that

$$g_n \in \mathcal{R}_\delta, \qquad h_n \in \mathcal{R}_\sigma, \qquad g_n \leq f_n \leq h_n, \qquad \int h_n \,\mathrm{d}\mu - \int g_n \,\mathrm{d}\mu < 2^{-n}\varepsilon.$$

Set $g = \sum_n g_n$, $h = \sum_n h_n$. Then, $g \leq f \leq h$ and

$$\begin{aligned} \int h \,\mathrm{d}\mu - \int g \,\mathrm{d}\mu &= \int \sum_n h_n \,\mathrm{d}\mu - \int \sum_n g_n \,\mathrm{d}\mu \\ &= \sum_n \left(\int h_n \,\mathrm{d}\mu - \int g_n \,\mathrm{d}\mu \right) \\ &< \sum_n 2^{-n}\varepsilon = \varepsilon. \end{aligned}$$

Furthermore, when n tends to infinity, $g_1 + \cdots + g_n \uparrow g$. As a result, when n is sufficiently large, the following inequality holds:

$$\int h \,\mathrm{d}\mu - \int (g_1 + \cdots + g_n) \,\mathrm{d}\mu < \varepsilon.$$

Since ε is arbitrary, this inequality, together with the relations

$$g_1 + \cdots + g_n \in \mathcal{R}_\delta, \qquad h \in \mathcal{R}_\sigma, \qquad g_1 + \cdots + g_n \leq f \leq h,$$

proves the desired identity.

It remains to prove that $\mathcal{R}$ is dense in $\mathcal{L}^1(\mu)$. To this end, taken an element f of $\mathcal{L}^1(\mu)$, a strictly positive real number ε, and chosen an element h of $\mathcal{R}_\sigma$, such that

$$f \leq h, \qquad \int (h - f) \,\mathrm{d}\mu < \varepsilon,$$

let us write h in the form $h = \sup_n f_n$, where (f_n) is an increasing sequence of elements of $\mathcal{R}$. Then, there exists an index n such that $\int (h - f_n) \,\mathrm{d}\mu < \varepsilon$, and therefore

$$\int |f - f_n| \,\mathrm{d}\mu \leq \int |f - h| \,\mathrm{d}\mu + \int |h - f_n| \,\mathrm{d}\mu < 2\varepsilon.$$

This is sufficient to complete the proof. □

5. Carathéodory-Hahn's extension theorem

The next example of application of Daniell's theory consists in the so-called Carathéodory-Hahn's extension theorem, which is the paramount tool which enables to define a measure, starting from an additive function on a suitable system of generators for the σ-field.

Before introducing the theorem, let us describe the properties of this suitable system of generators. To this end, let us give the following definition.

(5.1) Definition. Let E be a non-empty set. A class $\mathcal{A}$ of subsets of E is called a **ring** if the empty set, $\emptyset$, belongs to $\mathcal{A}$, and if it is stable for the operations of finite union, intersection, and set difference, that is to say

$$(A, B) \mapsto A \cup B, \qquad (A, B) \mapsto A \cap B, \qquad (A, B) \mapsto A \setminus B.$$

If a ring $\mathcal{A}$ contains the whole set E, we already know that it is then called a **field** (see Definition I(1.1)).

Let us observe immediately that the notion of 'additive' set functions can be defined on a ring in the exact same way as on a σ-field. Precisely, let us recall that a real valued function λ, defined on a ring $\mathcal{A}$, is **additive** if $\lambda(\emptyset) = 0$, and if, for any pair A, B of disjoint elements of $\mathcal{A}$,

$$\lambda(A \cup B) = \lambda(A) + \lambda(B).$$

The notion of 'countably additive' set function, instead, requires a little attention to the fact that the countable union of elements of a ring may not be an element of the ring. Hence, we shall say that λ is **countably additive** if, for every sequence (A_n) of pairwise disjoint sets in $\mathcal{A}$, such that $\bigcup_{n=0}^{\infty} A_n$ belongs to $\mathcal{A}$, we have

$$\lambda\left(\bigcup_{n=1}^{\infty} A_n\right) = \sum_{n=1}^{\infty} \lambda(A_n).$$

Observe that if λ is a positive additive function on a ring, then it is also monotone. Let us prove a useful lemma that aims to link a set function on a ring to a Daniell functional on a Riesz space.

(5.2) Lemma. *Let λ be a positive real additive function on a ring $\mathcal{A}$ defined on the non-empty set E. Let us denote by $\mathcal{R}$ the vector space generated by $\{I_A : A \in \mathcal{A}\}$. Then:*

(a) $\mathcal{R}$ is a Riesz space.

(b) There is one and only one monotone linear form L defined on $\mathcal{R}$, such that $L(I_A) = \lambda(A)$ for every element A of $\mathcal{A}$.

(c) If λ is countably additive, L is a Daniell functional.

Proof. Statement (a) is an obvious consequence of the stability under finite intersection. In fact, if A and B are elements of $\mathcal{A}$, then $I_A \wedge I_B = I_{A \cap B}$ and therefore $I_A \wedge I_B \in \mathcal{R}$. The uniqueness of statement (b) is also obvious, whereas the existence follows from the following two facts:

- Since $\mathcal{A}$ is a ring, every element f of $\mathcal{R}$ can be written in the form

$$f = \sum_{i \in I} a_i I_{A_i} \tag{5.3}$$

where I is a finite set, the a_i's are real numbers, and the A_i's are pairwise disjoint elements of $\mathcal{A}$.

• The sum $\sum_{i\in I} a_i\lambda(A_i)$ does not depend on the choice of the form (5.3) used to represent f. To see this, it suffices to observe that, thanks to the fact that λ is finitely additive, the value of this sum does not change when passing from a family of sets to a finer one.

Finally, to prove statement (c), suppose that λ is countably additive, and consider a decreasing sequence (f_n) of functions in $\mathcal{R}$, such that $\inf_n f_n = 0$. Let us set $A = \{f_1 > 0\}$ and let us denote by c the maximum value of f_1. Given a strictly positive real number ε, for every index n, we have

$$0 \leq f_n = f_n I_{\{f_n \leq \varepsilon\}} + f_n I_{\{f_n > \varepsilon\}} \leq \varepsilon I_A + c I_{\{f_n > \varepsilon\}}$$

and therefore, since the sets A and $\{f_n > \varepsilon\}$ both belong to the ring $\mathcal{A}$,

$$0 \leq L(f_n) \leq \varepsilon\lambda(A) + c\lambda\{f_n > \varepsilon\}. \tag{5.4}$$

On the other hand, since λ is countably additive, and since $\{f_n > \varepsilon\} \downarrow \emptyset$, because $f_n \downarrow 0$, one has $\lim_n \lambda\{f_n > \varepsilon\} = 0$. Taking the limit as n tends to infinity in (5.4) we get to

$$0 \leq \inf_n L(f_n) \leq \varepsilon\lambda(A).$$

Since ε is arbitrary, it follows that $\inf_n L(f_n) = 0$, and this proves that L is a Daniell functional. □

Using the theorem just proved, it is now easy to prove Carathéodory-Hahn's extension theorem.

(5.5) Theorem (Carathéodory-Hahn's extension theorem). *Let E be a non-empty set, $\mathcal{A}$ be a ring of subsets of E, such that E can be written as the increasing union of elements of $\mathcal{A}$, and let λ be a real positive countably additive function defined on $\mathcal{A}$. Denote by $\mathcal{E}$ the σ-field generated by $\mathcal{A}$.*

Then, there exists one and only one measure μ on $\mathcal{E}$ that extends λ.

Proof. The *uniqueness* of the measure follows from the fundamental criterion for the coincidence of two measures, because a ring with the additional property described in the hypothesis is a base. In order to prove the *existence*, consider the Riesz space $\mathcal{R}$ and the Daniell functional L defined in the previous lemma. Clearly, the σ-field generated by $\mathcal{R}$ coincides the σ-field $\mathcal{E}$ generated by $\mathcal{A}$. Furthermore, thanks to the hypothesis that the whole set E can be written as the increasing union of elements of $\mathcal{A}$, the constant 1 (i.e. the indicator function of the set E) belongs to $\mathcal{R}_\sigma$. Thus, it is possible to apply Daniell's representation theorem to the functional L. As a result, there exists a measure μ on $(E, \mathcal{E})$ which satisfies the relation (4.2). In particular, this gives, for every element A of the ring $\mathcal{A}$, the identity $L(I_A) = \int I_A \, d\mu$, which means $\lambda(A) = \mu(A)$, and this proves that μ is an extension of λ. □

6. Existence of the Borel-Lebesgue measure

Let us now use Carathéodory-Hahn's extension theorem to prove the existence of the Borel-Lebesgue measure on the real line. We shall do this in a slightly more general fashion, by proving the existence of a wider class of measures on the real line. Beforehand, let us introduce a new piece of terminology. Precisely, any measure defined on the Borel σ-field of a topological space shall be referred to as a **Borel measure**. The following theorem describes the properties of a Borel measure on the real line.

(6.1) Theorem.

(a) *Let μ be a Borel measure on $\mathbb{R}$. Then there is a function F, defined on $\mathbb{R}$, such that*

$$\mu\big((a,b]\big) = F(b) - F(a)$$

for any pair a, b of real numbers, such that $a < b$. This function F is right-continuous, increasing, and it is uniquely determined up to an additive constant.

(b) *Conversely, given a real function F, defined on $\mathbb{R}$, increasing, and right-continuous, there exists one and only one Borel measure μ on $\mathbb{R}$ such that*

$$\mu\big((a,b]\big) = F(b) - F(a)$$

for any pair a, b of real numbers, with $a < b$.

Proof. Let us start by proving statement (a). To this end, observe that a function F satisfying the required property is necessarily increasing, because a Borel measure is positive. Furthermore, such a function is also necessarily right-continuous, because, if (b_n) is a strictly decreasing sequence of real numbers, whose limit is a, then $(a, b_n] \downarrow \emptyset$, and therefore

$$F(b_n) - F(a) = \mu\big((a, b_n]\big) \downarrow 0.$$

To prove the existence of such a function F, let us set

$$F(0) = 0, \qquad F(x) = \begin{cases} \mu\big((0,x]\big) & \text{for } x > 0, \\ -\mu\big((x,0]\big) & \text{for } x < 0. \end{cases}$$

The fact that F is determined up to an additive constant is also obvious: In fact, if two real functions F_1 and F_2 both satisfy the same condition

$$\mu\big((a,b]\big) = F_1(b) - F_1(a) = F_2(b) - F_2(a),$$

their difference is zero over any interval, which means that their difference is a constant.

Let us now prove statement (b). To this end, denote by $\mathcal{A}$ the ring of all finite unions of bounded intervals of the form $(a, b]$, with $a < b$. The σ-field generated by $\mathcal{A}$ coincides with the Borel σ-field $\mathcal{B}(\mathbb{R})$. Moreover, $\mathbb{R}$ is the union of an increasing sequence of elements of $\mathcal{A}$. Thus, thanks to Carathéodory-Hahn's extension theorem, it suffices to prove that there is a positive real function λ, countably additive on $\mathcal{A}$, such that

$$\lambda\big((a, b]\big) = F(b) - F(a)$$

for any pair a, b of real numbers, such that $a < b$. Now, if, for any element A of the ring $\mathcal{A}$ we set

$$\lambda(A) = \textstyle\sum_{i\in I} \big[F(b_i) - F(a_i)\big],$$

where $\big((a_i, b_i]\big)_{i\in I}$ is any finite partition of A made of half-open intervals, this defines with no ambiguity a positive real function λ, additive on the ring $\mathcal{A}$. It remains to prove that λ is countably additive. To this end, let (A_n) be a sequence of pairwise disjoint elements of $\mathcal{A}$, assume that $A = \bigcup_{n=1}^{\infty} A_n$ also belongs to $\mathcal{A}$, and let us prove that $\lambda(A) = \sum_{n=1}^{\infty} \lambda(A_n)$. For this, observe that, taken a whole number m, thanks to the obvious inclusion $\bigcup_{n=1}^{m} A_n \subseteq A$, it follows that

$$\textstyle\sum_{n=1}^{m} \lambda(A_n) = \lambda\big(\bigcup_{n=1}^{m} A_n\big) \leq \lambda(A),$$

hence $\sum_{n=1}^{\infty} \lambda(A_n) \leq \lambda(A)$. To conclude that λ is countably additive, it remains to prove the opposite inequality. In order to show this, observe that, since F is right-continuous, taken a strictly positive real number ε, there always are B, B_n in $\mathcal{A}$ such that

$$B \subseteq \overline{B} \subseteq A, \quad \lambda(A) - \lambda(B) < \varepsilon, \quad A_n \subseteq B_n^{\circ} \subseteq B_n, \quad \lambda(B_n) - \lambda(A_n) < 2^{-n}\varepsilon.$$

Since $\overline{B}$ is the finite union of closed bounded intervals, it is a compact set, and since (A_n) is a covering of A, then (B_n°) is an open covering of $\overline{B}$. Thus, we can extract a finite sub-covering. In other words, there must be an index m such that $\overline{B} \subseteq \bigcup_{n=1}^{m} B_n^{\circ}$, and even more so $B \subseteq \bigcup_{n=1}^{m} B_n$. Hence,

$$\begin{aligned} \lambda(B) &\leq \lambda\big(\textstyle\bigcup_{n=1}^{m} B_n\big) \\ &= \textstyle\sum_{n=1}^{m} \lambda(B_n) \\ &\leq \textstyle\sum_{n=1}^{\infty} \lambda(B_n). \end{aligned}$$

Finally, from the fact that $\lambda(A) < \lambda(B) + \varepsilon$ and $\lambda(B_n) < \lambda(A_n) + 2^{-n}\varepsilon$, it follows that

$$\lambda(A) \leq \textstyle\sum_{n=1}^{\infty} \big[\lambda(A_n) + 2^{-n}\varepsilon\big] + \varepsilon = \sum_{n=1}^{\infty} \lambda(A_n) + 2\varepsilon.$$

Since ε is arbitrary, this proves that $\lambda(A) \leq \sum_{n=1}^{\infty} \lambda(A_n)$, and λ is thereby countably additive. We may now apply Carathéodory-Hahn's extension theorem and infer that there is one and only one measure μ on the σ-field generated by $\mathcal{A}$ that extends λ. In other words, there is a uniquely determined Borel measure on the real line, such that $\mu\big((a, b]\big) = F(b) - F(a)$ for any pair a, b of real numbers, such that $a < b$. □

We are finally ready to give the following general definition.

(6.2) Definition. Let F be a real function, defined on $\mathbb{R}$, increasing and right-continuous. The only Borel measure μ on $\mathbb{R}$ such that

$$\mu\big((a,b]\big) = F(b) - F(a) \qquad \text{for every } a, b \in \mathbb{R}, \text{ with } a < b,$$

is called the **Stieltjes measure** generated by F. In particular, the measure generated by the identity is the **Borel-Lebesgue measure** on $\mathbb{R}$.

When dealing with this type of measures, one usually adopts the following notation: The monotone Riesz space of all integrable functions with respect to this measure shall be denoted by $\mathcal{L}^1(F)$ and, if f is a real function belonging to this space, the integral of f with respect to the measure generated by F shall preferably be denoted by $\int f \, \mathrm{d}F$ or $\int f(x) \, \mathrm{d}F(x)$.

To complete this section, note that if μ is the measure generated by F, then

$$\mu(\mathbb{R}) = \sup_x F(x) - \inf_x F(x),$$

so the measure μ is finite if and only if the function F is bounded. Furthermore, for each real number x, we have

$$\mu\{c\} = F(c) - \lim_{x \to c,\, x < c} F(x)$$

and therefore the measure μ is diffuse (i.e. null on every singleton) if and only if F is continuous.

7. Carathéodory's outer measures

Let us now change point of view, and construct measures with the aid of the so-called Carathéodory's outer measure, after the Greek mathematician C. Carathéodory who first introduced them in 1914. The main idea behind the notion of outer measure is to define a measure-like set function on the class of all subsets of a non-empty set, which may not be countably additive, and then to restrict this to a suitable σ-field, where the function turns out to be a measure.

(7.1) Definition. A positive set function $\mathfrak{m}$, defined on the set $\mathcal{P}(E)$ of all subsets of a non-empty set E, shall be called an **outer measure** if

(O1) $\mathfrak{m}(\emptyset) = 0$,

(O2) $\mathfrak{m}$ is monotone, that is, $\mathfrak{m}(A) \leq \mathfrak{m}(B)$, whenever $A \subseteq B$,

(O3) $\mathfrak{m}$ is **countably sub-additive**, that is to say that, for any countable family $(A_i)_{i \in I}$ of subsets of E one has

$$\mathfrak{m}\left(\bigcup_{i \in I} A_i\right) \leq \sum_{i \in I} \mathfrak{m}(A_i).$$

Although in general this function need not be even finitely additive, it does happen that it is additive when it is restricted to a suitable σ-field on E. In order to see this, let us start by giving the following definition.

(7.2) Definition. Let $\mathfrak{m}$ be an outer measure defined on a non-empty set E. A subset A of E is said to be (**Carathéodory**) $\mathfrak{m}$**-measurable** if, for every subset X of E, one has

$$\mathfrak{m}(X \cap A) + \mathfrak{m}(X \cap A^c) = \mathfrak{m}(X). \tag{7.3}$$

The class of all $\mathfrak{m}$-measurable sets of E shall be denoted by $\mathfrak{M}_{\mathfrak{m}}$.

In other words, a set A is $\mathfrak{m}$-measurable if it splits every set according to the requirement of additivity of $\mathfrak{m}$. Let us also note at once that in general the measurability does not follow from the equality $\mathfrak{m}(A) + \mathfrak{m}(A^c) = \mathfrak{m}(E)$, even in the case of an outer measure with $\mathfrak{m}(E) < \infty$. To see this, it suffices to consider on the set $E = \{1, 2, 3\}$ the set function $\mathfrak{m}$ defined by:

$$\mathfrak{m}(\emptyset) = 0, \qquad \mathfrak{m}(E) = 2, \qquad \mathfrak{m}(A) = 1 \quad \text{for all other sets } A.$$

It is easy to check that $\mathfrak{m}$ is an outer measure on E. In addition, observe that, here, every subset $A \subseteq E$ obviously satisfies $\mathfrak{m}(A) + \mathfrak{m}(A^c) = \mathfrak{m}(E)$, but for $A = \{1\}$ and $X = \{1, 2\}$, the equality (7.3) does not hold (its left-hand side equals 2 and the right-hand side equals 1). It is easy to see that only the two sets $\emptyset$ and E are $\mathfrak{m}$-measurable.

Let us also observe that, if $\mathfrak{m}$ is sub-additive — in particular, if $\mathfrak{m}$ is an outer measure, in order to check whether a set is $\mathfrak{m}$-measurable, it suffices to show that

$$\mathfrak{m}(X) \geq \mathfrak{m}(X \cap A) + \mathfrak{m}(X \cap A^c)$$

for every subset X of E. Furthermore, observe that this inequality becomes trivially true when $\mathfrak{m}(X) = \infty$, and so we only need to establish it for sets X for which $\mathfrak{m}(X)$ is finite.

The next result establishes the main properties of the class $\mathfrak{M}_{\mathfrak{m}}$ constituted of all $\mathfrak{m}$-measurable sets on E. Before we may state it, however, let us give the following definition.

(7.4) Definition. A measure μ on a measurable space $(E, \mathcal{E})$ is **complete** if every subset of a μ-negligible set is measurable. In other words, the measure μ is complete if and only if the following condition holds:

$$[A \in \mathcal{E},\ \mu(A) = 0,\ B \subseteq A] \Rightarrow B \in \mathcal{E}$$

We can now prove the following result.

(7.5) Theorem. *Let $\mathfrak{m}$ be a positive numeric set function, defined on the class of all subsets of E, such that $\mathfrak{m}(\emptyset) = 0$. Then, the class $\mathfrak{M}_{\mathfrak{m}}$ is a σ-field, and the restriction of $\mathfrak{m}$ to $\mathfrak{M}_{\mathfrak{m}}$ is countably additive, hence a measure. In addition, the measure $\mathfrak{m}$, defined on $\mathfrak{M}_{\mathfrak{m}}$, is complete.*

Proof. It is obvious that $\emptyset \in \mathfrak{M}_{\mathfrak{m}}$ and that the class $\mathfrak{M}_{\mathfrak{m}}$ is closed under complementation. Suppose that the sets A and B both belong to $\mathfrak{M}_{\mathfrak{m}}$, and let X be any subset of E. Since A and B are both $\mathfrak{m}$-measurable, we have

$$\begin{aligned}\mathfrak{m}(X) &= \mathfrak{m}(X \cap A) + \mathfrak{m}(X \cap A^c)\\ &= \mathfrak{m}(X \cap A) + \mathfrak{m}(X \cap A^c \cap B) + \mathfrak{m}(X \cap A^c \cap B^c)\\ &= \mathfrak{m}(X \cap A) + \mathfrak{m}(X \cap A^c \cap B) + \mathfrak{m}\big(X \cap (A \cup B)^c\big).\end{aligned}$$

Since $X \cap A = X \cap (A \cup B) \cap A$, and A is $\mathfrak{m}$-measurable, one also has

$$\mathfrak{m}\big(X \cap (A \cup B)\big) = \mathfrak{m}(X \cap A) + \mathfrak{m}(X \cap A^c \cap B).$$

Hence, substituting in the previous equality, we obtain

$$\mathfrak{m}(X) = \mathfrak{m}\big(X \cap (A \cup B)\big) + \mathfrak{m}\big(X \cap (A \cup B)^c\big).$$

Thus, the set $A \cup B$ belongs to $\mathfrak{M}_{\mathfrak{m}}$ which is thereby stable under finite union. Then, let us show that, for every sequence (A_n) of disjoint elements of $\mathfrak{M}_{\mathfrak{m}}$, the set $A = \bigcup_n A_n$ is $\mathfrak{m}$-measurable. To this end, for every index n, let us set $B_n = \bigcup_{i=1}^n A_i$. Then, each B_n is $\mathfrak{m}$-measurable, and

$$\mathfrak{m}(X) = \mathfrak{m}(X \cap B_n) + \mathfrak{m}(X \cap B_n^c) \geq \mathfrak{m}(X \cap B_n) + \mathfrak{m}(X \cap A^c),$$

because $A^c \subseteq B_n^c$. Now, since $B_n \cap A_n = A_n$ and $B_n \cap A_n^c = B_{n-1}$, and thanks to the fact that A_n is $\mathfrak{m}$-measurable, we also have

$$\mathfrak{m}(X \cap B_n) = \mathfrak{m}(X \cap A_n) + \mathfrak{m}(X \cap B_{n-1}).$$

Hence, by induction, we find

$$\mathfrak{m}(X \cap B_n) = \textstyle\sum_{i=1}^n \mathfrak{m}(X \cap A_i),$$

and therefore

$$\begin{aligned}\mathfrak{m}(X) &\geq \mathfrak{m}(X \cap A^c) + \textstyle\sum_{i=1}^\infty \mathfrak{m}(X \cap A_i)\\ &\geq \mathfrak{m}(X \cap A) + \mathfrak{m}(X \cap A^c),\end{aligned}$$

because $X \cap A \subseteq \bigcup_{i=1}^\infty (X \cap A_i)$. Thus, A is $\mathfrak{m}$-measurable. So $\mathfrak{M}_{\mathfrak{m}}$ is also stable under pairwise disjoint union. In order to conclude that this class is a σ-field, it suffices to notice that, if (A_n) is a sequence of elements of $\mathfrak{M}_{\mathfrak{m}}$, such is the pairwise disjoint sequence (B_n) defined by

$$B_0 = A_0, \qquad B_{n+1} = A_{n+1} \setminus \textstyle\bigcup_{i=1}^n B_i,$$

thus the set $\bigcup_n A_n = \bigcup_n B_n$ is $\mathfrak{m}$-measurable. So, $\mathfrak{M}_{\mathfrak{m}}$ is a σ-field.

It remains to prove that the restriction of $\mathfrak{m}$ on $\mathfrak{M}_{\mathfrak{m}}$ is a measure. To do so, let us start by showing that $\mathfrak{m}$ is finitely additive on $\mathfrak{M}_{\mathfrak{m}}$. To this end, take two disjoint $\mathfrak{m}$-measurable set A and B. Then, by the fact that A is $\mathfrak{m}$-measurable, it follows that

$$\begin{aligned}\mathfrak{m}(A \cup B) &= \mathfrak{m}\big((A \cup B) \cap A\big) + \mathfrak{m}\big((A \cup B) \cap A^c\big) \\ &= \mathfrak{m}(A) + \mathfrak{m}(B).\end{aligned}$$

Furthermore, if (A_n) is a pairwise disjoint sequence of elements of $\mathfrak{M}_{\mathfrak{m}}$ and if $A = \bigcup_n A_n$, then

$$\mathfrak{m}(A) \geq \mathfrak{m}\big(\textstyle\bigcup_{i=1}^n A_i\big) = \sum_{i=1}^n \mathfrak{m}(A_i)$$

hence $\mathfrak{m}(A) \geq \sum_{i=1}^{\infty} \mathfrak{m}(A_i)$. On the other hand, $\mathfrak{m}(A) \leq \sum_{i=1}^{\infty} \mathfrak{m}(A_i)$, because $\mathfrak{m}$ is countably sub-additive, hence $\mathfrak{m}$ is countably additive and thus a measure on $\mathfrak{M}_{\mathfrak{m}}$ since it is positive and, by definition, $\mathfrak{m}(\emptyset) = 0$. To complete the proof, we need to prove that $\mathfrak{m}$ is also complete. To this end, let A be a subset of E, with $\mathfrak{m}(A) = 0$, and let X be any subset of E. Since $X \cap A \subseteq A$ and $\mathfrak{m}$ is monotone, we have $\mathfrak{m}(X \cap A) = 0$. On the other hand, since $X \cap A^c \subseteq X$, and since $\mathfrak{m}$ is sub-additive, we also have

$$\mathfrak{m}(X \cap A^c) \leq \mathfrak{m}(X) \leq \mathfrak{m}(X \cap A) + \mathfrak{m}(X \cap A^c) = \mathfrak{m}(X \cap A^c),$$

which yields $\mathfrak{m}(X) = \mathfrak{m}(X \cap A) + \mathfrak{m}(X \cap A^c)$, whence A is $\mathfrak{m}$-measurable. □

As a result of the previous theorem, every outer measure defines a σ-field, and its restriction to this σ-field is a measure. An interesting property of $\mathfrak{m}$-measurable sets is that the outer measure $\mathfrak{m}$ is countably additive even on more general sets, such as those of the form $X \cap A$, with X any subset of E and A is the union of a sequence of pairwise disjoint $\mathfrak{m}$-measurable sets. Precisely, let us prove the following theorem.

(7.6) Theorem. *Let E be a non-empty set, $\mathfrak{m}$ an outer measure on E, X a subset of E, and (A_n) a sequence of pairwise disjoint $\mathfrak{m}$-measurable sets in E. Set $A = \bigcup_n A_n$. Then,*

$$\mathfrak{m}(X \cap A) = \textstyle\sum_n \mathfrak{m}(X \cap A_n).$$

Proof. Let C_1 and C_2 be two disjoint $\mathfrak{m}$-measurable sets, and let X be any subset of E. Then, since in particular C_1 is $\mathfrak{m}$-measurable, and since C_1 and C_2 are disjoint,

$$\begin{aligned}\mathfrak{m}\big(X \cap (C_1 \cup C_2)\big) &= \mathfrak{m}\big(X \cap (C_1 \cup C_2) \cap C_1\big) + \mathfrak{m}\big(X \cap (C_1 \cup C_2) \cap C_1^c\big) \\ &= \mathfrak{m}(X \cap C_1) + \mathfrak{m}(X \cap C_2).\end{aligned}$$

Hence, it follows, by mathematical induction, that

$$\mathfrak{m}\big(X \cap \textstyle\bigcup_{i=1}^n A_i\big) = \sum_{i=1}^n \mathfrak{m}(X \cap A_i)$$

for every positive integer n. Now, let us set $B_n = \bigcup_{i=1}^n A_i$. By definition, this yields $\bigcup_{n=1}^\infty B_n = \bigcup_{i=1}^\infty A_i = A$, and, since $\mathfrak{M}_{\mathfrak{m}}$ is a σ-field, for every positive integer n, the sets B_n are $\mathfrak{m}$-measurable. Thus,

$$\begin{aligned}\mathfrak{m}(X \cap A) &= \mathfrak{m}(X \cap A \cap B_n) + \mathfrak{m}(X \cap A \cap B_n^c) \\ &\geq \mathfrak{m}(X \cap B_n) \\ &= \textstyle\sum_{i=1}^n \mathfrak{m}(X \cap A_i).\end{aligned}$$

Since this is true for every n, we obtain

$$\mathfrak{m}(X \cap A) \geq \textstyle\sum_{i=1}^n \mathfrak{m}(X \cap A_i).$$

To complete the proof, it suffices to observe that the opposite inequality follows from the fact that, since $\mathfrak{m}$ is an outer measure, in particular it is countably sub-additive. □

Let us conclude this section with a theorem that shows that the supremum of a family of outer measures is again an outer measure. We shall find this simple result useful in §9.

(7.7) Proposition. *Let $\mathfrak{H}$ be a family of outer measures on a non-empty set E. Then, the set function $\mathfrak{s}$ defined for every subset A of E as*

$$\mathfrak{s}(A) = \sup\Big\{\mathfrak{m}(A) : \mathfrak{m} \in \mathfrak{H}\Big\}$$

is an outer measure on E.

Proof. Clearly, one has $\mathfrak{s}(\emptyset) = 0$, because $\mathfrak{m}(\emptyset) = 0$ for every subset $\mathfrak{m}$ in $\mathfrak{H}$. Furthermore, if $A \subseteq B$, then

$$\begin{aligned}\mathfrak{s}(A) &= \sup\{\mathfrak{m}(A) : \mathfrak{m} \in \mathfrak{H}\} \\ &\leq \sup\{\mathfrak{m}(B) : \mathfrak{m} \in \mathfrak{H}\} \\ &= \mathfrak{s}(B),\end{aligned}$$

which shows that the set function $\mathfrak{s}$ is also monotone. Finally, if $(A_i)_{i\in I}$ is a countable family of subsets of E, then, for every $\mathfrak{m}$ in $\mathfrak{H}$, one has

$$\mathfrak{m}\left(\textstyle\bigcup_{i\in I} A_i\right) \leq \textstyle\sum_{i\in I} \mathfrak{m}(A_i) \leq \textstyle\sum_{i\in I} \mathfrak{s}(A_i),$$

and taking the supremum of all $\mathfrak{m}$ in $\mathfrak{H}$ yields

$$\mathfrak{s}\left(\textstyle\bigcup_{i\in I} A_i\right) \leq \textstyle\sum_{i\in I} \mathfrak{s}(A_i)$$

as required. □

8. Constructing outer measures on an abstract set

Unfortunately, in spite of the nice properties of an outer measure, it turns out that assigning one is virtually as complicated as assigning a measure. It is therefore essential to establish a criterion to define an outer measure, starting from simpler set functions. In fact, we are going to introduce two methods to construct outer measures. Precisely, the rest of this section is devoted to describe the so-called 'Method I', which is fairly easy and intuitive, while the next section shall be devoted to describe the so-called 'Method II', which requires some more sophisticated tools. Let us start by giving the following definition.

(8.1) Definition. Let us suppose that a non-empty set E is assigned, and let $\mathfrak{X}$ be a class of subsets of E such that $\varnothing \in \mathfrak{X}$. A **pre-measure** on $\mathfrak{X}$ is any positive numeric set function τ, defined on $\mathfrak{X}$, such that $\tau(\varnothing) = 0$.

Given a pre-measure τ on E, let us set

$$\tau^*(A) = \inf\left\{ \textstyle\sum_{k=1}^{\infty} \tau(X_k) : X_k \in \mathfrak{X},\, A \subseteq \bigcup_{k=1}^{\infty} X_k \right\},$$

where, in case of absence of such sets X_k, we shall set $\tau^*(A) = \infty$. We shall prove that τ^* is an outer measure on E. To this end, we need to show that $\tau^*(\varnothing) = 0$, that τ^* is monotone, and that τ^* is countably sub-additive. Now by definition, it is obvious, that $\tau^*(\varnothing) = 0$, and it is also clear, that τ^* is monotone, that is to say that if $B \subseteq A$, then $\tau^*(B) \le \tau^*(A)$. The following lemma proves that τ^* is also countably sub-additive, which means an outer measure.

(8.2) Lemma. *Let τ be a positive numeric set function on a class $\mathfrak{X}$ of subsets of E. Then the set function τ^* is countably sub-additive, that is,*

$$\tau^*\left(\textstyle\bigcup_{n=1}^{\infty} A_n\right) \le \textstyle\sum_{n=1}^{\infty} \tau^*(A_n)$$

for any sequence (A_n) of subsets of E.

Proof. If the series $\sum_n \tau^*(A_n)$ is divergent, there is nothing to prove, so assume that it is convergent, so in particular every $\tau^*(A_n)$ is finite. Then, take a strictly positive real number ε, and observe that, by definition of τ^*, for every index n, there is a sequence $(X_{n,k})$ of sets in $\mathfrak{X}$ such that

$$A_n \subseteq \textstyle\bigcup_k X_{n,k}, \qquad \textstyle\sum_k \tau(X_{n,k}) \le \tau^*(A_n) + 2^{-n}\varepsilon.$$

Then, $\bigcup_n A_n \subseteq \bigcup_n \bigcup_k X_{n,k}$ and therefore

$$\tau^*\left(\textstyle\bigcup_n A_n\right) \le \textstyle\sum_n \sum_k \tau(X_{n,k}) \le \sum_n \tau^*(A_n) + \varepsilon.$$

Since ε is an arbitrary strictly positive real number, this is enough to conclude that τ^* is countably sub-additive. □

We can now summarise how to construct a measure from a pre-measure.

(8.3) Definition. Let E be a non-empty set, $\mathfrak{X}$ a class of subsets of E with $\emptyset \in \mathfrak{X}$, and τ a pre-measure on $\mathfrak{X}$, which means a positive set function on $\mathfrak{X}$, such that $\tau(\emptyset) = 0$. Then the set function τ^* defined on $\mathcal{P}(E)$ by

$$\tau^*(A) = \inf\Big\{\sum_{k=1}^{\infty} \tau(X_k) : X_k \in \mathfrak{X},\ A \subseteq \bigcup_{k=1}^{\infty} X_k\Big\}$$

is an outer measure, which is called the **outer measure generated** by τ. The restriction of τ^* to the σ-field $\mathfrak{M}_{\tau^*}$ of all τ^*-measurable sets is a measure, which is called the **measure generated** by τ.

It should be emphasised that it is not claimed that τ^* extends τ. In general, in fact, this may be false. In addition, sets in the original family $\mathfrak{X}$ may even turn out to be non-measurable with respect to τ^*. (For example, take E to be the set $\mathbb{N}$, let $\mathfrak{X}$ be the class of all singletons and the whole set E, and let τ be defined as $\tau(n) = 2^{-n}$ and $\tau(E) = 2$. Then, $\tau^*(E) = 1$ and E is measurable with respect to τ^*.) On the other hand, if $\mathfrak{X}$ is the set of all parts of E, and $\mathfrak{m}$ is an outer measure on E, taking $\tau = \mathfrak{m}$, we can show that $\tau^* = \mathfrak{m}$ and therefore, the class $\mathfrak{M}_{\mathfrak{m}}$ of $\mathfrak{m}$-measurable sets of E may be smaller than $\mathfrak{X}$, unless in some trivial cases in which $\mathfrak{M}_{\mathfrak{m}}$ turns out to be the whole class of all subsets of E.

Let us now introduce an important class of outer measures.

(8.4) Definition. Let $\mathfrak{m}$ be an outer measure on a non-empty set E. We shall say that $\mathfrak{m}$ is **regular** if, for every subset A of E, it is always possible to find a $\mathfrak{m}$-measurable set B such that $A \subseteq B$ and $\mathfrak{m}(A) = \mathfrak{m}(B)$.

Regular outer measures have some very nice properties. For instance, if $\mathfrak{m}$ is a regular outer measure on E, and $\mathfrak{m}(E) < \infty$, then a set A is $\mathfrak{m}$-measurable (if and) only if

$$\mathfrak{m}(A) + \mathfrak{m}(A^c) = \mathfrak{m}(E)$$

To see this, let X be a subset of E, and, since $\mathfrak{m}$ is regular, let us consider an $\mathfrak{m}$-measurable subset Y of E such that $X \subseteq Y$ and $\mathfrak{m}(X) = \mathfrak{m}(Y)$. As we already mentioned, it suffices to show that $\mathfrak{m}(X) \geq \mathfrak{m}(X \cap A) + \mathfrak{m}(X \cap A^c)$. Let us also note that, by sub-additivity,

$$\mathfrak{m}(A \cap Y^c) + \mathfrak{m}(A^c \cap Y^c) \geq \mathfrak{m}(Y^c), \tag{8.5}$$

On the other hand, since Y is measurable, one has

$$\mathfrak{m}(A) = \mathfrak{m}(A \cap Y) + \mathfrak{m}(A \cap Y^c), \qquad \mathfrak{m}(A^c) = \mathfrak{m}(A^c \cap Y) + \mathfrak{m}(A^c \cap Y^c).$$

Putting these inequalities together, and combining with the sub-additivity of $\mathfrak{m}$, we get to

$$\begin{aligned}
\mathfrak{m}(E) &= \mathfrak{m}(A) + \mathfrak{m}(A^c)\\
&= \big(\mathfrak{m}(A \cap Y) + \mathfrak{m}(A \cap Y^c)\big) + \big(\mathfrak{m}(A^c \cap Y) + \mathfrak{m}(A^c \cap Y^c)\big)\\
&\geq \mathfrak{m}(Y) + \mathfrak{m}(Y^c)\\
&= \mathfrak{m}(E).
\end{aligned}$$

Thus, the inequality in the last chain must be an equality, and, subtracting it from (8.5), which is possible because $\mathfrak{m}$ is finite, we eventually get to the inequality $\mathfrak{m}(A \cap Y) + \mathfrak{m}(A^c \cap Y) \leq \mathfrak{m}(Y)$. Finally, this latter inequality, alongside the inclusion $X \subseteq Y$ and the fact that $\mathfrak{m}$ is monotone, yields

$$\mathfrak{m}(X \cap A) + \mathfrak{m}(X \cap A^c) \leq \mathfrak{m}(Y \cap A) + \mathfrak{m}(Y \cap A^c) \leq \mathfrak{m}(Y) = \mathfrak{m}(X).$$

The example discussed at the beginning of this section shows that this inequality is generally not sufficient to show that a set is $\mathfrak{m}$-measurable, henceforth that not all outer measures are regular. The following theorem shows a useful condition to ensure that an outer measure τ^* generated by a pre-measure τ be regular.

(8.6) Theorem. *Let E be a non-empty set. Suppose that $\mathfrak{X}$ is a ring on E, and let τ be an additive set function defined on $\mathfrak{X}$. Then, all sets in $\mathfrak{X}$ are τ^*-measurable, and the outer measure τ^* is regular. Also, if τ is countably additive, then τ^* coincides with τ on $\mathfrak{X}$, that is to say, the outer measure τ^* is an extension of τ.*

Proof. In order to prove that the element of the class $\mathfrak{X}$ are τ^*-measurable, let us take an element A of the ring $\mathfrak{X}$, and any subset X of E. It suffices to show that $\tau^*(X) \geq \tau^*(X \cap A) + \tau^*(X \cap A^c)$. To this end, observe that, by definition of τ^*, for any strictly positive real number ε, there is a sequence (X_n) of elements of $\mathfrak{X}$ whose union contains X, and such that

$$\tau^*(X) > \textstyle\sum_n \tau(X_n) + \varepsilon.$$

Since $\mathfrak{X}$ is a ring, every term in the two sequences $(X_n \cap A)$ and $(X_n \cap A^c)$ belongs to $\mathfrak{X}$; and since the unions $\bigcup_n (X_n \cap A)$ and $\bigcup_n (X_n \cap A^c)$ contain $X \cap A$ and $X \cap A^c$, respectively,

$$\begin{aligned}\tau^*(X) &> \textstyle\sum_n \tau(X_n) + \varepsilon \\ &= \textstyle\sum_n \tau(X_n \cap A) + \sum_n \tau(X_n \cap A^c) + \varepsilon \\ &\geq \tau^*(X \cap A) + \tau^*(X \cap A^c) + \varepsilon\end{aligned}$$

where, in the second step, we have used the fact that τ is additive. Since ε is arbitrary, this proves that $\tau^*(X) \geq \tau^*(X \cap A) + \tau^*(X \cap A^c)$, i.e. that A is τ^*-measurable.

Let us prove now that τ^* is regular, i.e. that for any subset A of E, there is a τ^*-measurable subset B of E such that $A \subseteq B$ and $\tau^*(A) = \tau^*(B)$. To this end, let A be any subset of E, and, for every whole number n, let us consider a sequence $(X_{n,k})$ of elements of $\mathfrak{X}$, such that

$$A \subseteq \textstyle\bigcup_k X_{n,k}, \qquad \tau^*(A) \leq \sum_k \tau(X_{n,k}) \leq \tau^*(A) + (1/n).$$

Set $B = \bigcap_n \bigcup_k X_{n,k}$. It is easy to check that $A \subseteq B$. Therefore, it only suffices to show that $\tau^*(B) = \tau^*(A)$. In fact, since $A \subseteq B$, it suffices to show that $\tau^*(B) \leq \tau^*(A)$. To this end, let us observe that, since all elements of the class $\mathfrak{X}$ are τ^*-measurable, then $\tau^*(X) \leq \tau(X)$ for all $X \in \mathfrak{X}$. To see this, it suffices to choose the covering (X_n) where $X_1 = X$ and $X_n = \emptyset$ for $n > 1$. Thus, by countable sub-additivity of τ^*, we have

$$\begin{aligned} \tau^*(B) &\leq \tau^*\left(\textstyle\bigcup_k X_{n,k}\right) \\ &\leq \textstyle\sum_k \tau^*(X_{n,k}) \\ &\leq \textstyle\sum_k \tau(X_{n,k}) \\ &\leq \tau^*(A) + (1/n) \end{aligned}$$

Thus, $\tau^*(B) \leq \tau^*(A)$, hence τ^* is a regular outer measure.

To complete the proof, assume that τ is countably additive and let us prove that $\tau^*(X) = \tau(X)$ for all $X \in \mathfrak{X}$. We have already mentioned the fact that $\tau^*(X) \leq \tau(X)$ whenever $X \in \mathfrak{X}$. To show that the opposite inequality holds, let (X_n) be a sequence of elements of $\mathfrak{X}$ such that $X \subseteq \bigcup_n X_n$. Then, since $X \in \mathfrak{X}$ and since $X_n \cap X \in \mathfrak{X}$, it turns out that $\bigcup_n X_n \cap X = X \in \mathfrak{X}$ and therefore

$$\begin{aligned} \tau(X) &= \tau(\textstyle\bigcup_n X_n \cap X) \\ &= \textstyle\sum_n \tau(X_n \cap X) \\ &\leq \textstyle\sum_n \tau(X_n) \end{aligned}$$

and taking the infimum over all possible sequences (X_n) yields the inequality $\tau(X) \leq \tau^*(X)$, which proves that the latter is in fact an identity, and this completes the proof of the theorem. □

(8.7) Remark. Let us stress again that, in the most general case of the construction of an outer measure τ^* from a pre-measure τ on a class $\mathfrak{X}$, the outer measure τ^* may differ from τ on $\mathfrak{X}$, and the equality between these two functions is only assured when τ is itself countably additive.

(8.8) Remark. Theorem (8.6) ensures that, if a positive countably additive set function τ is defined on a ring $\mathfrak{X}$ of subsets of a non-empty set E, this additive function τ can be extended to a regular outer measure τ^*, such that the σ-field $\mathfrak{M}_{\tau^*}$ always includes $\mathfrak{X}$ — hence the σ-field generated by the ring $\mathfrak{X}$, and that the restriction of τ^* to $\mathfrak{M}_{\tau^*}$ is a countably additive extension of τ. Observe how this corresponds exactly to the statement of Carathéodory-Hahn's extension theorem, whose proof, using the notion of outer measures rather than Daniell's theory, in fact resembles the original method employed by Carathéodory himself to prove this famous result.

(8.9) Example (Lebesgue measure). On the real line, $\mathbb{R}$, let us consider the pre-measure τ defined on the ring $\mathfrak{R}$ of all finite union of pairwise disjoint intervals by

$$\lambda(X) = \textstyle\sum_{k=1}^{m}(b_k - a_k) \qquad \text{for every subset } X = \textstyle\bigcup_{k=1}^{m}(a_k, b_k] \text{ in } \mathfrak{R},$$

By virtue of Theorem (8.6), this pre-measure extends to an outer measure, λ^*, whose restriction to the σ-field $\mathfrak{L}$ of all λ^*-measurable subsets of $\mathbb{R}$ is a complete measure, that is generally called the **Lebesgue measure** on $\mathbb{R}$. In other words, the Lebesgue measure is the measure defined by the outer measure

$$\lambda^*(A) = \inf\Big\{\textstyle\sum_{k=1}^{\infty} \lambda(I_k) : I_k \in \mathfrak{R},\ A \subseteq \bigcup_{k=1}^{\infty} I_k\Big\}.$$

Since $\mathfrak{R}$ is a base for the Borel σ-field $\mathcal{B}(\mathbb{R})$ (see I(5.5)) and since $\mathfrak{R} \subseteq \mathfrak{L}$, we also have $\mathcal{B}(\mathbb{R}) \subseteq \mathfrak{L}$.

It is not a simple task to show that the σ-field $\mathfrak{L}$ is actually strictly larger than the Borel σ-field $\mathcal{B}(\mathbb{R})$, that is to say, there are sets in $\mathbb{R}$ that belong to $\mathfrak{L}$ but that do not belong to $\mathcal{B}(\mathbb{R})$.

Let us also observe that the Lebesgue measure also coincides with the measure constructed with Method I starting from the pre-measure

$$\operatorname{diam}(X) = \max(X) - \min(X) \qquad \text{for every subset } X \text{ of } \mathbb{R},$$

defined on the set $\mathfrak{X}$ of all subsets X of $\mathbb{R}$ with $\operatorname{diam}(X) < \infty$. In other words, the Lebesgue measure coincides with the measure defined by restriction to the σ-field $\mathfrak{M}_{\mathfrak{l}}$ by the outer measure

$$\mathfrak{l}(A) = \inf\Big\{\textstyle\sum_{k=1}^{\infty} \operatorname{diam}(X_k) : X_k \in \mathfrak{X},\ A \subseteq \bigcup_{k=1}^{\infty} X_k\Big\}.$$

In fact, on one hand, it is completely obvious that $\mathfrak{l}(A) \leq \lambda^*(A)$, because the family of coverings $\mathfrak{X}$ that defines the measure $\mathfrak{l}$ is larger than the class of covering $\mathfrak{R}$ that defines λ. On the other hand, consider a covering (X_k) in $\mathfrak{X}$ such that $A \subseteq \bigcup_{k=1}^{\infty} X_k$, and observe that, for every strictly positive real number ε, from the fact that $\operatorname{diam}(X_k) < \infty$ it follows that there always exists an interval I_k in $\mathbb{R}$ containing X_k, such that

$$\lambda(I_k) \leq \operatorname{diam}(X_k) + 2^{-k}\varepsilon.$$

Thus, the family (I_k) is a family of elements of $\mathfrak{R}$ such that

$$\textstyle\sum_{k=1}^{\infty} \lambda(I_k) \leq \sum_{k=1}^{\infty} \operatorname{diam}(X_k) + \varepsilon,$$

whence, thanks to the fact that ε is chosen arbitrarily, yields $\lambda^*(A) \leq \mathfrak{l}(A)$.

Finally, by virtue of the fundamental criterion for the coincidence of two measures (see Theorem I(5.6)), observe that the Lebesgue measure coincides with the Borel-Lebesgue measure on the Borel σ-field $\mathcal{B}(\mathbb{R})$. Thus, the construction via Carathéodory-Hahn's theorem done in §6 and this construction of the Lebesgue measure, done with the aid of Carathéodory's outer measures, are similar but not identical. The main difference between this and the Borel-Lebesgue measure is that the latter is only defined on the Borel σ-field of $\mathbb{R}$, while the former is complete, and it is defined on a strictly larger σ-field, $\mathfrak{L}$, which is called the **Lebesgue σ-field**.

9. Constructing outer measures on a metric space

In the previous section, we have discussed 'Method I' to construct an outer measure starting from a pre-measure. In this section, we shall particularise our hypotheses, by adding to the set E some topological structure. Specifically, we are going to assume that E is a metric space, and let us denote by d its metric. With this additional hypothesis, if A and B are subsets of E, the distance between the sets A and B is defined by

$$d(A, B) = \inf \Big\{ d(a, b) : a \in A,\, b \in B \Big\}.$$

This is already sufficient to prove a very important and meaningful result, due to Carathéodory, that shows a condition on an outer measure to ensure that all Borel sets are measurable under this outer measure.

(9.1) Theorem (Carathéodory's measurability criterion). *Let E be a metric space with metric d, and $\mathfrak{m}$ be an outer measure on E. In order for every Borel set to be $\mathfrak{m}$-measurable, it is necessary and sufficient that the following condition holds:*

$$\mathfrak{m}(A \cup B) = \mathfrak{m}(A) + \mathfrak{m}(B) \qquad \textit{whenever } A, B \in \mathcal{B}(E), \textit{ and } d(A, B) > 0.$$

Proof. Let us start by proving that the condition is *necessary*. To this end, suppose that $\mathfrak{M}_\mathfrak{m}$ contains all closed sets and assume that A and B are two Borel sets with $d(A, B) = d_0 > 0$. Consider the sets

$$C_1 = \Big\{ x \in E : d(x, A) \leq \tfrac{1}{4} d_0 \Big\}, \qquad C_2 = \Big\{ x \in E : d(x, B) \leq \tfrac{1}{4} d_0 \Big\}.$$

These are closed, $A \subseteq C_1$ and $B \subseteq C_2$. Let us observe that, since C_1 and C_2 are supposed to be $\mathfrak{m}$-measurable, by virtue of Theorem (7.6), one has

$$\mathfrak{m}\big((A \cup B) \cap (C_1 \cup C_2)\big) = \mathfrak{m}\big((A \cup B) \cap C_1\big) + \mathfrak{m}\big((A \cup B) \cap C_2\big),$$

and since

$$(A \cup B) \cap C_1 = A, \quad (A \cup B) \cap C_2 = B, \quad (A \cup B) \cap (C_1 \cup C_2) = A \cup B,$$

this means that $\mathfrak{m}(A \cup B) = \mathfrak{m}(A) + \mathfrak{m}(B)$. This proves that, if the Borel sets are $\mathfrak{m}$-measurable, then the condition holds.

Conversely, let us now prove that the condition is also *sufficient*. In other words, suppose that the condition stated in the theorem holds, and let us prove that, then every Borel set is $\mathfrak{m}$-measurable. For this to happen, it suffices to show that the class of all closed sets is $\mathfrak{m}$-measurable. Thus, let C be a closed set. As we have already mentioned, it suffices to prove the inequality

$$\mathfrak{m}(A) \geq \mathfrak{m}(A \cap C) + \mathfrak{m}(A \cap C^c),$$

for every subset A of E. If $\mathfrak{m}(A) = \infty$, this is completely obvious, so let us assume that $\mathfrak{m}(A) < \infty$. Since C is a closed set, the sequence (C_n) defined for every positive integer n by

$$C_n = \Big\{x \in E : d(x, C) \leq 1/n\Big\},$$

is decreasing and its intersection is C.

It is clear that $d(A \cap C_n^c, A \cap C) \geq 1/n > 0$, which, thanks to the fact that, by hypothesis, $(A \cap C) \cup (A \cap C_n^c) \subseteq A$, yields

$$\mathfrak{m}(A \cap C) + \mathfrak{m}(A \cap C_n^c) = \mathfrak{m}\big((A \cap C) \cup (A \cap C_n^c)\big) \leq \mathfrak{m}(A).$$

To complete the proof, it suffices to show that the sequence $(\mathfrak{m}(A \cap C_n^c))$ converges to the number $\mathfrak{m}(A \cap C^c)$ when n approaches ∞. In fact, if this latter condition holds, then

$$\mathfrak{m}(A \cap C) + \mathfrak{m}(A \cap C^c) = \mathfrak{m}(A \cap C) + \lim_{n \to \infty} \mathfrak{m}(A \cap C_n^c) \leq \mathfrak{m}(A).$$

So, to prove that the sequence $(\mathfrak{m}(A \cap C_n^c))$ converges towards $\mathfrak{m}(A \cap C^c)$, for every positive integer k, consider the set

$$D_k = \Big\{x \in A : 1/(k+1) < d(x, C) < 1/k\Big\}.$$

It is obvious that $A \cap C^c = \bigcup_{k=n}^{\infty} D_k \cup (A \cap C_n^c)$. Hence,

$$\mathfrak{m}(A \cap C_n^c) \leq \mathfrak{m}(A \cap C^c) \leq \mathfrak{m}(A \cap C_n^c) + \textstyle\sum_{k=n}^{\infty} \mathfrak{m}(D_k).$$

Now, in order for the sequence $(\mathfrak{m}(A \cap C_n^c))$ to converge, it is sufficient to show that the series with general term $\mathfrak{m}(D_k)$ is convergent. Indeed, one has $d(D_k, D_h) > 0$ if $h > k + 2$. Thus, the hypothesis of the theorem ensures that $\mathfrak{m}(D_k \cup D_h) = \mathfrak{m}(D_k) + \mathfrak{m}(D_h)$ whenever $h > k + 2$, and, by induction, this finally gives

$$\textstyle\sum_{k=1}^{N} \mathfrak{m}(D_{2k}) = \mathfrak{m}\big(\bigcup_{k=1}^{N} D_{2k}\big) \leq \mathfrak{m}(A).$$

A similar identity obviously holds for odd numbers, and this ensures that the sequence $(\mathfrak{m}(A \cap C_n^c))$ converges, and therefore C is $\mathfrak{m}$-measurable. Now, since every closed set is $\mathfrak{m}$-measurable, and since the class of $\mathfrak{m}$-measurable sets is a σ-field, then all Borel sets are also $\mathfrak{m}$-measurable. □

The condition expressed by this criterion is so important to deserve a name.

(9.2) Definition. On a metric space E with metric d, an outer measure $\mathfrak{m}$ is called a **metric** outer measure if

$$\mathfrak{m}(A \cup B) = \mathfrak{m}(A) + \mathfrak{m}(B) \qquad \text{whenever } A, B \in \mathcal{B}(E), \text{ and } d(A, B) > 0.$$

With this new definition, Carathéodory's measurability criterion can be rephrased by saying that for every Borel set of E to be $\mathfrak{m}$-measurable it is necessary and sufficient that $\mathfrak{m}$ be a metric outer measure. In other words, the σ-field $\mathcal{B}(E)$ of all Borel sets in E is $\mathfrak{m}$-measurable if and only if $\mathfrak{m}$ is a metric outer measure.

As we have already mentioned, the aim of this section is to develop another method to construct a metric outer measure from a pre-measure, which shall be called 'Method II', by exploiting the metric properties of the underlying set E. To this end, let us start by recalling that, if E is a metric space, d denotes its metric, and A is a subset of E, the **diameter** of A is defined as

$$\operatorname{diam}(A) = \sup\Big\{ d(x,y) : x, y \in E \Big\},$$

with the convention that $\operatorname{diam}(\varnothing) = 0$.

(9.3) Theorem. *Let E be a metric space with metric d, $\mathfrak{X}$ a class of subsets of E such that $\varnothing \in \mathfrak{X}$, τ a pre-measure on $\mathfrak{X}$. i.e. a positive set function on $\mathfrak{X}$, such that $\tau(\varnothing) = 0$. For every subset A of E, let us set*

$$\mathfrak{h}_\delta(A) = \inf\Big\{ \textstyle\sum_{n=1}^{\infty} \tau(X_n) : X_n \in \mathfrak{X},\ \operatorname{diam}(X_n) \le \delta,\ A \subseteq \bigcup_{n=1}^{\infty} X_n \Big\}.$$

Then, the set function $\mathfrak{h}$ defined on $\mathcal{P}(E)$ by

$$\mathfrak{h}(A) = \sup\Big\{ \mathfrak{h}_\delta(A) : \delta > 0 \Big\} \qquad \text{for every subset } A \text{ of } E,$$

is a metric outer measure.

Proof. Let us start by showing that $\mathfrak{h}$ is indeed an outer measure. To this end, for each positive real number δ, let us denote by $\mathfrak{X}(\delta)$ the set of all elements X of $\mathfrak{X}$ such that $\operatorname{diam}(X) \le \delta$. Then, it follows immediately that the restriction of τ to $\mathfrak{X}(\delta)$ remains a pre-measure on E. Then, for every positive real number δ, we can write $\mathfrak{h}_\delta$ in the form

$$\mathfrak{h}_\delta(A) = \inf\Big\{ \textstyle\sum_{n=1}^{\infty} \tau(X_n) : X_n \in \mathfrak{X}(\delta),\ A \subseteq \bigcup_{n=1}^{\infty} X_n \Big\}.$$

which shows that $\mathfrak{h}_\delta$ is the outer measure generated by τ_δ (see (8.3)). By virtue of Proposition (7.7), the set function $\mathfrak{h}$ is an outer measure on E, for it is the supremum of a family of outer measures on E.

It remains to prove that $\mathfrak{h}$ is a metric outer measure. Since $\mathfrak{h}$ is an outer measure, hence sub-additive, it suffices to show that

$$\mathfrak{h}(A \cup B) \ge \mathfrak{h}(A) + \mathfrak{h}(B)$$

for any pair A and B of non-empty subsets of E such that $d(A,B) > 0$. In proving this, we may clearly suppose that $\mathfrak{h}(A \cup B)$ is finite. The idea of the proof consists in constructing, from a 'fine' cover for $A \cup B$, separate 'fine' covers A and B separately, and we shall do that using a process called 'filtering out'. To this end, let us observe that, thanks to the fact that $d(A, B) > 0$, we may choose a positive real number δ_0 such that

$$\inf\Big\{d(a,b) : a \in A,\ b \in B\Big\} \geq \delta_0.$$

Let ε be a fixed strictly positive real number, and let δ_1 and δ_2 be given with $0 < \delta_1 < \operatorname{diam}(E)$ and $0 < \delta_2 < \operatorname{diam}(E)$. Let us denote by δ the minimum of $\frac{1}{2}\delta_0$, δ_1, and δ_2. Since $\mathfrak{h}(A \cup B)$ is assumed to be finite, we have

$$\inf\Big\{\textstyle\sum_{n=1}^{\infty} \tau(X_n) : X_n \in \mathfrak{X},\ \operatorname{diam}(X_n) \leq \delta,\ A \cup B \subseteq \bigcup_{n=1}^{\infty} X_n\Big\} \leq \mathfrak{h}(A \cup B),$$

so we can choose a sequence (X_n) of sets in $\mathfrak{X}$ with

$$\operatorname{diam}(X_n) < \delta, \qquad A \cup B \subseteq \textstyle\bigcup_n X_n, \qquad \sum_n \tau(X_n) \leq \mathfrak{h}(A \cup B) + \varepsilon.$$

Now, if, for some index n, one had both $X_n \cap A \neq \emptyset$ and $X_n \cap B \neq \emptyset$, there would be two elements $a_0 \in X_n \cap A$ and $b_0 \in X_n \cap B$ such that

$$d(a_0, b_0) \leq \operatorname{diam}(X_n) \leq \delta \leq \tfrac{1}{2}\delta_0 \leq \tfrac{1}{2}\left(\inf_{a \in A,\, b \in B} d(a,b)\right) \leq \tfrac{1}{2} d(a_0, b_0),$$

with the contradictory implication $d(a_0, b_0) = 0$. This means that, for no index n can we have both $X_n \cap A \neq \emptyset$ and $X_n \cap B \neq \emptyset$. Thus, we may define two sequences (Y_n) and (Z_n) of elements of $\mathfrak{X}$ as follows:

$$\begin{array}{llll} Y_n = X_n & \text{if } X_n \cap A \neq \emptyset, \qquad & Z_n = X_n & \text{if } X_n \cap B \neq \emptyset, \\ Y_n = \emptyset & \text{if } X_n \cap A = \emptyset, \qquad & Z_n = \emptyset & \text{if } X_n \cap B = \emptyset. \end{array}$$

Then, clearly,

$$\begin{aligned} A &= A \cap \left(\textstyle\bigcup_n X_n\right) = \textstyle\bigcup_n (A \cap X_n) \subseteq \textstyle\bigcup_n Y_n \\ B &= B \cap \left(\textstyle\bigcup_n X_n\right) = \textstyle\bigcup_n (B \cap X_n) \subseteq \textstyle\bigcup_n Z_n \end{aligned}$$

Furthermore, the two sequences (Y_n) and (Z_n) of elements of $\mathfrak{X}$ are such that, for every index n, $\operatorname{diam}(Y_n) \leq \delta$ and $\operatorname{diam}(Z_n) \leq \delta$, and, by construction, one also has

$$\begin{array}{ll} \text{either} & \tau(Y_n) + \tau(Z_n) = \tau(\emptyset) + \tau(\emptyset) = 0, \\ \text{or} & \tau(Y_n) + \tau(Z_n) = \tau(X_n) + \tau(\emptyset) = \tau(X_n). \end{array}$$

In either case,

$$\tau(Y_n) + \tau(Z_n) \leq \tau(X_n).$$

Now, by construction, for every index n, one has

$$\begin{array}{lll} Y_n \in \mathfrak{X}, \qquad & \operatorname{diam}(Y_n) \leq \delta \leq \delta_1, \qquad & A \subseteq \bigcup_n Y_n, \\ Z_n \in \mathfrak{X}, \qquad & \operatorname{diam}(Z_n) \leq \delta \leq \delta_2, \qquad & B \subseteq \bigcup_n Z_n. \end{array}$$

Hence

$$\mathfrak{h}_{\delta_1}(A) \leq \sum_{n=1}^{\infty} \tau(Y_n), \qquad \mathfrak{h}_{\delta_2}(B) \leq \sum_{n=1}^{\infty} \tau(Z_n),$$

which yields

$$\begin{aligned} \mathfrak{h}_{\delta_1}(A) + \mathfrak{h}_{\delta_2}(B) &\leq \sum_n \big[\tau(Y_n) + \tau(Z_n)\big] \\ &\leq \sum_n \tau(X_n) \\ &\leq \mathfrak{h}(A \cup B) + \varepsilon. \end{aligned}$$

As this holds for all pairs δ_1 and δ_2, with $0 < \delta_{1,2} < \operatorname{diam}(E)$, it follows that

$$\mathfrak{h}(A) + \mathfrak{h}(B) \leq \mathfrak{h}(A \cup B) + \varepsilon.$$

As this holds for all positive real numbers ε, we finally have

$$\mathfrak{h}(A) + \mathfrak{h}(B) \leq \mathfrak{h}(A \cup B),$$

and that completes the proof. □

To sum up, if E is a metric space, given a family $\mathfrak{X}$ of subsets of E and a pre-measure τ on E, Theorem (9.3) shows that the set function $\mathfrak{h}$ constructed from τ is a metric outer measure. The restriction of $\mathfrak{h}$ to the σ-field $\mathfrak{M}_{\mathfrak{h}}$ of all the $\mathfrak{h}$-measurable sets of E, which contains the σ-field $\mathcal{B}(E)$ of all Borel sets of E, is a measure, which is said to be the measure constructed from the pre-measure τ using 'Method II'.

We shan't give any example of measures constructed using Method II right now, but we are going to hold up until Chapter XII, when we are going to use this result to define a whole family of measures on the Borel σ-field of a metric space that are going to be the focus of that whole chapter. Incidentally, we shall also see that Method II will enable us to define (for the third time) the Borel-Lebesgue measure, for this measure is going to be one of the measures in the aforementioned family.

10. Completion of a measure

Let μ be a measure on a measurable space $(E, \mathcal{E})$. If $\mathcal{E}$ is not the σ-field of all subsets of E, it may be useful to extend μ to a larger σ-field, especially if one can add sets whose measure is somewhat determined by the measure of a set already in the σ-field $\mathcal{E}$. For example, if $A \subseteq B \subseteq C$, and both A and C are in $\mathcal{E}$, but B is not, and if $\mu(A) = \mu(C)$, it would seem reasonable for B to also be measurable and to have the same measure, under μ, as A and C.

The goal of this section is to construct such an extension. The method used to extend this measure uses the notion of outer measure, and more precisely, the outer measure μ^* generated by μ as defined in (8.3). Let us start by proving the following result.

(10.1) Proposition. *Let $(E, \mathcal{E}, \mu)$ be a measure space. For every set A in $\mathcal{E}$, there exists a set B in $\mathcal{E}$, with $A \subseteq B$ such that $\mu^*(A) = \mu(B)$.*

Proof. Let (X_k) be a sequence of elements of $\mathcal{E}$ with $A \subseteq \bigcup_k X_k$, let us set $X = \bigcup_k X_k$, whence $\mu(X) \leq \sum_k \mu(X_k)$. It thus follows that

$$\mu^*(A) = \inf\{\mu(X) : X \in \mathcal{E},\, A \subseteq X\}.$$

As a result of this, for every strictly positive integer k, choosing an element X_k of $\mathcal{E}$ with $\mu(X_k) \leq \mu^*(A) + 1/k$ and setting $B = \bigcap_k X_k$, it follows that B is then an element of $\mathcal{E}$, with $A \subseteq B$, and $\mu^*(A) = \mu(B)$. □

In the same hypotheses and with the same notations of the previous proposition, a set B like the one found in the proposition is sometimes called a **measurable covering** of A. Conversely, if A is an element of $\mathcal{E}$, then a measurable covering of A is evidently A itself. If C is an element of $\mathcal{E}$, and the only subsets of C that are in $\mathcal{E}$ are $\emptyset$ and C itself (but C may contain more than one element), then C is a measurable covering of any subset A of C. If, on the other hand, there is a set A in $\mathcal{E}$ such that $A \subseteq B \subseteq C$, and $\mu(C \setminus A) = 0$, then $\mu^*(B \setminus A) = \mu^*(C \setminus B) = 0$.

Let us now extend the definition of negligible set given in Definition II(3.1).

(10.2) Definition. Let $(E, \mathcal{E}, \mu)$ be a measure space. Every subset A of E such that $\mu^*(A) = 0$ is called a **μ-negligible set** or a **negligible set** for μ, and the set of all negligible sets for μ shall be denoted by $\mathcal{N}(\mu)$.

In the same hypotheses of the previous definition, let us observe that, thanks to the sub-additivity of μ^* (see (8.2)), $\mathcal{N}(\mu)$ is stable under countable union. In fact, if (X_k) is an infinite sequence of elements of $\mathcal{N}(\mu)$, then

$$\mu^*(\textstyle\bigcup_k X_k) \leq \sum_k \mu^*(X_k) = 0.$$

In order to state the next result, let us recall that $A \triangle B$ denotes the 'symmetric difference' of A and B, that is to say the set $(A \setminus B) \cup (B \setminus A)$. It is also useful to give the following definition.

(10.3) Definition. Let $(E, \mathcal{E}, \mu)$ be a measure space. Two subsets A and B of E are said to be **almost equivalent**, or that one is **almost equivalent to** the other, **modulo** μ, if $A \triangle B$ is μ-negligible.

It is not difficult to recognise that the definition given above is an equivalence relation on the class $\mathcal{P}(E)$ of all subsets of E.

(10.4) Proposition. *In the same hypotheses as in Definition (10.3), the collection $\mathcal{E}_\mu$ of all subsets A of E for which there exists a set B which belongs to $\mathcal{E}$ and is almost equivalent to A modulo μ is a σ-field. In addition, $\mathcal{E}_\mu$ coincides with the σ-field generated by $\mathcal{E} \cup \mathcal{N}(\mu)$.*

Proof. It is entirely obvious that E and $\emptyset$ are elements of $\mathcal{E}_\mu$. In addition, if (A_n) and (B_n) are two infinite sequences of subsets of E, with B_n in $\mathcal{E}$ and $A_n \triangle B_n \in \mathcal{N}(\mu)$ for every positive integer n, then

$$\left(\bigcup_n A_n\right) \triangle \left(\bigcup_n B_n\right) \subseteq \bigcup_n (A_n \triangle B_n) \in \mathcal{N}(\mu).$$

Finally, if A and B are two subsets of E, with B in $\mathcal{E}$ and $A \triangle B \in \mathcal{N}(\mu)$, then $A^c \triangle B^c = A \triangle B \in \mathcal{N}(\mu)$. So, $\mathcal{E}_\mu$ is a σ-field.

To complete the proof, observe that, if A belongs to $\mathcal{E}_\mu$, then there exists a set B in $\mathcal{E}$ with $A \triangle B \in \mathcal{N}(\mu)$ and $B \in \mathcal{E}$, which yields

$$A \setminus B \in \mathcal{N}(\mu) \qquad \text{and} \qquad B \setminus A \in \mathcal{N}(\mu).$$

So, $A = (B \cup (A \setminus B)) \setminus (B \setminus A)$ lies in the σ-field generated by $\mathcal{E} \cup \mathcal{N}(\mu)$, which shows that $\mathcal{E}_\mu$ is included in the σ-field generated by $\mathcal{E} \cup \mathcal{N}(\mu)$. Conversely, observe that $\mathcal{E} \subseteq \mathcal{E}_\mu$ and $\mathcal{N}(\mu) \subseteq \mathcal{E}_\mu$, which yields $\sigma(\mathcal{E} \cup \mathcal{N}(\mu)) \subseteq \mathcal{E}_\mu$. □

Now, given a measure space $(E, \mathcal{E}, \mu)$, on the σ-field $\mathcal{E}_\mu$, which clearly extends $\mathcal{E}$, it is possible to extend μ to a measure $\bar{\mu}$ so that, whenever two sets are almost equivalent modulo μ, and μ is defined on one of them, then $\bar{\mu}$ will be defined and equal for both sets. Precisely, let us prove the following.

(10.5) Proposition. *Let $(E, \mathcal{E}, \mu)$ a measure space. For every element A of the σ-field $\mathcal{E}_\mu$, let B be an element of $\mathcal{E}$, with $A \triangle B \in \mathcal{N}(\mu)$, and let us define $\bar{\mu}(A) = \mu(B)$. Then, $\bar{\mu}$ is a well-defined measure on the σ-field $\mathcal{E}_\mu$ and it coincides with μ on $\mathcal{E}$.*

Proof. To show that $\bar{\mu}$ is well-defined, let A be an element of $\mathcal{E}_\mu$ and let B and C be two elements of $\mathcal{E}$ such that $A \triangle B \in \mathcal{N}(\mu)$ and $A \triangle C \in \mathcal{N}(\mu)$. Then $B \triangle C = (A \triangle B) \triangle (A \triangle C)$. Thus, $B \triangle C \in \mathcal{N}(\mu)$ which yields that $\mu(B) = \mu(C)$, hence $\bar{\mu}(A)$ is well-defined.

To show that $\bar{\mu}$ is countably additive, let (A_n) be an infinite sequence of pairwise disjoint elements of $\mathcal{E}_\mu$, and, for every index n, consider a set B_n in $\mathcal{E}$ with $C_n = A_n \triangle B_n \in \mathcal{N}(\mu)$, so that, by definition, $\bar{\mu}(A_n) = \mu(B_n)$. Observe that, if n_1 and n_2 are two distinct indices, then $B_{n_1} \cap B_{n_2} \subseteq C_{n_1} \cup C_{n_2}$, so $\mu(B_{n_1} \cap B_{n_2}) = 0$. In addition, let $D_n = B_n \setminus \bigcup_{i<n} B_i$ and set $B = \bigcup_n B_n$. Then, for every index n, $\mu(D_n) = \mu(B_n)$ and

$$\mu(B) = \sum_n \mu(D_n) = \sum_n \mu(B_n) = \sum_n \bar{\mu}(A_n).$$

Thus, if $A = \bigcup_n A_n$, then $A \triangle B \in \mathcal{N}(\mu)$, so $\bar{\mu}(A) = \mu(B) = \sum_n \bar{\mu}(A_n)$, and this shows that $\bar{\mu}$ is countably additive. Finally, it evident that $\bar{\mu}$ coincides with μ on $\mathcal{E}$, and this completes the proof. □

(10.6) Definition. In the same hypotheses as in Proposition (10.5), the measure $\bar{\mu}$, defined on the σ-field $\mathcal{E}_\mu$, is called the **completion** of μ. Clearly, $\bar{\mu}$ is a complete measure, in the sense of Definition (7.4).

In §8 we have seen that, starting from μ, by means of Theorem (8.6), we can construct a complete measure that extends μ on a σ-field $\mathfrak{M}_{\mu^*}$. Precisely, let us denote by $\tilde{\mu}$ the restriction of μ^* to $\mathfrak{M}_{\mu^*}$. It is interesting that, constructing an outer measure from μ or $\tilde{\mu}$ gives exactly the same result, as shown in the next lemma.

(10.7) Lemma. *Let τ be a countably additive positive set function on an algebra $\mathfrak{X}$ on a given set E. Then, $\tilde{\tau}^* = \tau^*$ on $\mathcal{P}(E)$.*

Proof. Since, by means of Theorem (8.6), on $\mathfrak{X}$ one has $\tilde{\tau} = \tau$, it is clear that $\tilde{\tau}^* \leq \tau^*$. Conversely, for every measurable subset A of E, by means of Proposition (10.1), there exists a set B in $\mathcal{E}$ which is included in A and such that $\tilde{\tau}^*(A) = \tilde{\tau}(B)$, and this guarantees that

$$\tau^*(A) \leq \tau^*(B) = \tilde{\tau}(B) = \tilde{\tau}^*(A). \qquad \square$$

In light of the construction done in this section, there are now two complete measures, namely $\bar{\mu}$ and $\tilde{\mu}$, defined on two σ-fields $\mathcal{E}_\mu$ and $\mathfrak{M}_{\mu^*}$ which both include $\mathcal{E}$ and that both coincide with μ on $\mathcal{E}$. It is just natural to wonder whether these two constructions lead to the same extension, or if, indeed, the two extensions $\bar{\mu}$ and $\tilde{\mu}$ are different. The answer is given in the following theorem.

(10.8) Theorem. *If $(E, \mathcal{E}, \mu)$ is a σ-finite measure space, then $\mathcal{E}_\mu = \mathfrak{M}_{\mu^*}$.*

Proof. Since, by Theorem (8.6), the restriction of μ^* to the σ-field $\mathfrak{M}_{\mu^*}$ is a complete measure that extends μ, one evidently has

$$\mathcal{E} \subseteq \mathfrak{M}_{\mu^*} \qquad \text{and} \qquad \mathcal{N}(\mu) \subseteq \mathfrak{M}_{\mu^*},$$

hence $\mathcal{E}_\mu \subseteq \mathfrak{M}_{\mu^*}$. Conversely, since the measure space is σ-finite, there exists a sequence (E_n) of elements of the σ-field $\mathcal{E}$ with finite measure under μ, and such that $E = \bigcup_n E_n$. Let A be any element of $\mathfrak{M}_{\mu^*}$. In order to show that A belongs to $\mathcal{E}_\mu$ it suffices to show that $A \bigtriangleup E_n \in \mathcal{N}(\mu)$ for every n. Restricting the measure to E_n, we may assume that μ is finite. Then, by virtue of Proposition (10.1), let B be a measurable covering for A. Then,

$$\mu^*(A) = \mu(B) = \mu^*(B) = \mu^*(A) + \mu^*(B \setminus A),$$

so $\mu^*(B \setminus A) = 0$ and $A = A \setminus (B \setminus A)$ belongs to $\mathcal{E}_\mu$, so $\mathfrak{M}_{\mu^*} \subseteq \mathcal{E}_\mu$. $\square$

(10.9) Example. Let E be a non-empty set. On the σ-field $\mathcal{E} = \{\emptyset, E\}$ define μ by $\mu(\emptyset) = 0$ and $\mu(E) = \infty$. Then $\mu^*(A) = \infty$ whenever A is a non-empty subset of E, whence all subsets of E are in $\mathfrak{M}_{\mu^*}$. On the other hand, $\mathcal{E}_\mu = \mathcal{E} = \{\emptyset, E\}$, which means that $\mathcal{E}_\mu$ is a smaller σ-field than $\mathfrak{M}_\mu$ as soon as E has more than one element.

Chapter IV

Kernels and Products of Measures

This chapter is devoted to proving the existence of the product of a family of measures on the product space of a family of measurable space. Precisely, let us consider a family of measurable spaces $\big((E_i, \mathcal{E}_i)\big)_{i\in I}$, let us set

$$E = \prod_{i\in I}, \qquad \mathcal{E} = \bigotimes_{i\in I} \mathcal{E}_i,$$

and, for any index i in the set of indices I, let us assign a measure μ_i on $(E_i, \mathcal{E}_i)$. In this framework, we wish to use the family of measures $(\mu_i)_{i\in I}$ to construct a measure μ on the product space $(E, \mathcal{E})$ which is 'compatible' with the given family, which means a measure μ such that

$$\mu(\prod_{i\in I} A_i) = \prod_{i\in I} \mu_i(A_i) \qquad \text{for all } A_i \in \mathcal{E}_i.$$

We shall see that this can be done in full generality, but it is going to require some work. In particular, if the family of measurable spaces is finite, one could effectively answer this question using Carathéodory-Hahn's extension theorem. A different approach, which is the one we are going to follow here, uses the beautiful and elegant notion of 'kernel', which shall be presented in detail in §1. After delving into the theory of kernels, we shall move onto §2 and define the product of two measures. In this section, we shall also see that, at its core, the famous 'Fubini's Theorem' is nothing but a trivial consequence of the properties of kernels. Finally, in §3 and §4, we shall focus on the product of an arbitrary family of measures. To this end, in §3, we shall start by discussing a very important theorem, due to C. Ionescu-Tulcea, that allows to construct a measure, on a product space, starting from a sequence of kernels. Using this result, which has many applications in probability theory, but that turns out to be handy in abstract measure theory as well, we shall then, throughout §4, construct the product measure for a sequence of measures, to then extend this construction to an arbitrary family of measures of probabilities.

DOI: 10.1201/9781003724766-IV

1. The notion of kernel

In the sequel, for the integral of a measurable real function f on a given measurable space, semi-integrable with respect to the measure μ, alongside the usual notations such as $\int f \, d\mu$ and $\int f(x) \, \mu(dx)$ or $\int \mu(dx) f(x)$, shall also deploy the notation $\langle \mu, f \rangle$. In fact, not only will the latter simplify some of the formulae and clarify some of the steps, but it will also highlight the 'dual' nature of the relationship between measures and functions. Anyway, whenever it is possible, we shall always endeavour to use the traditional notation as well, to give time to familiarise fully with this alternative notation.

(1.1) Definition. A **kernel** from a measurable space $(E, \mathcal{E})$ to another measurable space $(F, \mathcal{F})$ is a positive numeric function N, defined on the product set $E \times \mathcal{F}$, such that:

(K1) For every x in E, the map $B \mapsto N(x, B)$ is a measure on $\mathcal{F}$.

(K2) For every B in $\mathcal{F}$, the map $x \mapsto N(x, B)$ is $\mathcal{E}$-measurable.

A kernel from $(E, \mathcal{E})$ to $(E, \mathcal{E})$ shall preferably be called a kernel **on** $(E, \mathcal{E})$.

Observe immediately that a kernel N from $(E, \mathcal{E})$ to $(F, \mathcal{F})$, can also be thought of as a family

$$N = \big(N(x, \cdot)\big)_{x \in E}$$

of measures on $\mathcal{F}$, where the parameter x ranges over the set E, and with the additional property that, for every set B in $\mathcal{F}$, the function $x \mapsto N(x, B)$ is measurable on the measurable space $(E, \mathcal{E})$.

(1.2) Definition. In the same hypotheses as in Definition (1.1), if each of the measures $N(x, \cdot)$ is a probability measure, that is to say, if $N(x, F) = 1$ for every element x of E, then N is called a **Markov** kernel.

The next result shows that a kernel N from a measurable space $(E, \mathcal{E})$ to another measurable space $(F, \mathcal{F})$ acts on the class of measurable positive real functions on $(F, \mathcal{F})$ so that, every such function g is transformed into a measurable positive function Ng on $(E, \mathcal{E})$. In other words, a kernel defines a map $g \mapsto Ng$ from the class of positive real functions on $(F, \mathcal{F})$ to that of positive real functions on $(E, \mathcal{E})$.

Symmetrically, or, with a more accurate jargon, *dually*, a kernel N acts on the set of all measures on the σ-field $\mathcal{E}$ by associating, to any such measure μ, a measure μN on $\mathcal{F}$, thereby defining a map $\mu \mapsto \mu N$ from the class of all measures on $\mathcal{E}$ to the class of all measures on $\mathcal{F}$.

The next theorem shows the true elegance of kernels, which is the fact that these two maps are 'dual' of one another. In other words, if g is a positive real function on $(F, \mathcal{F})$ and μ a measure on $(E, \mathcal{E})$, then $\langle \mu N, g \rangle = \langle \mu, Ng \rangle$.

(1.3) Theorem. *Let N be a kernel from $(E, \mathcal{E})$ to $(F, \mathcal{F})$. Then,*

(a) *For every measurable positive real function g on $(F, \mathcal{F})$, the function Ng that associates to each element x of E the number $Ng(x)$ defined by*

$$Ng(x) = \int N(x, \mathrm{d}y)\, g(y)$$

is a measurable function on $(E, \mathcal{E})$.

(b) *For every measure μ on $(E, \mathcal{E})$, the function μN that associates to each element B of $\mathcal{F}$ the number $\mu N(B)$ defined by*

$$\mu N(B) = \langle \mu, NI_B \rangle = \int \mu(\mathrm{d}x)\, N(x, B)$$

is a measure on $(F, \mathcal{F})$. The measure μN is sometimes described by saying that it is a **mixture** *of the $N(x, \cdot)$, weighted with the measure μ.*

(c) *For every positive measurable function g defined on $(F, \mathcal{F})$, the following* duality condition *holds:*

$$\langle \mu N, g \rangle = \langle \mu, Ng \rangle = \int \mu(\mathrm{d}x) \int N(x, \mathrm{d}y)\, g(y).$$

Proof. (a) If B is an element of $\mathcal{F}$, then $NI_B(x) = \int_B N(x, \mathrm{d}y) = N(x, B)$, hence the function Ng is measurable when g is the indicator function of a measurable set. By linearity of the integral, therefore, Ng is measurable also when g is a simple function. When g is any positive measurable function, we know that g is the supremum of an increasing sequence (g_n) of positive simple functions and therefore, thanks to Beppo Levi's theorem, $g_n \uparrow g$ entails $Ng_n \uparrow Ng$ and therefore Ng is measurable because it is the supremum of a sequence of measurable functions.

(b) For every element B of $\mathcal{F}$ and every sequence (B_n) of pairwise disjoint elements of $\mathcal{F}$, whose union is B, the obvious relation $I_B = \sum_n I_{B_n}$ implies

$$NI_B = \sum_n NI_{B_n}$$

and, taking the integral with respect to μ of this identity and applying Beppo Levi's theorem, we get $\langle \mu, NI_B \rangle = \sum_n \langle \mu, NI_{B_n} \rangle$, which is the same as $\mu N(B) = \sum_n \mu N(B_n)$. This proves that μN is a measure on $\mathcal{F}$.

It remains to prove statement (c). This is obvious when g is the indicator function of an element B of $\mathcal{F}$, following from the very definition of μN. Then, the relation keeps holding when g is a simple function by linearity. Finally, in the general case, if g is a measurable positive real function, then g is the supremum of an increasing sequence (g_n) of simple functions, and since $\langle \mu N, g_n \rangle = \langle \mu, Ng_n \rangle$, the relation now follows by virtue of Beppo Levi's theorem, because $g_n \uparrow g \Rightarrow Ng_n \uparrow Ng$. □

It is worth noticing that, thanks to the relation $N(x,B) = NI_B(x)$, the kernel is uniquely determined by the map $g \mapsto Ng$. Symmetrically, the map $\mu \mapsto \mu N$ also determines the kernel. In fact, if ε_x denotes (for every x in E) Dirac's measure on x, that is to say the measure on $\mathcal{E}$ defined by

$$\varepsilon_x(A) = I_A(x) \qquad \text{for every } A \text{ in } \mathcal{E},$$

then $N(x,\,\cdot\,) = \varepsilon_x N$. In fact, if x is an element of E and B an element of $\mathcal{F}$, the duality relation can be written in the form

$$\langle \varepsilon_x N, I_B \rangle = \langle \varepsilon_x, NI_B \rangle = I_B(x) = N(x,B).$$

Notice further that, for every sequence (g_n), of positive measurable functions on $(F,\mathcal{F})$, by means of Beppo Levi's theorem and the theorem of series integration, the following properties hold:

$$(1.4) \qquad g_n \uparrow g \Rightarrow Ng_n \uparrow Ng, \qquad g = \textstyle\sum_n g_n \Rightarrow Ng = \textstyle\sum_n Ng_n.$$

If N is a kernel from $(E,\mathcal{E})$ to $(F,\mathcal{F})$, the duality condition stated in assertion (c) of Theorem (1.3), is not just limited to positive measurable functions on $(F,\mathcal{F})$, but it also extends to all semi-integrable functions with respect to the measures $N(x,\,\cdot\,)$. To see this, let us start by observing that, if g is any measurable function on $(F,\mathcal{F})$ (not necessarily positive), then Ng is well-defined and measurable on $(E,\mathcal{E})$. In fact, the sets A_- and A_+ of all elements x of E, such that g is lower and upper semi-integrable with respect to the measure $N(x,\,\cdot\,)$, respectively, is an element of $\mathcal{E}$. In fact, these two sets can be written in the form

$$A_- = \{Ng^- < \infty\}, \qquad A_+ = \{Ng^+ < \infty\}$$

and they are thereby measurable because N is a kernel. Thus, if one denotes by Ng the function that is zero outside of the set $A = A_+ \cup A_-$ and that, for every point x of this set, assumes the value

$$\langle N(x,\,\cdot\,), g \rangle = \textstyle\int N(x,\mathrm{d}y)\, g(y),$$

it turns out that Ng is measurable on $(E,\mathcal{E})$. In fact,

$$Ng = I_A Ng^+ - I_A Ng^-.$$

We can now prove the following.

(1.5) Theorem. *Let N be a kernel from $(E,\mathcal{E})$ to $(F,\mathcal{F})$, let μ be a measure on $(E,\mathcal{E})$, and g a measurable real function on $(F,\mathcal{F})$ which is upper (resp. lower) semi-integrable with respect to the measure μN. Then,*

(a) *$Ng^\pm$ is integrable with respect to μ. Hence, with respect to μ, the complement of the set $A_\pm = \{Ng^\pm < \infty\}$ is negligible.*

(b) *Ng is upper (resp. lower) semi-integrable with respect to μ.*

(c) *The duality formula $\langle \mu N, g\rangle = \langle \mu, Ng\rangle$ holds.*

Proof. By duality, for the two positive functions g^+ and g^-, one has

$$\langle \mu N, g^+\rangle = \langle \mu, Ng^+\rangle, \qquad \langle \mu N, g^-\rangle = \langle \mu, Ng^-\rangle.$$

From these relations (in which one of the left-hand sides is finite by hypothesis) assertion (a) follows. This, in turn, implies assertion (b), thanks to the decomposition $Ng = I_A Ng^+ - I_A Ng^-$. Finally, since the complement of the set A is μ-negligible, because g is semi-integrable with respect to the measure μN, the previous relations can be written in the form

$$\langle \mu N, g^+\rangle = \langle \mu, Ng^+\rangle, \qquad \langle \mu N, g^-\rangle = \langle \mu, Ng^-\rangle,$$

and, by linearity, one also gets assertion (c). □

If $(E, \mathcal{E})$ and $(F, \mathcal{F})$ are two measurable space, we have seen that a kernel induces a linear transformation $g \mapsto Ng$ from the set of the positive real functions on $(F, \mathcal{F})$ to the set of the positive real functions on $(E, \mathcal{E})$ such that the conditions (1.4) hold. Conversely, if L is a linear map that associates to every measurable positive real function g on $(F, \mathcal{F})$ a measurable positive real function $L(g)$ on $(E, \mathcal{E})$, we may wonder under which conditions we can say that the map L is induced by a kernel. The answer is the following.

(1.6) Theorem. *Let L be a map, defined on the class of all positive measurable functions on $(F, \mathcal{F})$, taking values in the class of all positive measurable functions on $(E, \mathcal{E})$, with the following properties:*

(a) *$L(\alpha g) = \alpha L(g)$ for every g and every positive real number α,*

(b) *$L(\sum_n g_n) = \sum_n L(g_n)$ for every sequence (g_n).*

Then there is a kernel N such that $Ng = L(g)$.

Proof. Assume that the map L satisfies the two conditions expressed in the statement of the theorem. For every element x of E, and every set B in $\mathcal{F}$, set $N(x, B) = L(I_B)(x)$. Obviously, the function $x \mapsto N(x, B)$ is measurable for any $B \in \mathcal{F}$. On the other hand, for any x in E, the function $B \mapsto N(x, B)$ turns out to be a measure on $\mathcal{F}$. Therefore, the family of these measures is, by definition, a kernel from $(E, \mathcal{E})$ to $(F, \mathcal{F})$, and one has $Ng = L(g)$ for every positive measurable function g on $(F, \mathcal{F})$. □

It is time to show a few examples of kernels.

(1.7) Example. Let f be a positive measurable function on an assigned measurable space $(E, \mathcal{E})$. For every x in E, let us set

$$N(x, \cdot) = f(x)\varepsilon_x.$$

It is not difficult to show that,

$$N(x, A) = f(x)I_A(x) \qquad \text{for every } A \text{ in } \mathcal{E},$$

which is measurable because it is product of the two measurable functions f and I_A. Thus, N is a kernel on $(E, \mathcal{E})$, and, if g is a positive measurable function on $(E, \mathcal{E})$, then $Ng = fg$. So this kernel acts on the positive measurable functions on $(E, \mathcal{E})$ by multiplication by f. To see how this kernel acts on a measure, let μ be a measure on $(E, \mathcal{E})$, and observe that

$$\langle \mu N, g \rangle = \langle \mu, Ng \rangle = \textstyle\int fg \, \mathrm{d}\mu.$$

So $\mu N = f \cdot \mu$ is the measure with density f.

(1.8) Example. Let f be a measurable map from $(E, \mathcal{E})$ to $(F, \mathcal{F})$, and, for every element x in E, let us set

$$N(x, \cdot\,) = \varepsilon_{f(x)}.$$

It is not difficult to recognise that, if g is a positive measurable function on $(F, \mathcal{F})$, the transform of g under N is $Ng = g \circ f$, which is measurable because such are f and g. Thus, N is a kernel from $(E, \mathcal{E})$ to $(F, \mathcal{F})$, which acts on the set of all semi-integrable measurable functions on $(F, \mathcal{F})$ by composition with f. In order to see how this kernel acts on a measure μ, defined on $(E, \mathcal{E})$, observe that

$$\langle \mu N, g \rangle = \langle \mu, Ng \rangle = \textstyle\int g \circ f \, \mathrm{d}\mu.$$

Hence, the transform μN, of μ under N, coincides with the image $f(\mu)$ of the measure μ under f, and the kernel N acts on the measures on $(E, \mathcal{E})$ by taking the image under f.

(1.9) Example. As a special case of the previous example, let $(E, \mathcal{E})$ be a measurable space, and let us consider the family I defined, for every x in E, by $I(x, \cdot\,) = \varepsilon_x$. It is clear that I is a kernel on $(E, \mathcal{E})$ such that, for every positive real function g and every measure μ on $(E, \mathcal{E})$, one has

$$Ig = g \qquad \text{and} \qquad \mu I = \mu.$$

For this reason, this kernel is called the **identity** kernel of $(E, \mathcal{E})$.

(1.10) Example. Let ν be a positive σ-finite measure on $(F, \mathcal{F})$ and let n be a positive real function defined on $E \times F$ and measurable with respect to $\mathcal{E} \otimes \mathcal{F}$. One may then defined a kernel N from $(E, \mathcal{E})$ to $(F, \mathcal{F})$ by setting

$$N(x, B) = \textstyle\int_B n(x, y) \, \nu(\mathrm{d}y).$$

This is called an **integral kernel** with base ν.

The transform of a positive real function g on $(F, \mathcal{F})$ under N, and the transform of a measure μ on $(E, \mathcal{E})$ under N are, respectively,

$$Ng(x) = \int n(x,y)g(y)\,\nu(\mathrm{d}y), \qquad \mu N(B) = \int \mu(\mathrm{d}x) \int_B n(x,y)\,\nu(\mathrm{d}y).$$

Further, if $n = a \otimes b$, where a and b are two measurable functions on $(E, \mathcal{E})$ and $(F, \mathcal{F})$, respectively, it is not difficult to recognise that

$$Ng = a \int bg\,\mathrm{d}\nu = \langle \nu, bg \rangle a, \qquad \mu N = (b \int a\,\mathrm{d}\mu) \cdot \nu = (\langle \mu, a \rangle b) \cdot \nu.$$

In this case, we shall write $N = a \otimes b\nu$, and simply $N = a \otimes \nu$, when $b = 1$.

(1.11) Example. As a special case of integral kernel, assume that E and F are two countable sets, that the σ-fields $\mathcal{E}$ and $\mathcal{F}$ are the classes all subsets of E and F, respectively, and suppose that ν is the counting measure on F. Then, if g is a positive real function defined on F, one has $Ng(x) = \sum_{y \in F} n(x,y)g(y)$. On the other hand, since $N(x, \{y\}) = n(x,y)$, one may well mix the two notational devices and simply write

$$Ng(x) = \textstyle\sum_{y \in F} N(x,y)g(y).$$

Thus, in order to define a kernel from $(E, \mathcal{P}(E))$ to $(F, \mathcal{P}(F))$, it suffices to assign the numbers $N(x,y)$ for every pair (x,y) ranging over $E \times F$. The collection of these numbers may be thought as an infinite matrix indexed by the elements of E and F, respectively. Observe that, in this special case, measures on $(E, \mathcal{P}(E))$ and functions on $(F, \mathcal{P}(F))$ may be represented as row- and column-vectors, respectively, and the maps $g \mapsto Ng$ and $\mu \mapsto \mu N$ coincide with the usual operations on vectors defined by matrices.

If N is a kernel from $(E, \mathcal{E})$ to $(F, \mathcal{F})$ and M is a kernel from $(F, \mathcal{F})$ to $(G, \mathcal{G})$, it is easily checked that the map $g \mapsto NMg$, obtained composing the transformation induced by N with the transformation induced by M, satisfies conditions (a) and (b) of Theorem (1.6), thus it is in turn induced by a kernel from $(E, \mathcal{E})$ to $(G, \mathcal{G})$, which is called the **composite kernel** of N and M. In particular,

$$NM(x,B) = \int N(x,\mathrm{d}y)\,M(y,B) \qquad \text{for all } x \in E \text{ and } B \in \mathcal{G}.$$

Also, if μ is a measure on $(E, \mathcal{E})$, then the transform of μ under NM is given by $\mu(NM) = (\mu N)M$, that is to say it coincides with the transform of μN under M. Furthermore, it is easily seen that the composition of kernels is an associative operation. In fact, if g is a measurable positive real function, then $((LM)N)g = (LM)(Ng) = L(M(Ng)) = L(MNg) = (L(MN))g$. As a result of this, we may now define the *power* of a kernel N on a measurable space $(E, \mathcal{E})$. Precisely, for $n = 0$, let us set $N^0 = I$, the identity kernel, defined by $I(x, \cdot\,) = \varepsilon_x$, and, for every integer n inductively set

$$N^n g = N(N^{n-1}g) = N^{n-1}(Ng).$$

Let us complete this section by observing that when a family N of measures on a measurable space $(F, \mathcal{F})$ is given, in order to establish if N is a kernel from $(E, \mathcal{E})$ to $(F, \mathcal{F})$, it might be complicated to check whether or not all functions of the form Ng are measurable with respect to $\mathcal{E}$. Even though it suffices to only show that every function of the form NI_B, with $B \in \mathcal{F}$, is measurable with respect to $\mathcal{E}$, this may still remain too complicated a task to perform. However, if $\mathcal{J}$ is a base for the σ-field $\mathcal{F}$, it turns out that it is sufficient to only show that every function of the form NI_B, with $B \in \mathcal{J}$, is $\mathcal{E}$-measurable in order to guarantee that N is a kernel.

Precisely, let us prove the following.

(1.12) Theorem. *Let $(E, \mathcal{E})$ and $(F, \mathcal{F})$ be two measurable spaces, and*

$$N = \big(N(x, \cdot\,)\big)_{x \in E}$$

a family of measures on $(F, \mathcal{F})$, with the index x ranging in E. Let $\mathcal{J}$ be a base for the σ-field $\mathcal{F}$. Suppose that, for every B in $\mathcal{J}$, the function NI_B is finite and measurable on $(E, \mathcal{E})$. Then, N is a kernel between $(E, \mathcal{E})$ and $(F, \mathcal{F})$.

Proof. Let us denote by $\mathcal{L}$ the class of all numeric functions g, that belong to $\bigcap_{x \in E} \mathcal{L}^1(N(x, \cdot\,))$, such that the function Ng is measurable on $(E, \mathcal{E})$. The class $\mathcal{L}$ is a monotone vector space which contains the indicator functions of all sets in $\mathcal{J}$. It follows from Theorem I(6.8), that $\mathcal{L}$ contains the indicator functions of all sets of the form $B \cap (B_1 \cup \cdots \cup B_n)$, with B an element of $\mathcal{F}$ and $n \geq 1$. In other words, for every element B in $\mathcal{F}$, each of the functions $NI_{B \cap (B_1 \cup \cdots \cup B_n)}$ is measurable on $(E, \mathcal{E})$. Thus, such is

$$NI_B = \sup\nolimits_n NI_{B \cap (B_1 \cup \cdots \cup B_n)},$$

and this proves that N is a kernel. □

The next result looks just like a simple application of Theorem (1.12), but it is pivotal to construct the product measure. In fact, if $(E, \mathcal{E})$ and $(F, \mathcal{F})$ are two measurable spaces, it is the following result that allows to create a measure on the space $(E \times F, \mathcal{E} \otimes \mathcal{F})$ once a measure μ on $(E, \mathcal{E})$ and a kernel N from $(E, \mathcal{E})$ to $(F, \mathcal{F})$ are given. Indeed, this is the first result of this sort, and its importance is paramount.

(1.13) Theorem. *Let $(E, \mathcal{E})$ and $(F, \mathcal{F})$ be two measurable spaces, and N a kernel from $(E, \mathcal{E})$ to $(F, \mathcal{F})$. Let us denote by $\widetilde{N}(x, \cdot\,)$ the image under the measurable map $y \mapsto (x, y)$ from $(F, \mathcal{F})$ to $(E \times F, \mathcal{E} \otimes \mathcal{F})$. In other words, let us set, for every set C of the σ-field $\mathcal{E} \otimes \mathcal{F}$,*

$$\widetilde{N}(x, C) = N\big(x, C(x)\big),$$

where $C(x)$ denotes the set $\{y \in F : (x, y) \in C\}$. Then, the family of measures

$$\widetilde{N} = \big(\widetilde{N}(x, \cdot\,)\big)_{x \in E}$$

is a kernel, from $(E, \mathcal{E})$ to $(E \times F, \mathcal{E} \otimes \mathcal{F})$.

Proof. For every set A in $\mathcal{E}$ and every set B in $\mathcal{F}$, one has

$$\widetilde{N} I_{A\times B} = I_A N I_B,$$

so the function $\widetilde{N} I_{A\times B}$ is finite and measurable on $(E,\mathcal{E})$. Further, the class

$$\mathcal{H} = \{A\times B : A\in\mathcal{E},\, B\in\mathcal{F}\}$$

is a base for the σ-field $\mathcal{E}\otimes\mathcal{F}$, and the result follows from Theorem (1.12). □

Let us see how the kernel defined in the previous theorem acts on functions and measures. To this end, given a kernel N from $(E,\mathcal{E})$ to $(F,\mathcal{F})$, let us consider the corresponding kernel $\widetilde{N}$ from $(E,\mathcal{E})$ to $(E\times F,\mathcal{E}\otimes\mathcal{F})$ as defined in Theorem (1.13). Taking a measurable positive function h on the product space $(E\times F,\mathcal{E}\otimes\mathcal{F})$, it is not difficult to see that

$$\widetilde{N} h(x) = \int N(x,\mathrm{d}y) h(x,y). \tag{1.14}$$

In fact, the class $\mathcal{M}$ of all measurable real functions h on $(E\times F,\mathcal{E}\otimes\mathcal{F})$ such that the equality (1.14) holds is a monotone class which contains the indicator functions of all sets of the form $A\times B$, with $A\in\mathcal{E}$ and $B\in\mathcal{F}$. Since the class of all this special type of sets is a base for $\mathcal{E}\otimes\mathcal{F}$, Theorem I(6.8) yields that the class $\mathcal{M}$ contains the class of all bounded measurable real function h on $(E\times F,\mathcal{E}\otimes\mathcal{F})$. In particular, $\mathcal{M}$ contains all simple functions and therefore all positive measurable functions, as they are supremum of a sequence of simple functions.

Let us now see how the kernel $\widetilde{N}$ acts on a measure μ on $(E,\mathcal{E})$. To this end, let h be a positive measurable function on $(E\times F,\mathcal{E}\otimes\mathcal{F})$ and let us apply the duality relation:

$$\langle \mu\widetilde{N}, h\rangle = \langle \mu, \widetilde{N} h\rangle = \int \mu(\mathrm{d}x) \int N(x,\mathrm{d}y)\, h(x,y).$$

So, if h is the indicator function of a set C in $\mathcal{E}\otimes\mathcal{F}$, one has

$$\mu\widetilde{N}(C) = \int \mu(\mathrm{d}x) \int N(x,\mathrm{d}y)\, I_C(x,y) = \int \mu(\mathrm{d}x)\, N\big(x, C(x)\big).$$

Thus, the kernel $\widetilde{N}$ associates to each measure on the space $(E,\mathcal{E})$, a measure on the product space $(E\times F,\mathcal{E}\otimes\mathcal{F})$ which is the *mixture* of all the measures $N(x,\cdot)$ on $(F,\mathcal{F})$, weighted with the given measure on $(E,\mathcal{E})$.

(1.15) Definition. In the same hypotheses and with the same notations as in Theorem (1.14), the kernel $\widetilde{N}$ shall be called the **lifting** of N. In addition, we shall denote by $\mu\otimes N$ the transform of μ under the kernel $\widetilde{N}$, and we shall call it the **product** of μ and N.

(1.16) Remark. Thanks to the duality relation, the measure $\mu \otimes N$ is the only measure on the product space $(E \times F, \mathcal{E} \otimes \mathcal{F})$ described by

$$\langle \mu \otimes N, h \rangle = \int \mu(\mathrm{d}x) \int N(x, \mathrm{d}y)\, h(x,y)$$

for every semi-integrable real function h on $(E \times F, \mathcal{E} \otimes \mathcal{F})$. In particular, taking h of the form $I_{A \times B} = I_A \otimes I_B$, the previous relation may be written, in equivalent fashion, as

$$(\mu \otimes N)(A \times B) = \int_A \mu(\mathrm{d}x)\, N(x, B)$$

for every A in $\mathcal{E}$, and every B in $\mathcal{F}$.

(1.17) Remark. Observe that $\mu \otimes N$ can be written in the form

$$\mu \otimes N = \int \mu(\mathrm{d}x) \big(\varepsilon_x \otimes N(x, \cdot\,) \big),$$

where the symbol $\otimes$ on the right-hand side denotes the usual product of measures. In fact, if A is an element of $\mathcal{E}$ and B is an element of $\mathcal{F}$, then

$$\begin{aligned}(\mu \otimes N)(A \times B) &= \int_A N(x, B)\, \mu(\mathrm{d}x) \\ &= \int I_A(x) N(x, B)\, \mu(\mathrm{d}x) \\ &= \int \varepsilon_x(A) N(x, B)\, \mu(\mathrm{d}x).\end{aligned}$$

2. The product of two measures

Let $(E, \mathcal{E}, \mu)$ and $(F, \mathcal{F}, \nu)$ be two measure spaces. We wish to define on this product space the product measure of μ and ν, which means a measure λ on $(E \times F, \mathcal{E} \otimes \mathcal{F})$ such that

(2.1) $$\lambda(A \times B) = \mu(A)\nu(B) \qquad \text{for all } A \in \mathcal{E} \text{ and } B \in \mathcal{F}.$$

The easiest way to do this would be to observe that the function λ above is defined on the class

$$\mathcal{H} = \{ A \times B : A \in \mathcal{E},\ B \in \mathcal{F} \}$$

which is in turn a base for the σ-field $\mathcal{E} \otimes \mathcal{F}$. This class can also be extended to a ring, taking the set $\mathcal{H}_s$ of all finite unions of elements of $\mathcal{H}$, and the measure λ can be extended to this ring with no fuss. Then, applying Carathéodory-Hahn's extension theorem, one could infer that there exists one and only one measure on the product space $(E \times F, \mathcal{E} \otimes \mathcal{F})$ that extends λ on $\mathcal{H}_s$, hence on $\mathcal{H}$ as we require. However, this construction hides many important properties of the product measure, which we would now have to carefully prove. On the other hand, using the theory of kernels, these properties are in fact trivial. Precisely, let us prove the following theorem.

(2.2) Theorem. *Given two measure spaces $(E, \mathcal{E}, \mu)$ and $(F, \mathcal{F}, \nu)$, such that the measures μ and ν are both σ-finite, there exists a uniquely determined measure on the measurable space $(E \times F, \mathcal{E} \otimes \mathcal{F})$ that, on every set of the form $A \times B$, with $A \in \mathcal{E}$, $\mu(A) < \infty$, and $B \in \mathcal{F}$, $\nu(B) < \infty$, takes the value $\mu(A)\nu(B)$.*

Proof. Let us set

$$\mathcal{I} = \{A \in \mathcal{E} : \mu(A) < \infty\}, \qquad \mathcal{J} = \{B \in \mathcal{F} : \nu(B) < \infty\}.$$

It is clear that $\mathcal{I}$ is a base for $\mathcal{E}$ and $\mathcal{J}$ is a base for $\mathcal{F}$. Let us denote by N the kernel, from $(E, \mathcal{E})$ to $(F, \mathcal{F})$, defined setting $N(x, \cdot) = \nu$ for every x in E. It is easy to show that the measure $\mu \otimes N$ satisfies the condition

$$\mu \otimes N(A \times B) = \mu(A)\nu(B) \qquad \text{for } A \in \mathcal{E},\ B \in \mathcal{F}.$$

To complete the proof, it suffices to show that this measure is unique. To this end, it suffices to observe that the set

$$\mathcal{H} = \{A \times B : A \in \mathcal{E},\ B \in \mathcal{F}\}$$

is a base for $\mathcal{E} \otimes \mathcal{F}$, and since the measures μ and ν are supposed to be σ-finite, the conclusion follows from the fundamental criterion for the coincidence of two measures. □

(2.3) Definition. In the same hypotheses as in Theorem (2.2), the only measure on the product space $(E \times F, \mathcal{E} \otimes \mathcal{F})$ such that, for every set of the form $A \times B$, with $A \in \mathcal{E}$ and $B \in \mathcal{F}$, takes the value $\mu(A)\nu(B)$ is called the **product measure** of μ and ν, and we shall denote it by $\mu \otimes \nu$. In other words, the measure $\mu \otimes \nu$ is the only measure on $(E \times F, \mathcal{E} \otimes \mathcal{F})$ such that

$$(\mu \otimes \nu)(A \times B) = \mu(A)\nu(B) \qquad \text{for all } A \in \mathcal{E} \text{ and } B \in \mathcal{F}.$$

Once this measure has been defined, the first thing to do would be to understand how one can 'integrate' with respect to it. Precisely, given a positive real function h on $E \times F$, measurable with respect to the σ-field $\mathcal{E} \otimes \mathcal{F}$, one would like to relate the integral $\langle \mu \otimes \nu, h \rangle$ to integrals with respect to the two measures μ and ν. This is the content of Fubini's theorem, named after the Italian mathematician Guido Fubini (1879–1943), and sometimes also referred to as Fubini-Tonelli's theorem, in recognition of the important contributions given to the subject by the Italian mathematician Leonida Tonelli (1885–1946).

(2.4) Theorem (Fubini's theorem for positive numeric functions). *Assigned two measure spaces $(E, \mathcal{E}, \mu)$ and $(F, \mathcal{F}, \nu)$, with the measures μ and ν both σ-finite, let h be a measurable positive numeric function on $(E \times F, \mathcal{E} \otimes \mathcal{F})$. Then the function*

$$x \mapsto \int \nu(\mathrm{d}y)\, h(x, y)$$

is measurable on $(E, \mathcal{E})$ and

$$\langle \mu \otimes \nu, h \rangle = \int \mu(\mathrm{d}x) \int \nu(\mathrm{d}y)\, h(x, y).$$

Proof. Let N be the trivial kernel defined by $N(x,\cdot) = \nu$ for every $x \in E$ so that $\mu \otimes \nu = \mu\widetilde{N}$. If h is a positive function on $E \times F$, measurable on $\mathcal{E} \otimes \mathcal{F}$, then, by definition of kernel, the function $\widetilde{N}h(x) = \int \nu(\mathrm{d}y)\, h(x,y)$ is measurable on $(E, \mathcal{E})$. Furthermore, since $\mu \otimes \nu = \mu \otimes N$, thanks to the duality relation, one also has

$$\langle \mu \otimes \nu, h \rangle = \langle \mu, \widetilde{N}h \rangle = \int \mu(\mathrm{d}x) \int \nu(\mathrm{d}y)\, h(x,y),$$

and this proves the theorem. □

Starting from Fubini's theorem for positive numeric functions, an analogous result for semi-integrable functions also holds. Precisely, since a function h is semi-integrable if either the integral of h^+ or the integral h^- is finite, using the decomposition $h = h^+ - h^-$, and applying Theorem (1.5) to $\widetilde{N}$, where N is the kernel defined by $N(x,\cdot) = \nu$, yields the following more general form of Fubini's theorem.

(2.5) Theorem (Fubini's theorem for semi-integrable functions). *Assigned two measure spaces $(E, \mathcal{E}, \mu)$ and $(F, \mathcal{F}, \nu)$, with the measures μ and ν both σ-finite, let h be a numeric function on $(E \times F, \mathcal{E} \otimes \mathcal{F})$ which is upper (resp. lower) semi-integrable with respect to the measure $\mu \otimes \nu$. Then the function*

$$x \mapsto \int \nu(\mathrm{d}y)\, h^{\pm}(x,y)$$

is integrable with respect to μ, and therefore the complement of the set A, consisting of all elements x of E such that $h(x,\cdot)$ is upper (resp. lower) semi-integrable with respect to ν, is negligible with respect to μ. Thus, the function $x \mapsto \int \nu(\mathrm{d}y)\, h(x,y)$ is upper (resp. lower) semi-integrable and

$$\langle \mu \otimes \nu, h \rangle = \int \mu(\mathrm{d}x) \int \nu(\mathrm{d}y)\, h(x,y).$$

(2.6) Remark. On a product space $(E \times F, \mathcal{E} \otimes \mathcal{F})$, when integrating an integrable real function f with respect to a measure λ, which is the product of the measures μ and ν defined on $\mathcal{E}$ and $\mathcal{F}$, respectively, beside the notation $\langle \lambda, f \rangle$, which shows the duality, and the original notation $\int f\, \mathrm{d}\lambda$, we generally adopt the convention to utilise a 'double integral' sign such as

$$\iint f(x,y)\, \lambda(\mathrm{d}x, \mathrm{d}y), \qquad \iint \lambda(\mathrm{d}x, \mathrm{d}y)\, f(x,y), \qquad \iint f(x,y)\, \mathrm{d}\lambda(x,y),$$

or similar ones, to emphasise the fact that the measure λ is a product measure, and it splits into two measures, each called the **marginal** of λ.

(2.7) Remark. An important consequence of Fubini's theorem is the following. Let us assume that two measure spaces $(E, \mathcal{E}, \mu)$ and $(F, \mathcal{F}, \nu)$ are assigned, that the measures μ and ν are both σ-finite, and that f and g are two measurable real functions on $(E, \mathcal{E})$ and $(F, \mathcal{F})$, semi-integrable with respect to μ and ν, respectively. Then,

$$\langle \mu \otimes \nu, f \otimes g \rangle = \langle \mu, f \rangle \langle \nu, g \rangle.$$

In fact, applying Fubini's theorem to the function $h = f \otimes g$, we obtain

$$\begin{aligned}\langle \mu \otimes \nu, f \otimes g \rangle &= \int \mu(\mathrm{d}x) \int \nu(\mathrm{d}y)\, f(x) g(x) \\ &= \int \mu(\mathrm{d}x)\, f(x) \int \nu(\mathrm{d}y)\, g(y) \\ &= \langle \mu, f \rangle \langle \nu, g \rangle.\end{aligned}$$

Another important result is the following.

(2.8) Theorem. *Given two measure spaces $(E, \mathcal{E}, \mu)$ and $(F, \mathcal{F}, \nu)$, with the measures μ and ν both σ-finite, the measure $\mu \otimes \nu$ coincides with the image of $\nu \otimes \mu$ under the measurable map φ, from $(F \times E, \mathcal{F} \otimes \mathcal{E})$ to $(E \times F, \mathcal{E} \otimes \mathcal{F})$, defined by $\varphi(y, x) = (x, y)$. Hence, if h is a measurable real function defined on $(E \times F, \mathcal{E} \otimes \mathcal{F})$, semi-integrable with respect to $\mu \otimes \nu$, then*

$$\langle \mu \otimes \nu, h \rangle = \langle \nu \otimes \mu, h \circ \varphi \rangle.$$

Proof. The image of $\nu \otimes \mu$ under φ is a measure on $\mathcal{E} \otimes \mathcal{F}$ with the property that, on every set of the form $A \times B$, with $A \in \mathcal{E}$ and $B \in \mathcal{F}$, takes the value $\mu(A)\nu(B)$. Hence, this measure must coincide with the measure $\mu \otimes \nu$, and the first statement is proved. Finally, the second statement follows by definition of image of a measure under a map. □

Rewriting the following statement with the more usual integral notation, one finally gets the following result on the invertibility of the order of integration, which is going to be extremely useful to calculate multiple integrals in the next paragraph.

(2.9) Corollary (Invertibility of the order of integration). *Assigned two measure spaces $(E, \mathcal{E}, \mu)$ and $(F, \mathcal{F}, \nu)$, where the measures μ and ν are both σ-finite, let h be a measurable real function on $(E \times F, \mathcal{E} \otimes \mathcal{F})$, semi-integrable with respect to $\mu \otimes \nu$, then*

$$\int \mu(\mathrm{d}x) \int \nu(\mathrm{d}y)\, h(x, y) = \int \nu(\mathrm{d}y) \int \mu(\mathrm{d}x)\, h(x, y).$$

The product measure is particularly important when the measures are probabilities. To see why, assume that $(E, \mathcal{E}, \mu)$ and $(F, \mathcal{F}, \nu)$ are two probability space (i.e. assume that $\mu(E) = \nu(F) = 1$) and let us denote by ξ_E and ξ_F the canonic projections of $E \times F$ on E and F, respectively. Let us also consider the probability space

$$(E \times F, \mathcal{E} \otimes \mathcal{F}, \mu \otimes \nu).$$

Then, the image of the product measure $\mu \otimes \nu$ under the projections ξ_E and ξ_F coincides with μ and ν, respectively. In fact, if A is an element of $\mathcal{E}$ and B is an element of $\mathcal{F}$, then

$$\langle \xi_E(\mu \otimes \nu), I_A \rangle = \langle \mu \otimes \nu, I_A \otimes I_F \rangle = \mu(A),$$

and similarly

$$\langle \xi_F(\mu \otimes \nu), I_B \rangle = \langle \mu \otimes \nu, I_E \otimes I_B \rangle = \nu(B).$$

Let us see, however, that not all measures on the product space $(E \times F, \mathcal{E} \otimes \mathcal{F})$ are product measures. To see this, let μ be a probability measure on $(E, \mathcal{E})$ and let N be a non-trivial Markov kernel from $(E, \mathcal{E})$ to $(F, \mathcal{F})$ and let us take $\lambda = \mu \otimes N$. Then,

$$\lambda(A \times B) = \int_A \mu(\mathrm{d}x) \int N(x, B) \qquad \text{for every } A \in \mathcal{E} \text{ and } B \in \mathcal{F}.$$

This formula shows that, if ξ_E and ξ_F are the projections of $E \times F$ onto E and F, respectively, and if A is an element of $\mathcal{E}$, then

$$\langle \xi_E(\lambda), I_A \rangle = \langle \lambda, I_A \otimes I_F \rangle = \int_A \mu(\mathrm{d}x) \int N(x, F) = \mu(A),$$

so one has $\mu = \xi_E(\lambda)$. Similarly, if B is an element of $\mathcal{F}$, then

$$\langle \xi_F(\lambda), I_B \rangle = \langle \lambda, I_E \otimes I_B \rangle = \int \mu(\mathrm{d}x) \int N(x, B),$$

so one has $\nu = \xi_F(\lambda) = \int \mu(\mathrm{d}x) \int N(x, \cdot\,)$. On the other hand, it is clear that, in this case, $\lambda(A \times B)$ is not the product $\mu(A)\nu(A)$, and therefore λ is not the product measure $\mu \otimes \nu$.

Let us complete this section by showing that the result of existence of a product measure can be easily extended to a *finite* family of measure spaces. Precisely, let us prove the following.

(2.10) Theorem. *Let $\big((E_i, \mathcal{E}_i, \mu_i)\big)_{i \in I}$ a finite family of measure spaces, such that each of the measures μ_i is σ-finite. Then, there exists a unique measure μ on the σ-field $\bigotimes_{i \in I} \mathcal{E}_i$ such that, for every finite family $(A_i)_{i \in I}$, with $A_i \in \mathcal{E}_i$, for every index i, the following condition holds:*

$$\mu\big(\textstyle\prod_{i \in I} A_i\big) = \prod_{i \in I} \mu_i(A_i).$$

Proof. The *uniqueness* follows from the fact that the class of all sets of the form $\prod_{i \in I} A_i$, with $(A_i)_{i \in I}$ finite family, such that $A_i \in \mathcal{E}_i$ and $\mu_i(A_i) < \infty$ for every index i, forms a base for the σ-field $\bigotimes_{i \in I} \mathcal{E}_i$.

The *existence* can be proved by induction on the cardinal of the index set I. Precisely, the statement is obvious when the cardinal of I is 1. Thus, let us assume that the statement is true when the cardinal of I is n and let us prove it when the cardinal of I is $n+1$. To this end, let us choose any index j in I, and denote by μ the corresponding measure given by the family of the family $(\mu_i)_{i \in I \setminus \{j\}}$, which exists by inductive hypothesis. Then, consider the product measure $\mu \otimes \mu_j$. This is a measure defined on the σ-field

$$\Big(\textstyle\bigotimes_{i \in I \setminus \{j\}} \mathcal{E}_i\Big) \otimes \mathcal{E}_j.$$

In order to obtain the required measure corresponding to the whole family $(\mu_i)_{i\in I}$, it suffices to take the image of the measure $\mu \otimes \mu_j$ under the inverse of the one-to-one map that, to each element $(x_i)_{i\in I}$ of the product space $\prod_{i\in I} E_i$, associates the pair $\big((x_i)_{i\in I\setminus\{j\}}, x_j\big)$. □

(2.11) Definition. With the same hypotheses as in Theorem (2.10), the only measure μ which is defined in the statement of the theorem is called the **product measure** of the family $(\mu_i)_{i\in I}$ and shall be denoted by $\bigotimes_{i\in I} \mu_i$. In particular, when the finite family of measures is a finite sequence $(\mu_1, \dots, \mu_n)$, instead of $\bigotimes_{i=1}^n \mu_i$ one would preferably write $\mu_1 \otimes \cdots \otimes \mu_n$.

3. Ionescu-Tulcea's theorem

The aim of this and the next section is to prove the existence of the product of an arbitrary family of measures. To this end, we shall utilise a fundamental result, borrowed from probability theory, that goes by the name of 'Ionescu-Tulcea's theorem', in honour of the Romanian-American mathematician Cassius Ionescu-Tulcea (1923–2021). This theorem, that is usually stated in the language of probability, guarantees the existence of a probability measure on the product space of a sequence of measurable spaces, and therefore it may also be thought as a result of measure theory.

(3.1) Theorem (Ionescu-Tulcea). *Let $((E_n, \mathcal{E}_n))_{n\geq 0}$ be a sequence of measurable spaces, μ be a probability measure on the measurable space $(E_0, \mathcal{E}_0)$, and $(N_n)_{n\geq 0}$ a sequence of Markov kernels, such that, for every index n,*

N_n is a kernel from $(E_0 \times \cdots \times E_n, \mathcal{E}_0 \otimes \cdots \otimes \mathcal{E}_n)$ into $(E_{n+1}, \mathcal{E}_{n+1})$.

Let us also set

$$F = \prod_n E_n, \qquad \mathcal{F} = \bigotimes_n \mathcal{E}_n,$$

and let j_n be the projection of F onto $E_0 \times \cdots \times E_n$. Then, there exists a unique probability measure ν on $(F, \mathcal{F})$ such that

$$j_0(\nu) = \mu, \qquad j_{n+1}(\nu) = j_n(\nu) \otimes N_n \qquad \text{for every } n \geq 0.$$

The necessity of such a measure is to ensure that a certain class of probabilistic models, given in terms of a special sequences of kernels, do actually exist. Nevertheless, this result turns out to be quite useful in the abstract framework of measure theory as well. For example, we shall utilise this result to prove the existence of the product of a sequence of probability measures, and further applications of shall be seen in Chapter X as well.

Proof. Let us introduce some preparatory notation. To this end, in the hypotheses of the theorem, let us denote by ξ_n the projection with index n: for every index n, set

$$\xi_n(x) = x_n \qquad \text{if } x = (x_0, x_1, \dots).$$

It is clear that ξ_n is a measurable map from $(F, \mathcal{F})$ onto $(E_n, \mathcal{E}_n)$. In terms of these projections, the map j_n of F onto $E_0 \times \cdots \times E_n$ can be written as $j_n(x) = \big(\xi_0(x), \ldots, \xi_n(x)\big)$. Again, it is clear that j_n is a measurable map from $(F, \mathcal{F})$ onto $(E_0 \times \cdots \times E_n,\ \mathcal{E}_0 \otimes \cdots \otimes \mathcal{E}_n)$. Finally, also set

$$\mathcal{F}_n = \sigma(j_n) = \sigma(\xi_0, \ldots, \xi_n).$$

By definition, this is the smallest σ-field on F that makes measurable all the projections $\xi_0, \ldots, \xi_n$. Moreover, the map j_n is also $\mathcal{F}_n$-measurable, and it is readily shown that, for every index n, one has $\mathcal{F}_n \subseteq \mathcal{F}_{n+1}$ and $\mathcal{F} = \bigvee_n \mathcal{F}_n$.

Let us observe that, if g is a real function defined on F, measurable with respect to the σ-field $\mathcal{F}_{n+1}$, then, by means of Doob's measurability criterion, there exists a uniquely determined measurable real function f, defined on the product space $(E_0 \times \cdots \times E_{n+1},\ \mathcal{E}_0 \otimes \cdots \otimes \mathcal{E}_{n+1})$, such that $g = f \circ j_{n+1}$. Hence, setting

$$K_n g(x) = \int N_n\big(j_n(x), \mathrm{d}y\big)\, f\big(j_n(x), y\big) \tag{3.2}$$

it is easy to show that, for every positive index n, K_n is a Markov kernel from $(F, \mathcal{F}_n)$ to $(F, \mathcal{F}_{n+1})$ such that $K_n g = g$ whenever g is $\mathcal{F}_n$-measurable. In fact, if g is a measurable positive real function on $(F, \mathcal{F}_{n+1})$, and α is a positive real number, then $K_n(\alpha g) = \alpha K_n g$ and, similarly, if (g_n) is any sequence of measurable positive real functions on $(F, \mathcal{F}_{n+1})$, then $K_n(\sum_k g_k) = \sum_k K_n g_k$. Hence, Theorem (1.6) ensures that K_n is indeed a kernel. Furthermore, if g is also $\mathcal{F}_n$-measurable, then $g = f \circ j_n$ and therefore

$$K_n g(x) = \int N_n\big(j_n(x), \mathrm{d}y\big) f\big(j_n(x)\big) = f\big(j_n(x)\big) = g(x).$$

Using the kernels (K_n), we shall now construct a sequence of probability measures (ν_n) such that, for every index n, ν_n is defined on the σ-field $\mathcal{F}_n$, and we shall see that this sequence can be extended to a probability measure ν on $\mathcal{F}$ which turns out to have the desired properties that

$$j_0(\nu) = \mu \qquad \text{and} \qquad j_{n+1}(\nu) = j_n(\nu) \otimes N_n.$$

To this end, when $n = 0$, observe that the map $A \mapsto \{j_0 \in A\}$ is a one-to-one correspondence between the elements of $\mathcal{E}_0$ and the elements of $\mathcal{F}_0$. Thus, the function $\nu_0\{j_0 \in A\} = \mu(A)$ is well-defined on the σ-field $\mathcal{F}_0$, and it is a probability measure.

Then, for any positive integer n, define, by induction, $\nu_{n+1} = \nu_n K_n$, and, since K_n is a Markov kernel from $(F, \mathcal{F}_n)$ to $(F, \mathcal{F}_{n+1})$, the transform of ν_n under K_n is a probability measure on $\mathcal{F}_{n+1}$. Furthermore, if A is in $\mathcal{F}_n$, then the function I_A is $\mathcal{F}_n$-measurable and therefore, by the properties of K_n,

$$\begin{aligned}\nu_{n+1}(A) &= \langle \nu_{n+1}, I_A\rangle \\ &= \langle \nu_n K_n, I_A\rangle \\ &= \langle \nu_n, K_n I_A\rangle \\ &= \langle \nu_n, I_A\rangle \\ &= \nu_n(A),\end{aligned}$$

so the probability measure ν_{n+1} is an extension of ν_n.

It is thereby constructed a sequence (ν_n) of probability measures, such that, for every index n, ν_n is a probability measure on $\mathcal{F}_n$, and ν_{n+1} extends ν_n.

Let us now set $\mathcal{G} = \bigcup_n \mathcal{F}_n$ and denote by ν_∞ the function, defined on $\mathcal{G}$, that coincides with ν_n on $\mathcal{F}_n$. This function is well-defined, and it is additive on $\mathcal{G}$. It suffices to prove that ν_∞ is also countably additive to be able to infer, by virtue of Carathéodory-Hahn's extension theorem, the existence of a measure ν, defined on the σ-field $\mathcal{F} = \sigma(\mathcal{G})$, that extends the function ν_∞. Furthermore, let us observe that, if such a measure ν exists, then it coincides with ν_n on $\mathcal{F}_n$, and $j_{n+1}(\nu) = \nu_n \otimes N_n$, as it is readily shown integrating (3.2) with respect to the measure ν, and observing that its right-hand side is measurable with respect to the σ-field $\mathcal{F}_n$.

In order to prove that ν_∞ is countably additive, it suffices to show that, for every sequence (A_k) of elements of the field $\mathcal{G}$, if $A_k \downarrow \emptyset$, then $\nu_\infty(A_k) \downarrow 0$. To this end, let us begin by proving the following lemma.

(3.3) Lemma. *In the same hypotheses and with the same notations of the theorem, suppose, for every index n, that ζ_n is a measurable function, defined on $(F, \mathcal{F}_n)$ with values in $[0,1]$. Let us suppose that (ζ_n) pointwise converges towards zero, and that the following relation holds:*

$$\zeta_n = K_n \zeta_{n+1} \qquad \textit{for every } n \geq 0.$$

Then, $\zeta_0 = 0$.

Proof. For every n, let us set $\zeta_n = f_n(\xi_0, \ldots, \xi_n)$. The hypothesis on the ζ_n implies that, for every n and every $(x_0, \ldots, x_n)$ in $E_0 \times \cdots \times E_n$,

$$\begin{aligned}f_n(x_0, \ldots, x_n) &= \textstyle\int N_n(x_0, \ldots, x_n, \mathrm{d}y)\, f_{n+1}(x_0, \ldots, x_n, y) \\ &\leq \sup_{y \in E_{n+1}} f_{n+1}(x_0, \ldots, x_n, y).\end{aligned}$$

Thus, if we assume by contradiction that there is an element x_0 in E_0 such that $f_0(x_0) > 0$, then, chosen a real number $0 < \varepsilon < f_0(x_0)$, it is possible to inductively construct from x_0 a sequence $(x_n)_{n\geq 0}$ that satisfies $f_n(x_0, \ldots, x_n) > \varepsilon$ for every positive index n. In fact, assuming that there exists a finite sequence $x_0, x_1, \ldots, x_n$, with $x_1 \in E_1, \ldots, x_n \in E_n$, such that $f_n(x_0, \ldots, x_n) > \varepsilon$, then, by means of the previous inequality, it is also possible to find an element x_{n+1} in E_{n+1} such that $f_{n+1}(x_0, \ldots, x_n, x_{n+1}) > \varepsilon$. Since this contradicts the hypothesis that the sequence (ζ_n) converges towards zero, the lemma follows. □

To complete the proof of Ionescu-Tulcea's theorem, let us introduce a further notation: For every element A of the field $\mathcal{G}$ and every positive integer n, denote by $L_n(A)$ the measurable real function defined by

$$L_n(A) = K_n \cdots K_m I_A,$$

where m is an integer, greater than or equal to n, sufficiently large that the right-hand side is meaningful, that is to say, such that A belongs to $\mathcal{F}_{m+1}$. Thanks to the properties of the kernels K_j, this definition is not ambiguous, which means that it does not depend on the choice of m. Furthermore, by definition of ν_n, it follows that

$$\begin{aligned}\big\langle \nu_n, L_n(A)\big\rangle &= \langle \nu_n, K_n \cdots K_m I_A\rangle \\ &= \langle \nu_n K_n \cdots K_m, I_A\rangle \\ &= \langle \nu_{m+1}, I_A\rangle \\ &= \nu_\infty(A).\end{aligned}$$

The following three properties also hold:

(a) $L_n(A) = I_A$ for n sufficiently big.

(b) $L_n(A) = K_n L_{n+1}(A)$.

(c) $L_n(A) \leq L_n(B)$ when $A \subseteq B$.

We are now ready to complete the proof of the theorem, showing that, for every sequence (A_k) of elements of $\mathcal{G}$, if $A_k \downarrow \emptyset$, then $\nu_\infty(A_k) \downarrow 0$. Let (A_k) be such a sequence, and set

$$\alpha = \inf_k \nu_\infty(A_k) = \lim_k \nu_\infty(A_k).$$

It suffices to show that α is zero. To this end, for every n, let us denote by ζ_n the infimum of the sequence $(L_n(A_k))$. In other words, set

$$\zeta_n = \inf_k L_n(A_k) = \lim_k L_n(A_k).$$

The sequence of measurable functions (ζ_n) converges towards zero. In fact, by the definition of ζ_n as an infimum, for every index k, one has $\zeta_n \leq L_n(A_k)$ and this gives

$$\limsup_n \zeta_n \leq \limsup_n L_n(A_k) = I_{A_k}$$

(where the last equality follows form property (a) above). This is sufficient to conclude that ζ_n indeed converges towards zero, because $I_{A_k} \downarrow 0$, thanks to the hypothesis that $A_k \downarrow \emptyset$. Furthermore, as a result of property (b), one also has

$$L_n(A_k) = K_n L_{n+1}(A_k),$$

hence, as k approaches infinity, and thanks to the fact that K_n is a kernel, one gets $\zeta_n = K_n\zeta_{n+1}$, and the previous lemma yields that therefore $\zeta_0 = 0$. On the other hand, one also has

$$\big\langle \nu_0, L_0(A_k)\big\rangle = \nu_\infty(A_k),$$

and, taking the limit as $k \to \infty$, one gets $\langle \nu_0, \zeta_0\rangle = \alpha$, thus $\alpha = 0$, and this completes the proof. □

Let us now draw some useful conclusions from this theorem. Let us start with the following simple result.

(3.4) Corollary. *In the same hypotheses as of Theorem (3.1), for every element x of E_0, let us denote by $Q(x, \cdot\,)$ the only probability measure on $(F, \mathcal{F})$ corresponding to the measure ε_x on the space $(E_0, \mathcal{E}_0)$.*

Then, $Q = (Q(x, \cdot\,))_{x\in E_0}$ is a Markov kernel from $(E_0, \mathcal{E}_0)$ to $(F, \mathcal{F})$.

Furthermore, for every probability measure μ on $(E_0, \mathcal{E}_0)$, the only probability measure ν on $(F, \mathcal{F})$ described by the theorem is given by $\nu = \mu Q$.

Proof. By definition,

$$\big\langle Q(x, \cdot\,),\, f \circ \xi_0\big\rangle = \langle \varepsilon_x, f\rangle = f(x) \tag{3.5}$$

for every bounded measurable function f on $(E_0, \mathcal{E}_0)$, and also

$$\big\langle Q(x, \cdot\,),\, h\big\rangle = \big\langle Q(x, \cdot\,)K_n,\, h\big\rangle = \big\langle Q(x, \cdot\,),\, K_n h\big\rangle \tag{3.6}$$

for every bounded measurable function h on $(F, \mathcal{F}_{n+1})$.

From these two properties it is easy to show by induction that every function of the form $x \mapsto \langle Q(x, \cdot\,), h\rangle$ with h bounded and measurable on $(F, \mathcal{F}_n)$ is measurable on $(E_0, \mathcal{E}_0)$. This is sufficient to prove that Q is a kernel.

Let us now take a probability measure μ on the measurable space $(E_0, \mathcal{E}_0)$, and let us observe that μQ is the only measure on $(F, \mathcal{F})$ such that

$$j_0(\mu Q) = \mu \quad \text{and} \quad j_{n+1}(\mu Q) = j_n(\mu Q) \otimes N_n \quad \text{for every index } n.$$

To see this, integrating the two relations (3.5) and (3.6) with respect to the probability measure μ gives

$$\big\langle \mu Q,\, f \circ \xi_0\big\rangle = \langle \mu, f\rangle, \qquad \big\langle \mu Q,\, h\big\rangle = \big\langle \mu Q,\, K_n h\big\rangle.$$

These two equalities show that μQ is the only probability measure, on the measurable space $(F, \mathcal{F})$ such that

$$j_0(\mu Q) = \xi_0(\mu Q) = \mu, \quad j_{n+1}(\mu Q) = j_n(\mu Q) \otimes N_n, \quad \text{for every index } n.$$

This completes the proof. □

(3.7) Definition. The Markov kernel Q defined in Corollary (3.4) shall be called **Ionescu-Tulcea's kernel** associated to the sequence $(N_n)_{n\geq 0}$.

Using this terminology, Ionescu-Tulcea's theorem can be rephrased as follows.

(3.8) Corollary. *Let $((E_n, \mathcal{E}_n))_{n\geq 0}$ be a sequence of measurable spaces, μ a probability measure on the measurable space $(E_0, \mathcal{E}_0)$, and $(N_n)_{n\geq 0}$ a sequence of Markov kernels, such that, for every positive integer n,*

$$N_n \text{ is a kernel from } (E_0 \times \cdots \times E_n,\ \mathcal{E}_0 \otimes \cdots \otimes \mathcal{E}_n) \text{ into } (E_{n+1}, \mathcal{E}_{n+1}),$$

Let us also set

$$F = \prod_n E_n, \qquad \mathcal{F} = \bigotimes_n \mathcal{E}_n,$$

let j_n be the projection of F onto $E_0 \times \cdots \times E_n$, ν a probability measure on $(F, \mathcal{F})$, and Q Ionescu-Tulcea's kernel associated to the sequence $(N_n)_{n\geq 0}$. Then, the following conditions are equivalent:

(a) *For each index n,*

$$j_0(\nu) = \mu, \qquad j_{n+1}(\nu) = j_n(\nu) \otimes N_n \qquad \text{for every } n \geq 0.$$

(b) $\nu = \mu Q$.

4. Infinite products of measures

Given an arbitrary family $((E_i, \mathcal{E}_i, \mu_i))_{i\in I}$ of probability spaces, the aim of this section is to prove that there exists one and only one product measure of the family $(\mu_i)_{i\in I}$. As we have already mentioned, this is a probability measure defined on the measurable space

$$\left(\prod_{i\in I} E_i,\ \bigotimes_{i\in I} \mathcal{E}_i\right),$$

that assigns to every *measurable rectangle*, that is to say, every set of the form $\prod_{i\in I} A_i$, with $A_i \in \mathcal{E}_i$ for every i, and with $A_i \neq E_i$ only for a finite number of indices i, the value $\prod_{i\in I} \mu_i(A_i)$.

It is clear that two product measures of the same family $(\mu_i)_{i\in I}$ are necessarily identical because they coincide on the class of all measurable rectangles, which forms a base for the product σ-field. It remains to prove the existence. To this end, let us begin with some simple observations.

(4.1) Remark. If ν is a product measure for the family $(\mu_i)_{i\in I}$, then, for every non-empty subset H of I, the subfamily $(\mu_i)_{i\in H}$ also admits a product measure: Precisely, it admits as a product measure the image of ν under the canonical projection from $\prod_{i\in I} E_i$ to $\prod_{i\in H} E_i$. (It suffices to observe that the inverse image, under this projection, of a generic rectangle of the form $\prod_{i\in H} A_i$ is the rectangle $\prod_{i\in I} A_i$, with $A_i = E_i$ for every element i of $I \setminus H$.)

(4.2) Remark. Let $\mathcal{H}$ be a set of non-empty subsets of I, such that every finite subset of I is contained in an element of $\mathcal{H}$. Then, for a measure ν on the σ-field $\bigotimes_{i\in I}\mathcal{E}_i$ to be a product measure of the family $(\mu_i)_{i\in I}$, it is necessary and sufficient that, for every element H of $\mathcal{H}$, the image of ν under the canonical projection from $\prod_{i\in I} E_i$ to $\prod_{i\in H} E_i$ is a product measure of the subfamily $(\mu_i)_{i\in H}$.

(4.3) Remark. If ν is a product measure for the family $(\mu_i)_{i\in I}$, and if σ is a bijective map from a set J to the set I, then the family $(\mu_{\sigma(j)})_{j\in J}$ also admits a product measure: Precisely, it admits as a product measure the image of ν under the map from the space $\prod_{i\in I} E_i$ to the space $\prod_{j\in J} E_{\sigma(j)}$ that associates to each element $(x_i)_{i\in I}$ of the first space the element $(x_{\sigma(j)})_{j\in J}$ of the second space.

Keeping these preliminaries into account, we are now able to prove a first partial result.

(4.4) Lemma. *Every countable family of probability measures admits a product measure.*

Proof. Thanks to Remark (4.3), we can further reduce to the case of an infinite sequence. Hence, consider a sequence $((E_n, \mathcal{E}_n, \mu_n))_{n\geq 0}$ of probability spaces. For every positive integer n, denote by N_n the trivial Markov kernel, relative to the pair of measurable spaces

$$(E_0 \times \cdots \times E_n, \mathcal{E}_0 \otimes \cdots \otimes \mathcal{E}_n), \quad (E_{n+1}, \mathcal{E}_{n+1}),$$

defined as follows:

$$N_n(x_0, \dots, x_n, \cdot) = \mu_{n+1}.$$

Thanks to Ionescu-Tulcea's theorem (see (3.1)), setting

$$F = \prod\nolimits_{n\geq 0} E_n, \qquad \mathcal{F} = \bigotimes\nolimits_{n\geq 0} \mathcal{E}_n,$$

and denoting by j_n the projection of F onto $E_0 \times \cdots \times E_n$, there exists a unique probability measure ν on $(F, \mathcal{F})$ such that, for every index n, we have

$$j_0(\nu) = \mu_0, \qquad j_{n+1}(\nu) = j_n(\nu) \otimes N_n \qquad \text{for every } n \geq 0.$$

We are going to show that the measure ν coincides with the product measure of the sequence $(\mu_n)_{n\geq 0}$. To this end, thanks to the content of Remark (4.2), it will suffice to prove that, for every positive integer n, the probability measure $\nu_n = j_n(\nu)$ is the product of the finite sequence $(\mu_0, \dots, \mu_n)$. This is verified by induction. For $n = 0$, the thesis is obvious. Assuming the thesis is true for a certain positive integer n, fix, for every index i between 0 and $n+1$, an element A_i of $\mathcal{E}_i$. Then, since the function $N_n I_{A_{n+1}}$, that is to say the function $(x_0, \dots, x_n) \mapsto N_n(x_0, \dots, x_n, A_{n+1})$, simply coincides with the constant

$\mu_{n+1}(A_{n+1})$, then

$$\begin{aligned}\nu_{n+1}(A_0 \times \cdots \times A_{n+1}) &= \langle j_{n+1}(\nu), I_{A_0\times\cdots\times A_n} \otimes I_{A_{n+1}}\rangle \\ &= \langle \nu_n \otimes N_n, I_{A_0\times\cdots\times A_n} \otimes I_{A_{n+1}}\rangle \\ &= \int_{A_0\times\cdots\times A_n} N_n I_{A_{n+1}}\, \mathrm{d}\nu_n \\ &= \nu_n(A_0 \times \cdots \times A_n) \cdot \mu_{n+1}(A_{n+1}) \\ &= \prod_{i=0}^{n+1} \mu_i(A_i)\end{aligned}$$

(where the final equality is due to the inductive hypothesis). The lemma is thus proved. □

We are now able to prove the more general result.

(4.5) Theorem. *Let $((E_i, \mathcal{E}_i, \mu_i))_{i\in I}$ be an arbitrary family of probability spaces. Then, on the product space*

$$\left(\prod_{i\in I} E_i, \bigotimes_{i\in I} \mathcal{E}_i\right),$$

there exists one (and only one) product measure of the family $(\mu_i)_{i\in I}$.

Proof. Let

$$E = \prod_{i\in I} E_i, \qquad \mathcal{E} = \bigotimes_{i\in I} \mathcal{E}_i$$

and, for every i, denote by ξ_i the canonical projection of index i (from E to E_i). Denote also by $\mathcal{H}$ the set of countable and non-empty subsets of I and, for every such subset H, denote by ν_H the product measure of the family $(\mu_i)_{i\in H}$ (existing by virtue of Lemma (4.4)). Finally, denote by $\mathcal{F}_H$ the σ-field generated by the ξ_i with $i \in H$, that is to say, by the projection map j_H of E onto the partial product $\prod_{i\in H} E_i$.

Every element of $\mathcal{F}_H$ can be written, in a unique way, in the form $\{j_H \in A\}$, with A an element of the σ-field $\bigotimes_{i\in H} \mathcal{E}_i$. Moreover, the map

$$A \mapsto \{j_H \in A\}$$

from $\bigotimes_{i\in H} \mathcal{E}_i$ to $\mathcal{F}_H$ is an isomorphism of Boolean σ-fields. Therefore, there exists on $\mathcal{F}_H$ a unique probability measure ϱ_H such that

$$\varrho_H\{j_H \in A\} = \nu_H(A)$$

for every element A of $\bigotimes_{i\in H} \mathcal{E}_i$, which means that $j_H(\varrho_H) = \nu_H$.

Let H, K be two elements of $\mathcal{H}$, with $H \subset K$. Then $\mathcal{F}_H \subset \mathcal{F}_K$. Moreover, if we denote by $\pi_{H,K}$ the canonical projection from $\prod_{i\in K} E_i$ to $\prod_{i\in H} E_i$, we can write j_H in the form:

$$j_H = \pi_{H,K} \circ j_K. \tag{4.6}$$

Thanks to Remark (4.1), the product measure ν_H coincides with the image of ν_K under the projection $\pi_{H,K}$:

$$\nu_H = \pi_{H,K}(\nu_K).$$

This relation can be written in the equivalent form

$$j_H(\varrho_H) = \pi_{H,K}\big(j_K(\varrho_K)\big) = (\pi_{H,K} \circ j_K)(\varrho_K),$$

or equivalently, thanks to (4.6), in the form

$$j_H(\varrho_H) = j_H(\varrho_K).$$

It thus means that, for every element A of $\bigotimes_{i \in H} \mathcal{E}_i$, we have

$$\varrho_K\{j_H \in A\} = \varrho_H\{j_H \in A\},$$

that is to say, the measure ϱ_K is an extension of ϱ_H.

Let us prove that the σ-field $\mathcal{E}$ coincides with the union of the σ-fields of the form $\mathcal{F}_H$, with H an element of $\mathcal{H}$. To this end, denoted by $\mathcal{G}$ this union, let us observe that it is contained in $\mathcal{E}$ and contains the class of sets of the form $\{\xi_i \in A\}$ with $i \in I$ and $A \in \mathcal{E}_i$. Thus, everything is now reduced to proving that $\mathcal{G}$ is a σ-field. But for this, it suffices to observe that, however one assigns a sequence $(A_n)_{n \geq 1}$ of elements of $\mathcal{G}$, if, for each index n, one chooses an element H_n of $\mathcal{H}$ such that $A_n \in \mathcal{F}_{H_n}$, and denotes by H the union of the H_n, then $A_n \in \mathcal{F}_H$ for every n.

Let us now observe that, if A is a fixed element of the σ-field $\mathcal{E}$, the measures ϱ_H corresponding to the elements H of $\mathcal{H}$ such that A belongs to $\mathcal{F}_H$ all assume the same value on A. Indeed, if ϱ_H, ϱ_K are two such measures, then, since the measure $\varrho_{H \cup K}$ extends both, it follows that $\varrho_H(A) = \varrho_{H \cup K}(A) = \varrho_K(A)$. There thus exists, on the σ-field $\mathcal{E}$, a function ϱ that simultaneously extends all the ϱ_H. It is immediately verified that ϱ is a measure. Indeed, given a sequence $(A_n)_{n \geq 1}$ of elements of $\mathcal{E}$, pairwise disjoint, one can find an element H of $\mathcal{H}$ such that all the A_n belong to the σ-field $\mathcal{F}_H$. Then, since ϱ coincides with ϱ_H on $\mathcal{F}_H$, we have

$$\varrho\Big(\bigcup_{n \geq 1} A_n\Big) = \sum_{n \geq 1} \varrho(A_n).$$

To complete the proof of the theorem, it suffices to observe that the measure ϱ is, thanks to Remark (4.2), the product measure for the family $(\mu_i)_{i \in I}$. Indeed, for every element H of $\mathcal{H}$, since ϱ extends ϱ_H, the image of ϱ under the canonical projection j_H, is the product measure ν_H. □

Chapter V

Riesz Spaces and Signed Measures

In Chapter I, §5 we have introduced the notion of Riesz space as a special subspace of the vector space $\mathbb{R}^E$ of all functions defined on a non-empty set E. In this chapter, we are going to extend the definition of Riesz space to the framework of a general vector space. In fact, in measure theory, one of the most recurring algebraic structures is indeed that of a Riesz space, which, roughly speaking, is a vector space with a partial order, 'compatible' with the vector structure, and stable under supremum and infimum calculated using this order. Due to its algebraic nature, the present chapter may look a little different from the rest of this book, for it shan't speak about measures and measurable spaces for quite a while. In fact, in §1, we shall delve into the abstract definition of Riesz spaces. This is going to be a rather lengthy list of definitions and observations aimed to describe the structure of abstract Riesz spaces. In §2, we shall continue developing the structure of Riesz spaces, describing particularly remarkable subsets of a Riesz space, namely Riesz subspaces, ideals, and bands, while in §3 we shall discuss the notion of duality in the context of a Riesz space, and we shall prove Riesz decomposition theorem.

Once the purely algebraic content has been fully developed, and hopefully digested by the reader, however, it is finally going to be the time to gather the fruit of this abstract labour, by deploying our full set of new shiny algebraic tools to the study of the set of all measures on an assigned measurable space, which is going to be a rather anti-climatic Riesz space. This is the content of §4. The lengthy discussion on Riesz spaces shall also empower us to unlock important properties of this space. In §5 and §6, we shall then focus on the set of all measures with base a given measure, showing an original proof of the famous 'Radon-Nikodým Theorem' which uses solely tools from measure theory, and we shall utilise this theorem to construct the *Lebesgue decomposition* from which we shall derive the *Hahn decomposition* as a simple consequence.

DOI: 10.1201/9781003724766-V

1. Partially ordered vector spaces and Riesz spaces

(1.1) Definition. Let V be a vector space over the field of real numbers. An order relation on V, denoted by $\leq$, is said to be **compatible** with the vector structure of V, if the following two conditions hold:

(i) (*Translation invariance*) For every triple x, y, z of elements of V, the relation $x \leq y$ is equivalent to $x + z \leq y + z$.

(ii) For every element x of V and every strictly positive real number, the relation $x \geq 0$ is equivalent to $cx \geq 0$.

The vector space V, endowed with an order structure compatible with its vector structure, is then called a **partially ordered vector space**. Every element x in V that satisfies the relation $x \geq 0$ is called a **positive** element of V, and the set of these special elements shall be denoted by V_+.

(1.2) Remark. Let us recall that a subset U of a vector space V is called a **cone** if the relation $x \in U$ implies $cx \in U$ for every positive real number c. If V is a partially ordered vector space, it is not difficult to recognise that the set V_+ of all positive elements of V is indeed a cone, and it is generally referred to as the **positive cone** of V. Furthermore, this cone fully describes the order relation on V. In fact, for every pair x, y of elements of V, the relation $x \leq y$ is equivalent to $y - x \in V_+$. The sum of two positive elements is positive, and the only element of V that is positive together with its opposite is the zero of V.

The notions of supremum and infimum are well-established for real numbers. In the sequel, we shall extend these definitions to the case of a partially ordered vector space. This, however, needs to be done with some caution. Specifically, if V is a partially ordered vector space, and U is a subset of V, there may, or may not, exists an element x in V with the property that

$$\text{for every } v \in V,\ x \leq v \text{ if and only if } u \leq v \text{ for every } u \in U,$$

that is, $x \leq v$ if and only if v is an upper bound for U. Clearly, there may exist at most one such x, and, if one such 'least upper bound' exists, it shall be called the **supremum** of U and denoted by $\sup U$. Similarly, the **infimum** of U, provided that it exists, is the one 'greatest lower bound' of U, that is to say, the one element of V such that

$$\text{for every } v \in V,\ v \leq x \text{ if and only if } v \leq u \text{ for every } u \in U.$$

The following identities are straightforward

$$\begin{aligned} \inf U &= \sup\{v \in V : v \leq u \text{ for every } u \in U\} \\ \sup U &= \inf\{v \in V : u \leq v \text{ for every } u \in U\} \\ \inf V &= \sup \varnothing, \qquad \sup V = \inf \varnothing, \end{aligned}$$

and these equalities being true in the sense that, if one side is defined, so is the other, and they are then equal. In general, it is not guaranteed that a partially ordered vector space does have a supremum or an infimum. For this reason, it is necessary to introduce a slightly more stable structure, which is named after the Hungarian mathematician Frigyes Riesz (1880–1956) who first defined them in 1928.

(1.3) Definition. A **Riesz space** or **vector lattice** is a partially ordered vector space V such that $\sup\{x, y\}$ and $\inf\{x, y\}$ exist for all elements x and y of V. In a Riesz space, we shall write $x \vee y$ and $x \wedge y$ for $\sup\{x, y\}$ and $\inf\{x, y\}$, respectively. Furthermore, if V is a Riesz space, we shall write

$$x^+ = x \vee 0, \qquad x^- = (-x) \vee 0, \qquad |x| = x \vee (-x)$$

for any $x \in V$.

If V is a Riesz space, the maps $(x, y) \mapsto x \vee y$ and $(x, y) \mapsto x \wedge y$ may be thought as two additional operations on the Riesz space, which are compatible with the vector operations of addition and multiplication by a scalar. In other words, the two operations $\wedge$ and $\vee$ enjoy a number of algebraic properties, the most important of which are summarised in the following.

(1.4) Lemma. *Let V be a Riesz space, x, y and z three elements of V, and α a real number. Then*

(a) $-(x \wedge y) = (-x) \vee (-y)$.

(b) $(x \vee y) + z = (x + z) \vee (y + z)$.

(c) $(x \wedge y) + z = (x + z) \wedge (y + z)$.

(d) $x + y = x \vee y + x \wedge y$.

(e) *If $\alpha \geq 0$, then $\alpha x \vee \alpha y = \alpha(x \vee y)$.*

(f) *If $\alpha \geq 0$, then $\alpha x \wedge \alpha y = \alpha(x \wedge y)$.*

Proof. Let us start by observing that, by rearranging the inequalities in the definition of supremum and infimum, (a) follows immediately. Similarly, (b) and (c) are an immediate consequence of (1.1)(i). In order to prove (d), it suffices to observe that (b) yields $x \vee y - (x + y) = (-x) \vee (-y)$. Thus, applying (c) gives $-(x \wedge y) = (-x) \vee (-y) = x \vee y - (x + y)$, hence the desired equality. Finally, (e) and (f) are an immediate consequence of (1.1)(ii), applied to the definitions of supremum and infimum, respectively. □

Starting from the next result, we are going to show in more detail the main features of the vectors x^+, x^-, and $|x|$, and we are going to establish relations that that are going to look familiar, for they are already well-known in the special case of subspaces of $\mathbb{R}^E$. As we shall see in the next result, even though, for a general Riesz space, the partial order relation is not linked to the order relation in $\mathbb{R}$, apart from (1.1)(ii), this is sufficient to create a very strong structure that mimic that of real functions numbers under many aspects.

(1.5) Proposition. *Let V be a Riesz space. Then, for every element x of V, the following relations hold:*

$$x = x^+ - x^-, \qquad |x| = x^+ + x^-.$$

Proof. To prove the first equality, observe that, thanks to (1.4)(b), one has $x \vee 0 - x = 0 \vee (-x)$, which shows that $x^+ - x = x^-$, which is equivalent to the first equality. Now, applying again (1.4)(b), gives $x \vee (-x) + x = 2x \vee 0$, that is to say, $|x| + x = 2x^+$. The latter equality, thanks to the relation $x = x^+ - x^-$, is equivalent to the second equality. □

Using the vectors x^+, x^-, and $|x|$ are also sufficient to establish whether a vector space is a Riesz space, as shown in the following proposition.

(1.6) Proposition. *Let V be a partially ordered vector space. Then, the following conditions are equivalent:*

(a) *V is a Riesz space.*

(b) $x \in V \Rightarrow x^+ \in V$.

(c) $x \in V \Rightarrow x^- \in V$.

(d) $x \in V \Rightarrow |x| \in V$.

Proof. It is perfectly obvious that, if V is a Riesz space, then the three conditions (b), (c), and (d) hold. Furthermore, the relations

$$x^- = x^+ - x, \qquad x^+ = \tfrac{1}{2}\big(x + |x|\big)$$

show that (b), (c), and (d) are equivalent, hence it suffices to prove (b) $\Rightarrow$ (a). To this end, assume that condition (b) holds, and observe that

$$x \vee y = y + (x - y)^+, \qquad x \wedge y = -(-x) \vee (-y).$$

From the first of these equalities, it turns of that the supremum of every pair x, y of elements of V lies in V, and therefore, from the second equality, the infimum must also lie in V. □

(1.7) Definition. In a Riesz space V, two elements x and y are **mutually singular**, (or **disjoint**, or **alien**) and we shall write $x \perp y$, whenever the equality $|x| \wedge |y| = 0$ holds. Observe that, thanks to (1.4)(d), this equality can also be written in the form $|x| + |y| = |x| \vee |y|$.

(1.8) Remark. If x and y are two positive elements of the Riesz space V, then, for them to be mutually singular, it is necessary and sufficient that the relation $x = x \vee y - y$, that is to say $x = (x - y)^+$, holds. In particular, if x is an arbitrary element of V, the relation $x^+ = (x^+ - x^-)^+$ entails that x^+ and x^- are mutually singular.

(1.9) Proposition. *Let V be a Riesz space, and y_1 and y_2 two elements of V_+. Then, for every element x of the positive cone V_+, that is bounded from above by $y_1 + y_2$, it is possible to decompose x into the sum of two elements of V_+, the first of which is bounded from above by y_1, and the second by y_2. More precisely, it is possible to choose, for the first term in this decomposition, the element $x \wedge y_1$.*

Proof. It is clear that the element $x \wedge y_1$ is bounded from above by y_1. Since it is also bounded from above by x, the difference $x - x \wedge y_1$ belongs to V_+. The proposition is proved once it is proved that this difference is bounded from above by y_2, that is to say, that the inequality $x - x \wedge y_1 \leq y_2$ holds. In an equivalent form, however, this inequality can be written as

$$x - y_2 \leq x \wedge y_1$$

which holds, for both the inequalities $x - y_2 \leq x$ and $x - y_2 \leq y_1$ hold. □

(1.10) Corollary. *Let V be a Riesz space, and let x, y_1 and y_2 be three elements of V_+. Then,*

$$x \wedge (y_1 + y_2) \leq x \wedge y_1 + x \wedge y_2.$$

As a result, the set of all elements of V which are mutually singular with respect to a given element of V renders a vector subspace of V.

Proof. The left-hand side of the inequality is obviously bounded from above by both x and $y_1 + y_2$. It suffices to observe that, thanks to the previous proposition, it is possible to express $x \wedge (y_1 + y_2)$ as the sum of two elements of V_+ (each necessarily bounded from above by x), of which the first is also bounded from above by y_1 (hence by $x \wedge y_1$), and the second by y_2 (hence by $x \wedge y_2$). □

2. Riesz subspaces, ideals, and bands

(2.1) Definition. If V is a Riesz space, a **Riesz subspace** of V is any vector subspace U of V that is stable under the lattice operations

$$(x, y) \mapsto x \vee y, \qquad (x, y) \mapsto x \wedge y.$$

Observe immediately that, thanks to Proposition (1.6), if V is a Riesz space, and U is a vector subspace of V, to establish whether U is a Riesz subspace of V, it suffices to check whether the set U is stable under one (hence each) of the following operations

$$x \mapsto x^+, \qquad x \mapsto x^-, \qquad x \mapsto |x|.$$

Before moving any further it is important to notice that the lattice operations, and these three operations, may be misleading, because they may operate in a different manner depending on whether they are calculated in one set or in another. In particular, in the definition of Riesz subspace, the lattice operations are supposed to be calculate with respect to the subspace and not to the ambient space. To clarify this point, the following remark should suffice.

(2.2) Remark. Let U be a vector subspace of a Riesz space V. Then,

(a) If U is a Riesz subspace of V, then it is clear that U, equipped with the partial order relation inherited by V, is itself a Riesz space and, in addition, for every pair of elements of U, the supremum and infimum calculated in U coincide with those calculated in V.

(b) It may happen that U be a Riesz space with respect to the partial order relation inherited by V, without U being a Riesz subspace of V. For instance, take the Riesz space $V = \mathbb{R}^{[0,1]}$ of all real functions defined on the unit interval $[0, 1]$, with its natural order, and the subspace U of all affine linear functions defined on $[0, 1]$, that is to say all functions of the form $y = mx + c$ with m and c real numbers.

The previous remark shows that the bare notion of Riesz subspace, without any additional hypothesis, is generally too weak, and unstable. The next definition solves this issue introducing a property that ensures that the issue highlighted in Remark (2.2) cannot occur.

(2.3) Definition. Let U be a vector subspace of a Riesz space V. We shall say that U is **solid** in V if, for any pair x, y of elements of V,

$$0 \leq x \leq y \in U \Rightarrow x \in U.$$

The next result shows that, when a vector subspace of a Riesz space is solid, the situation shown in Remark (2.2)(b) does not happen.

(2.4) Proposition. *Let V be a Riesz space and U a vector subspace of V, solid in V. The following conditions are equivalent:*

(a) *The set U is a Riesz subspace of V — i.e., for every element x of U, the supremum $\sup_V\{x, 0\}$ of the pair $\{x, 0\}$, calculated in V belongs to U.*

(b) *The set U, with the partial order inherited by V is a Riesz space — i.e., for every element x of U, there exists a supremum $\sup_U\{x, 0\}$ of the pair $\{x, 0\}$, calculated in U.*

Proof. The implication (a) $\Rightarrow$ (b) follows from Remark (2.2)(a). Thus, it suffices to prove the implication (b) $\Rightarrow$ (a). To this end, observe that, if condition (b) holds, then, thanks to the fact that U is solid in V, for every element x of U, the relation

$$0 \leq \sup_V\{x, 0\} \leq \sup_U\{x, 0\} \in U$$

implies that $\sup_V\{x, 0\} \in U$, hence it coincides with $\sup_U\{x, 0\}$. □

(2.5) Definition. If V is a Riesz space, every Riesz subspace of V that is solid in V is called an **ideal** of V.

(2.6) Proposition. *Let U be an ideal of a Riesz space V. Then, for every element x of V, the following conditions are equivalent:*

(a) *x belongs to U.*

(b) *$|x|$ belongs to U.*

(c) *x^+ and x^- belong to U.*

Proof. The implication (a) $\Rightarrow$ (b) follows from the fact that U is a Riesz subspace of V. The implication (b) $\Rightarrow$ (c) follows from the fact that U is solid in V. Finally, the implication (c) $\Rightarrow$ (a) follows immediately from the equality $x = x^+ - x^-$. □

We have established that, for a general partially ordered vector space, it is not guaranteed that the supremum and the infimum of a subset exist. For a Riesz space, instead, this is guaranteed for a set with two elements, hence inductively for a finite set of elements. In the sequel, we are now going to study those Riesz spaces for which the supremum and the infimum exist for any subset that has an upper bound. Let us start by giving to this special type of Riesz space a name, which has been selected to honour the German mathematician Richard Dedekind (1831–1916).

(2.7) Definition. A Riesz space V is **Dedekind complete** if every non-empty subset of V which has an upper bound in V does have a least upper bound in V.

For example, if E is a non-empty set, the vector space $\mathbb{R}^E$ of all real functions defined on E (equipped with the usual order relation) is Dedekind complete. Alongside this simple and uninteresting example, however, we also know already of some more useful Dedekind complete Riesz spaces. For instance, if, in addition, $\mathcal{E}$ is an assigned σ-field on E, we have already established that the class $\mathcal{M}(E, \mathcal{E})$ of all measurable functions on $(E, \mathcal{E})$ is another example of Dedekind complete Riesz space. Finally, if μ is a measure on $(E, \mathcal{E})$, the space $\mathcal{L}^1(\mu)$ is yet another example of Dedekind complete Riesz space.

(2.8) Proposition. *Let V be a Riesz space. For V to be Dedekind complete, it is (necessary and) sufficient that every non-empty subset of V_+, with an upper bound in V, filtering for the order on V, has a least upper bound in V.*

Proof. Assume that the condition expressed in the proposition holds, and let us consider a subset H of V that has an upper bound in V, and let us prove that it does have a least upper bound in V. With no loss of generality, substituting, if necessary, H with a translation of H, we may assume that H has 0 as its element. Let us consider the set K of all the suprema of all finite subsets of H which have 0 as their element. It is clear that the supremum of K

coincides with the supremum of H. On the other hand, K is, by construction, included in V_+ and filtering for the order on V, hence it has a supremum in V. Thus, H also has a supremum in V (identical to the supremum of K). □

(2.9) Definition. Let V be a Dedekind complete Riesz space, and U an ideal of V. We shall say that U is a **band** of V if, for every non-empty subset H of U, that has an upper bound in V, the supremum of H, calculated in V, belongs to U.

For instance, in the Riesz space $\mathbb{R}^E$, where E is a non-empty set, let A be a non-empty subset of E. Then the subspace of $\mathbb{R}^E$ of all functions that vanish on A is a band. On the contrary, the subset of $\mathbb{R}^E$ of all bounded functions is an ideal, but it is generally not a band. (It is a band only when the set E is finite.)

With the same argument as in Proposition (2.8), the following holds.

(2.10) Proposition. *Let V be a Dedekind complete Riesz space, and U an ideal of V. For U to be a band in V it is (necessary and) sufficient that, for every non-empty subset H of U_+ that has an upper bound in V, and that is filtering for the order on V, the supremum of H calculated in V belongs to U.*

We are also going to use this elementary result.

(2.11) Proposition. *The intersection of any set of Riesz subspaces is a Riesz subspace. The intersection of any set of ideals is an ideal. The intersection of any set of bands is a band.*

3. Duality and Riesz decomposition theorem

Let us recall that, if V is a Riesz space, we have already said that two of its elements, x and y are mutually singular if $|x| \wedge |y| = 0$, and we have introduced the notation $x \perp y$ to denote it. If A and B are two parts of V, we shall also write $A \perp B$ to say that every element of A is mutually singular to any element of the other. In other words, we shall write $A \perp B$ to say that $x \perp y$ for every $x \in A$ and $y \in B$.

(3.1) Definition. If V is a Riesz space, and A is a part of V, we shall denote by $A^\perp$, and we shall call it the **dual** of A, the set $\{x \in V : \{x\} \perp A\}$ of all elements of V that are mutually singular to all elements of A.

Note that, if A and B are two subsets of a Riesz space V, the relation $A \perp B$ is equivalent to the two relations $A \subseteq B^\perp$ and $B \subseteq A^\perp$. In particular, the relation $A \perp A^\perp$ (which is trivially true for any subset A of V) can also be written, in an equivalent fashion, as $A \subseteq A^{\perp\perp}$.

(3.2) Proposition. *If V is a Dedekind complete Riesz space, and A is a subset of V, then the set $A^\perp$ is a band in V.*

Proof. Note that $A^\perp = \bigcap_{y \in A} \{y\}^\perp$. Since the intersection of bands is a band, we may assume, with no loss of generality, $A = \{y\}$ with $y \in V$. In addition, we may also assume that $y \in V_+$. Thus, all the positive elements of $A^\perp$ are those elements x in V_+ such that $x + y = x \vee y$. Furthermore, we already know that $A^\perp$ is a vector subspace of V (see (1.10)). It is evident that $A^\perp$ is solid in V. It is thus sufficient to prove that, if H is a non-empty subset of $A^\perp$, included in V_+, with an upper bound in V, and if z denotes its least upper bound calculated in V, then one has $z \in A^\perp$, that is to say $z + y = z \vee y$. To this end, since the relation $x + y = x \vee y$ is true for any element x in H, it suffices to take the supremum of this relation, when x ranges in H. □

We are now finally able to prove the following important result.

(3.3) Theorem (Riesz's decomposition theorem). *Given a Dedekind complete Riesz space V and a band U of V, let us consider the band $U^\perp$ of all elements of V that are mutually singular to every element of U.*

Then, every element z of V can be decomposed, in a unique fashion, into the sum of two elements of V, of which the first belongs to U, and the second to $U^\perp$. In other words, $V = U \oplus U^\perp$ —that is to say, V can be decomposed in the direct sum of the two subspaces U and $U^\perp$.

In addition, if z is a positive element of V, then the elements of this decomposition are both positive: More precisely, setting $H_z = \{u \in U_+ : u \leq z\}$, the first element of the decomposition of z is the maximum element of H_z.

Proof. The *uniqueness* of the decomposition is ensured by $U \cap U^\perp = \{0\}$.

Thus, it suffices to prove the *existence* of such a decomposition. To this end, thanks to the relation $z = z^+ - z^-$, we may reduce the problem to the case of a positive z. Observe that the set H_z is a non-empty subset of U_+ (because 0 belongs to it), and that z is an upper bound of it. Let us denote by x the least upper bound of H_z (calculated in V) and, taking into account the fact that U is a band, we find that $x \in U_+$, and $x \leq z$. Thus, we can see that x belongs to the set H_z, hence it is its maximum element. It remains to prove that, setting $y = z - x$, one has $u \wedge y = 0$, for every element u of U_+. To this end, take u in U_+ and observe that $x + u \wedge y$ is an element of U_+ which is bounded from above by $x + y = z$ (hence an element of H_z), and it is an upper bound of x. Since x is the maximum, it follows that $u \wedge y = 0$, as desired. □

(3.4) Corollary. *In the hypotheses of the previous theorem, the band $U^{\perp\perp}$, of all elements of V which are mutually singular to any element of $U^\perp$, coincides with U.*

Proof. Since the inclusion $U \subseteq U^{\perp\perp}$ is obvious, it suffices to prove the opposite inclusion. To this end, let z be an element of $U^{\perp\perp}$, and, by virtue of Theorem (3.3), let (x, y) be an element of $U \times U^\perp$ such that $z = x + y$. Then, thanks to the aforementioned inclusion $U \subseteq U^{\perp\perp}$, one can readily see that (y, x) is an element of $U^{\perp} \times U^{\perp\perp}$ such that $y + x = z$. On the other hand, since $V = U^{\perp} \oplus U^{\perp\perp}$, one finally gets $(y, x) = (0, z)$ hence $z = x \in U$. □

(3.5) Corollary. *Let V be a Dedekind complete Riesz space and A a subset of V. Let us denote by U the band generated by A, that is to say the smallest band (with respect to the inclusion) which contains A. Then, U coincides with the band $A^{\perp\perp}$ (of all elements of V that are mutually singular to all elements of $A^{\perp}$).*

Proof. For every element x of V, for the band $\{x\}^{\perp}$ to include A, it is necessary and sufficient that it includes the band U, generated by A. Thus,

$$A^{\perp} = \{x : A \subseteq \{x\}^{\perp}\} = \{x : U \subseteq \{x\}^{\perp}\} = U^{\perp}.$$

Thanks to (3.4), it follows that $A^{\perp\perp} = U^{\perp\perp} = U$. □

4. The Riesz space of real measures

We can now utilise the theory of Riesz spaces to study a special, and especially important, example of Dedekind complete Riesz space, that is of interest in measure theory. To this end, recall that, if $(E, \mathcal{E})$ is a measurable space, a *set function* is any numeric function defined on the σ-field $\mathcal{E}$. We are now going to introduce a slight generalisation of the notion of measure on a measurable set, and we shall see that a measure turns out to be an element of the positive cone of the space of these generalised measures. This generalisation of the concept of measure is capable of modelling objects like electric charges, which behave like measures, but that can accept both positive and negative values.

(4.1) Definition. On a measurable space $(E, \mathcal{E})$, a **real measure** is any real valued set function φ, defined on the σ-field $\mathcal{E}$ of all measurable sets of E, such that, for any sequence (A_n) of pairwise disjoint measurable sets, the series $\sum_n \varphi(A_n)$ is absolutely convergent and its sum is $\varphi\left(\bigcup_n A_n\right)$.

Let us observe immediately that, in an equivalent fashion, one could say that, on a measurable space $(E, \mathcal{E})$, a real valued set function φ, defined on the σ-field $\mathcal{E}$, which is additive on $\mathcal{E}$, is a real measure if it satisfies one (hence both) of the following conditions:

(a) *For every decreasing sequence (A_n) of elements of $\mathcal{E}$, whose intersection is empty, the real sequence $(\varphi(A_n))$ converges towards zero, as $n \to \infty$.*

(b) *For every monotone sequence (A_n) of elements of $\mathcal{E}$, if A denotes its limit, the real sequence $(\varphi(A_n))$ converges towards the number $\varphi(A)$, as $n \to \infty$.*

Furthermore, it is clear that the linear combination (with real coefficients) of real measures on a measurable space $(E, \mathcal{E})$ is again a real measure on $(E, \mathcal{E})$.

In other words, the set $\mathfrak{M}(E,\mathcal{E})$ of all real measures on $(E,\mathcal{E})$ is a vector space. In the sequel, this vector space shall always be considered as equipped with the order relation induced by the natural order relation among real functions, that is to say, we shall convene to say that, given two real measures φ_1 and φ_2, the relation $\varphi_1 \leq \varphi_2$ in $\mathfrak{M}(E,\mathcal{E})$ is equivalent to

$$\varphi_1(A) \leq \varphi_2(A) \qquad \text{for every element } A \text{ of } \mathcal{E}.$$

This order relation is compatible with the vector structure of $\mathfrak{M}(E,\mathcal{E})$, and this means that this space is, in fact, a partially ordered vector space, whose cone of positive elements, $\mathfrak{M}_+(E,\mathcal{E})$, coincides with the set of all positive finite measures on $(E,\mathcal{E})$, whose zero is the measure identically zero. The following theorem shows that this space is a Riesz space. We shall see in section 8 that this space is, in fact, a Dedekind complete Riesz space.

(4.2) Theorem. *The partially ordered vector space $\mathfrak{M}(E,\mathcal{E})$ of all real measures on an assigned measurable space $(E,\mathcal{E})$ is a Riesz space. For every element φ of $\mathfrak{M}(E,\mathcal{E})$, the element φ^+ coincides with the set function μ defined on $\mathcal{E}$ by:*

$$\mu(B) = \sup\{\varphi(A) : A \in \mathcal{E},\ A \subseteq B\}.$$

Proof. It is evident that the set function μ is positive (for $\mu(\emptyset) = 0$) and increasing. Furthermore, this function is clearly finite. It remains to prove that it is also countably additive. To this end, consider a sequence (B_n) of pairwise disjoint elements of $\mathcal{E}$, and let us denote by B its union. For every element A of $\mathcal{E}$, with $A \subseteq B$, on has

$$\varphi(A) = \textstyle\sum_n \varphi(A \cap B_n) \leq \sum_n \mu(B_n)$$

The fact that A is chosen arbitrarily yields $\mu(B) \leq \sum_n \mu(B_n)$. On the other hand, for any strictly positive real number ε, choosing, for every index n, an element A_n of $\mathcal{E}$, such that

$$A_n \subseteq B_n, \qquad \mu(B_n) \leq \varphi(A_n) + 2^{-n}\varepsilon,$$

and setting $A = \bigcup_n A_n$, one gets

$$\textstyle\sum_n \mu(B_n) \leq \sum_n \big[\varphi(A_n) + 2^{-n}\varepsilon\big] = \varphi(A) + \varepsilon \leq \mu(A) + \varepsilon.$$

Since ε is chosen arbitrarily, it follows that $\sum_n \mu(B_n) \leq \mu(B)$.

This proves that the function μ is countably additive, whence a measure. In addition, this measure is obviously an upper bound for φ. Now, let μ' be any other measure which is an upper bound for φ. Then, for any element A of $\mathcal{E}$, with $A \subseteq B$, one has $\varphi(A) \leq \mu'(A) \leq \mu'(B)$. Since A is chosen arbitrarily, this yields $\mu(B) \leq \mu'(B)$. Hence, μ is the least upper bound of φ, that is to say, it is the supremum of the pair $\{\varphi, 0\}$ in the partially ordered vector space $\mathfrak{M}(E,\mathcal{E})$. This is sufficient to infer that this vector space is a Riesz space, and that the element φ^+ (i.e. $\varphi \vee 0$) coincides with the measure μ. $\square$

In light of the previous theorem, the next definition is more a terminology adaptation of the corresponding notions in the theory of Riesz spaces, to the case of measures. To start off, let us just remark that, in the sequel, with the term 'measure' we shall refer indistinguishably to a measure according to Definition I(3.1) or as a positive real measure, i.e. an element of the positive cone of $\mathfrak{M}(E, \mathcal{E})$. These two notions, in fact, coincide. Hence, the phrases 'a positive real measure' and 'a measure' mean the exact same thing. Furthermore, let us give the following definition.

(4.3) Definition. Let φ be an element of the Riesz space $\mathfrak{M}(E, \mathcal{E})$, of all real measures on a measurable space $(E, \mathcal{E})$. The elements φ^+, φ^-, and $|\varphi|$ shall be referred to as the **positive variation**, **negative variation**, and **total variation** of φ. Furthermore, for every function f that belongs to the space $\mathcal{L}^1(|\varphi|)$, that is to say to $\mathcal{L}^1(\varphi^+) \cap \mathcal{L}^1(\varphi^-)$, we shall set

$$\int f \, \mathrm{d}\varphi = \int f \, \mathrm{d}\varphi^+ - \int f \, \mathrm{d}\varphi^-.$$

Let us complete this section with a simple remark.

(4.4) Notation. The Riesz space of all real measures defined on a measurable space $(E, \mathcal{E})$ shall be denoted by $\mathfrak{M}(E, \mathcal{E})$. When E is a topological space, and $\mathcal{E} = \mathcal{B}(E)$ is the Borel σ-field associated to its topology, this notation may be simplified to $\mathfrak{M}(E)$, meaning the Riesz space of all real measures on the Borel measurable space $(E, \mathcal{B}(E))$. As a general rule, whenever a measurable space is intended with respect to its Borel σ-field, the reference to the σ-field shall always be omitted.

5. Real measures defined by a density

Let us now extend the definitions and results that we had previously introduced, in Chapter II, §5, only in the special case of positive measures. As we shall see, the same properties already known apply for real measures, but, in the end, thanks to the decomposition $x = x^+ - x^-$, valid in every Riesz space, the proofs of the results can be re-conducted to the case of elements of the positive cone in the various Riesz space under exam (either $\mathfrak{M}$ or $\mathcal{L}^1(\mu)$).

(5.1) Definition. Let μ be a σ-finite measure on an assigned measurable space $(E, \mathcal{E})$. If g is a function, not necessarily positive, belonging to $\mathcal{L}^1(\mu)$, we shall denote by $g \cdot \mu$ the real measure ν defined by

$$\nu(A) = \int_A g^+ \, \mathrm{d}\mu - \int_A g^- \, \mathrm{d}\mu = \int_A g \, \mathrm{d}\mu \qquad \text{for every element } A \text{ of } \mathcal{E}.$$

Every measure of the form $g \cdot \mu$, as g runs over $\mathcal{L}^1(\mu)$, is referred to as a measure **with base** μ.

(5.2) Proposition. *In the same hypotheses as in Definition (5.1), denote by $\mathfrak{M}$ the Riesz space of all real measures on $(E, \mathcal{E})$. Then, the map $g \mapsto g \cdot \mu$, from $\mathcal{L}^1(\mu)$ to $\mathfrak{M}$ is linear. Furthermore, for the real measure $g \cdot \mu$ to be positive it is necessary and sufficient that the function g is positive almost everywhere with respect to μ.*

Proof. The first assertion is obvious. To prove the second, let us observe that from the already known results for positive measures (see II(5.4)), for the measure $g \cdot \mu$ to be positive, that is to say, $g^+ \cdot \mu \geq g^- \cdot \mu$, it is necessary and sufficient that g^+ be greater than g^- almost everywhere with respect to μ. On the other hand, this latter condition is equivalent to say that g is positive almost everywhere with respect to μ. □

(5.3) Corollary. *In the same hypotheses as in Definition (5.1), let g_1 and g_2 be two elements of $\mathcal{L}^1(\mu)$. In order to have $g_1 \cdot \mu \leq g_2 \cdot \mu$ it is necessary and sufficient that the function g_1 is smaller than g_2 μ-almost everywhere.*

In particular, the measures $g_1 \cdot \mu$ and $g_2 \cdot \mu$ are identical if and only if the functions g_1 and g_2 are equivalent modulo μ.

(5.4) Definition. In the same hypotheses as in Definition (5.1), if ν is a real measure with base μ, any of the functions g, belonging to $\mathcal{L}^1(\mu)$, such that $\nu = g \cdot \mu$, shall be called a **version of the density** of ν with respect to μ. The set of all these versions is an element of the quotient space $L^1(\mu)$ which is referred to as the **density** of ν with respect to μ.

6. The Radon-Nikodým theorem

In the sequel, we shall assume that a measurable space $(E, \mathcal{E})$ is assigned and that μ is a σ-finite positive real measure on this space.

(6.1) Definition. A real measure ν is said to be **absolutely continuous** with respect to μ, and we shall write

$$\nu \ll \mu,$$

if every negligible set with respect to μ is also negligible with respect to ν, that is to say, if $\mu(A) = 0 \Rightarrow \nu(A) = 0$, for every element A of $\mathcal{E}$.

Let us observe immediately that if ν is a real measure with base μ, then ν is also absolutely continuous with respect to μ. The ultimate aim of this section is to show that the converse is true, that is to say, that all real measures absolutely continuous with respect to μ are with base μ. To achieve this, it is necessary to first prove a few preliminary results that describe in more detail the properties of the Riesz space $\mathfrak{M}(E, \mathcal{E})$ of all real measures on $(E, \mathcal{E})$.

(6.2) Lemma. *Let $(f_n \cdot \mu)$ be an increasing sequence of measures, all bounded from above by a given σ-finite measure ν, and set $f = \sup_n f_n$. Then, $f \cdot \mu$ is a measure on $(E, \mathcal{E})$, bounded from above by ν, which is also the supremum, in the Riesz space $\mathfrak{M}(E, \mathcal{E})$, of the sequence $(f_n \cdot \mu)$.*

Proof. By modifying, if needed, the function f_n on a negligible set with respect to μ, we can assume that the sequence (f_n) is increasing. Then, by means of Beppo Levi's theorem, we have

$$\int_A f \, \mathrm{d}\mu = \sup_n \int_A f_n \, \mathrm{d}\mu \leq \nu(A).$$

This proves that the measure $f \cdot \mu$ is bounded from above by the measure ν, which is σ-finite, thus $f \cdot \mu$ is also σ-finite. In particular, it is obvious that $f \cdot \mu$ is greater than all of the $f_n \cdot \mu$. Conversely, if λ is a measure that is greater than all the $f_n \cdot \mu$, then, for every measurable set A, one has

$$\lambda(A) \geq \sup_n \int_A f_n \, \mathrm{d}\mu = \int_A f \, \mathrm{d}\mu,$$

hence $\lambda \geq f \cdot \mu$. □

(6.3) Corollary. *Every σ-finite measure ν is the supremum, in the Riesz space $\mathfrak{M}(E, \mathcal{E})$, of a sequence of increasing finite measures.*

Proof. Since ν is σ-finite, the whole space E is the union of an increasing sequence (B_n) of measurable sets, with finite measure under ν. The measure ν is the supremum of the increasing sequence $(I_{B_n} \cdot \nu)$ of finite measures. □

We are now ready to prove the following result that establishes a sufficient condition for a measure to be with base a pre-assigned measure. This is also the core result of the section, and the remaining ones, which perhaps may seem more meaningful, all depend on it.

(6.4) Theorem. *Assume that on the measurable space $(E, \mathcal{E})$, the real measures μ and ν are both finite, positive and that $\nu \leq \mu$. Then, ν is with base μ.*

Proof. With no loss of generality, we may assume that ν is neither the measure constantly 0, nor equal to the measure μ. In fact, in each of these cases, the conclusion follows trivially, by taking $f = 0$ and $f = 1$, respectively.

Let us denote by $\mathcal{H}$ the set of all positive functions f in $\mathcal{L}^1(\mu)$ such that

$$\int_A f \, \mathrm{d}\mu \leq \nu(A) \qquad \text{for every element } A \text{ of } \mathcal{E}.$$

This set is not empty, because at least the function identically 0 lies in it. Furthermore, if f_1 and f_2 are two elements of $\mathcal{H}$, then $f_1 \vee f_2$ is also in $\mathcal{H}$. In fact, given the two functions f_1 and f_2, and setting $H = \{f_1 \geq f_2\}$, it is easy to recognise that $f_1 \vee f_2 = f_1$ on H and $f_1 \vee f_2 = f_2$ on H^c. Thus,

$$\begin{aligned}\int_A (f_1 \vee f_2) \, \mathrm{d}\mu &= \int_{A \cap H} f_1 \, \mathrm{d}\mu + \int_{A \cap H^c} f_2 \, \mathrm{d}\mu \\ &\leq \nu(A \cap H) + \nu(A \cap H^c) \\ &= \nu(A).\end{aligned}$$

In other words, the set $\mathcal{H}$ is stable under finite suprema. Let us set

$$\alpha = \sup\Big\{\textstyle\int f \,\mathrm{d}\mu : f \in \mathcal{H}\Big\}.$$

The number α is necessarily finite, because $\alpha \leq \nu(E)$ and $\nu(E)$ is finite.

Consider any sequence (f_n) of elements of $\mathcal{H}$, such that $\int f_n \,\mathrm{d}\mu \to \alpha$. Since $\mathcal{H}$ is stable under finite suprema, setting $g_n = f_1 \vee \cdots \vee f_n$, the sequence (g_n) is increasing, belongs to $\mathcal{H}$, and $\int g_n \,\mathrm{d}\mu \uparrow \alpha$. Thus, by virtue of Beppo Levi's theorem, the sequence (g_n) converges towards an element f of $\mathcal{H}$ such that $\int f \,\mathrm{d}\mu = \alpha$.

We wish to prove that $\nu(A) = \int_A f \,\mathrm{d}\mu$ for every element A in $\mathcal{E}$. This is done by contradiction. Precisely, assume that there exists a set B in $\mathcal{E}$ such that

$$\textstyle\int_B f \,\mathrm{d}\mu < \nu(B).$$

In other words, if we denote by λ the measure $\nu - f \cdot \mu$, the latter hypothesis can be rephrased by assuming that the measure λ is not identically zero.

Let us prove that there exist a positive real number ε and a measurable set D, included in B, with $\mu(D) > 0$, such that

$$\varepsilon\mu(A) \leq \lambda(A) \qquad \text{for every } A \in \mathcal{E} \text{ with } A \subseteq D.$$

To this end, let ε be a strictly positive real number with $\varepsilon\mu(B) < \lambda(B)$, and let us denote by $\mathcal{G}$ the family of all measurable subsets X of B such that $\lambda(X) \leq \varepsilon\mu(X) \leq \lambda(B)$. Let us observe that, if $\mathcal{G}$ is empty, then the set B satisfies the desired property, so we may assume, with no loss of generality, that the class $\mathcal{G}$ is not empty. In these hypotheses, by induction, it is possible to construct an increasing sequence (C_n) of elements of $\mathcal{G}$, such that

$$C_0 = \emptyset, \quad \varepsilon\mu(C_n) \geq \sup\big\{\lambda(X) : C_{n-1} \subseteq X,\, X \in \mathcal{G}\big\} - 1/n \quad \text{for } n > 1.$$

Set $C = \bigcup_n C_n$ and $D = B \setminus C$, and observe that

$$\lambda(C) = \sup_n \lambda(C_n) \leq \sup_n \big[\varepsilon\mu(C_n) + 1/n\big] = \varepsilon\mu(C),$$

which means that C belongs to $\mathcal{G}$, and

$$\varepsilon\mu(D) = \varepsilon\mu(B) - \varepsilon\mu(C) < \lambda(B) - \lambda(C) = \lambda(D),$$

thereby ensuring that $\mu(D) > 0$. Let us show that D is the set desired. To this end, observe that, if A is a measurable subset of D, such that $\lambda(A) \leq \varepsilon\mu(A)$, then A belongs to $\mathcal{G}$, and since the two sets A and C are disjoint, $A \cup C$ also belongs to $\mathcal{G}$, and $C_n \subseteq C \subseteq A \cup C$, for every index n. Thus,

$$\mu(A \cup C) \leq \sup\big\{\mu(X) : C_{n-1} \subseteq X,\, X \in \mathcal{G}\big\} \leq \mu(C_n) + 1/n.$$

Taking the limit as n approaches ∞, this gives $\mu(A \cup C) = \mu(C)$, which yields $\mu(A) = 0$. On the other hand, since $\lambda \leq \mu$, one also has $\lambda(A) = 0$.

Now, we may finally prove that $f + \varepsilon I_D \in \mathcal{H}$. To this end, observe that, for any element A of $\mathcal{E}$, one has

$$\begin{aligned} \textstyle\int_A (f + \varepsilon I_D)\,\mathrm{d}\mu &= \textstyle\int_A f\,\mathrm{d}\mu + \varepsilon\mu(A \cap D) \\ &= \textstyle\int_{A\cap D} f\,\mathrm{d}\mu + \varepsilon\mu(A \cap D) + \int_{A\cap D^c} f\,\mathrm{d}\mu \\ &\leq \textstyle\int_{A\cap D} f\,\mathrm{d}\mu + \lambda(A \cap D) + \int_{A\cap D^c} f\,\mathrm{d}\mu \\ &= \textstyle\int_{A\cap D} f\,\mathrm{d}\mu + \nu(A \cap D) - \int_{A\cap D} f\,\mathrm{d}\mu + \int_{A\cap D^c} f\,\mathrm{d}\mu \\ &\leq \nu(A \cap D) + \nu(A \cap D^c) = \nu(A). \end{aligned}$$

On the other hand,

$$\textstyle\int (f + \varepsilon I_D)\,\mathrm{d}\mu = \int f\,\mathrm{d}\mu + \varepsilon\mu(D) = \alpha + \varepsilon\mu(D) > \alpha,$$

which is a contradiction. □

(6.5) Corollary. *If μ is a measure that is greater than both ν_1 and ν_2 (for example $\mu = \nu_1 + \nu_2$), and if f_1, f_2 are two densities of ν_1, ν_2, respectively, with respect to μ, then*

$$\nu_1 \wedge \nu_2 = (f_1 \wedge f_2) \cdot \mu,$$

where $\nu_1 \wedge \nu_2$ is the infimum, in $\mathfrak{M}(E, \mathcal{E})$, of ν_1 and ν_2.

Proof. Every measure that is bounded from above by both ν_1 and ν_2 is also bounded from above by μ and therefore it is of the form $f \cdot \mu$ with f function such that

$$f \leq f_1 \pmod{\mu}, \qquad f \leq f_2 \pmod{\mu}.$$

On the other hand, this relations are summarised as $f \leq f_1 \wedge f_2$ modulo μ, which in turn is equivalent to the inequality $f \cdot \mu \leq (f_1 \wedge f_2) \cdot \mu$ in $\mathfrak{M}$. This proves that $(f_1 \wedge f_2) \cdot \mu$ is the infimum of the pair ν_1, ν_2. □

(6.6) Theorem (Lebesgue decomposition). *Let μ and ν be two positive finite measures on $(E, \mathcal{E})$. Then,*

(a) *The measure $\nu_\mathrm{a} = \sup_n(\nu \wedge (n\mu))$ is the supremum, in $\mathfrak{M}(E, \mathcal{E})$, of all measures with base μ, that are bounded from above by ν.*

(b) *There is a function $h \in \mathcal{L}^1(\mu)$ such that*

$$\nu_\mathrm{a}(A) = \textstyle\int h\,\mathrm{d}\mu \qquad \text{for every } A \text{ in } \mathcal{E},$$

and this function is essentially unique modulo μ.

(c) *Setting $\nu_\mathrm{s} = \nu - \nu_\mathrm{a}$, the real measure ν_s is positive and $\nu_\mathrm{a} \perp \nu_\mathrm{s}$. Furthermore, the pair of measures $(\nu_\mathrm{a}, \nu_\mathrm{s})$ defined on $\mathcal{E}$ in uniquely determined by μ and enjoys the following properties:*

$$\nu = \nu_\mathrm{a} + \nu_\mathrm{s}, \qquad \nu_\mathrm{a} \ll \mu, \qquad \nu_\mathrm{s} \perp \mu.$$

Proof. Let us start by proving that ν_{a} is the supremum of all measures λ that are at the same time with base μ and bounded from above by ν. To this end, let us set $\pi = \lambda + \mu$ and observe that, by means of Theorem (6.4), one has

$$\lambda = f \cdot \pi, \qquad \mu = g \cdot \pi.$$

For every positive integer n, Corollary (6.5) yields $\lambda \wedge (n\mu) = (f \wedge (ng)) \cdot \pi$; whence, from the identity $\lambda = \sup_n \big(\lambda \wedge (n\mu)\big)$, applying Lemma (6.2), it then follows $f = \sup_n f \wedge (ng)$. Thus,

$$\begin{aligned} \lambda = f \cdot \pi &= \sup_n \big(f \wedge (ng)\big) \cdot \pi \\ &= \sup_n \big((f \cdot \pi) \wedge (ng \cdot \pi)\big) \\ &= \sup_n \big(\lambda \wedge (n\mu)\big) \\ &\leq \sup_n \big(\nu \wedge (n\mu)\big) = \nu_{\mathrm{a}}. \end{aligned}$$

Furthermore, by Theorem (6.4), for every positive integer n, the measure

$$\nu \wedge (n\mu)$$

is with base $n\mu$, hence ultimately on μ. This means that there exists a function g_n in $\mathcal{L}^1(\mu)$ such that $\nu \wedge (n\mu) = g_n \cdot \mu$. Moreover, by virtue of Lemma (6.2), this yields $\nu_{\mathrm{a}} = \sup_n(\nu \wedge (n\mu)) = \sup_n(g_n \cdot \mu) = (\sup_n g_n) \cdot \mu$, hence ν_{a} is a measure with base μ, and since ν_{a} is also bounded from above by ν, it follows that ν_{a} is the supremum of all positive real measures with base μ, bounded from above by ν, and that there exists a function $h = \sup_n g_n$ in $\mathcal{L}^1(\mu)$ such that $\nu_{\mathrm{a}} = h \cdot \mu$, proving at the same time (a) and (b).

To prove (c), it suffices to show that the two real measures $\nu_{\mathrm{s}} = \nu - \nu_{\mathrm{a}}$ and μ are mutually singular. In fact, if this has been proved, the identity $\nu_{\mathrm{a}} = h \cdot \mu$ will then show that ν_{s} and ν_{a} must also be mutually singular. Furthermore, the maximality of ν_{a} ensures that the pair $(\nu_{\mathrm{a}}, \nu_{\mathrm{s}})$ is unique, and the condition $\nu_{\mathrm{a}} \leq \nu$ ensure that ν_{s} is also a positive real measure. Thus, let us show that $\nu_{\mathrm{s}} \wedge \mu = 0$. This is indeed very simple: In fact, from the obvious equality $\nu_{\mathrm{s}} \wedge \mu \leq \nu_{\mathrm{s}}$, it follows that

$$\nu_{\mathrm{a}} + \nu_{\mathrm{s}} \wedge \mu \leq \nu_{\mathrm{a}} + \nu_{\mathrm{s}} = \nu,$$

and since $\nu_{\mathrm{s}} \wedge \mu \leq \mu$, it follows that $\nu_{\mathrm{s}} \wedge \mu$ is with base μ. Thus, $\nu_{\mathrm{a}} + \nu_{\mathrm{s}} \wedge \mu$ is with base μ, and bounded from above by ν. By maximality of ν_{a}, it must then be $\nu_{\mathrm{a}} = \nu_{\mathrm{a}} + \nu_{\mathrm{s}} \wedge \mu$, whence $\nu_{\mathrm{s}} \wedge \mu = 0$. □

The following simple corollary is in fact an incredibly important result in its own right, and it is usually referred to as the 'Radon-Nikodým theorem', after the Austrian mathematician Johann Radon (1887–1956), who proved this theorem for the special case where the underlying space is $\mathbb{R}^n$ in 1913, and the Polish mathematician Otto M. Nikodým (1887–1974), who proved the general case in 1930.

(6.7) Corollary (Radon-Nikodým). *If a measure ν is absolutely continuous with respect to μ, then ν is with base μ. In other words, if $\nu \ll \mu$, then there exists a positive function $h \in \mathcal{L}^1(\mu)$ such that $\nu = f \cdot \mu$.*

Proof. If $\nu \ll \mu$, then $\nu = \nu_a$, and the result follows from Theorem (6.6). □

7. A characterisation of mutually singular measures

On an assigned measurable space $(E, \mathcal{E})$, we have already encountered a few examples of mutually singular measures. For instance, given a real measure φ on $(E, \mathcal{E})$, we know that its positive and negative variations, that is to say the measures φ^+ and φ^-, are mutually singular, and Lebesgue decomposition theorem states that, if μ and ν are two finite positive real measures on $(E, \mathcal{E})$, then $\nu = \nu_a + \nu_s$ and the measures ν_a and ν_s are mutually singular.

On the other hand, in accordance with the Riesz space structure of the space $\mathfrak{M}(E, \mathcal{E})$ of all real measures on $(E, \mathcal{E})$, we have defined two measures μ and ν to be mutually singular if

$$|\mu| \wedge |\nu| = 0.$$

Whilst legitimate, this definition remains rather obscure and hides the more intuitive idea behind the concept of mutual singularity. In this section, we are going to introduce a new concept, and we shall prove that this new definition leads to a rather intuitive characterisation of mutually singular measures.

(7.1) Definition. If there is a set $B \in \mathcal{E}$ such that $\nu(A) = \nu(A \cap B)$ for every $A \in \mathcal{E}$, we say that ν is **concentrated on** B or, equivalently, that B **carries** ν. This is equivalent to saying that $\nu(A) = 0$ whenever $A \cap B = \emptyset$.

Let us observe immediately that the condition expressed by the previous definition can be rephrased, in an equivalent fashion, by saying that a real measure ν is concentrated on a set B if it coincides with its trace onto the set B or, equivalently, if $\nu = I_B \cdot \nu$.

The following result shows a characterisation of mutually singular measures in terms of the sets were there measures are concentrated.

(7.2) Proposition. *In order for two measures μ and ν on the measurable space $(E, \mathcal{E})$ to be mutually singular it is necessary and sufficient that they are concentrated on two disjoint elements of $\mathcal{E}$.*

Proof. Let us set $\lambda = \mu + \nu$. Then, by means of Theorem (6.4), there exist two positive functions g and h, measurable on $(E, \mathcal{E})$, with values in $[0, 1]$, such that

$$\mu = g \cdot \lambda, \qquad \nu = h \cdot \lambda.$$

Assume that μ and ν are mutually singular. Thus, thanks to Corollary (6.5),

$$(g \wedge h) \cdot \lambda = \mu \wedge \nu = 0,$$

hence $g \wedge h$ is negligible with respect to λ. Thus, substituting, if necessary, the functions g and h with two equivalent functions modulo λ, we may assume that $g \wedge h = 0$. This means that the two measurable sets $\{g > 0\}$ and $\{h > 0\}$ are disjoint and the measures μ and ν are concentrated on the former and the latter, respectively.

Conversely, let us assume that A and B are two disjoint subsets of $\mathcal{E}$, such that μ is concentrated on A and ν is concentrated on B. Then,

$$\mu = I_A \cdot \mu = (I_A g) \cdot \lambda, \qquad \nu = I_B \cdot \nu = (I_B h) \cdot \lambda,$$

hence $\mu \wedge \nu = \big((I_A g) \wedge (I_B h)\big) \cdot \lambda = 0$. □

As an example of application of this result, we may prove very quickly the so-called 'Hahn's decomposition theorem' of a real measure, which named after the Austrian mathematician Hans Hahn (1879–1934).

(7.3) Corollary (Hahn decomposition). *If φ is a real measure on a measurable space $(E, \mathcal{E})$, then there exist two disjoint measurable sets A and B whose union is E, such that:*

(a) *$\varphi(X) \geq 0$ for every element X of $\mathcal{E}$ with $X \subseteq A$,*

(b) *$\varphi(X) \leq 0$ for every element X of $\mathcal{E}$ with $X \subseteq B$.*

Proof. We already know that, since the set of all real measures is a Riesz space, for a real measure φ, one has $\varphi = \varphi^+ - \varphi^-$, where φ^+ and φ^- are two mutually singular positive real measures. By virtue of Proposition (7.2), there exist two disjoint sets A and B in $\mathcal{E}$ such that the measure φ^+ is concentrated on A and the measure φ^- is concentrated on B. In order to complete the proof, it suffices to show that $A \cup B = E$. In fact, if the set $C = (A \cup B)^c = A^c \cap B^c$ were not empty, then $\varphi(C) = \varphi^+(C) - \varphi^-(C) = 0 + 0 = 0$, so the conclusion follows taking the set $A \cup C$ in lieu of A. □

8. Further properties of the space of real measures

(8.1) Theorem. *Assume that, on an assigned measurable space $(E, \mathcal{E})$, we are given a σ-finite measure μ. Let us denote by $\mathfrak{M}$ the Riesz space of all real measures on $(E, \mathcal{E})$ and by $\mathfrak{M}_\mu$ the vector subspace of $\mathfrak{M}$ of all the real measures with base μ. Then,*

(a) *$\mathfrak{M}_\mu$ is an ideal of $\mathfrak{M}$.*

(b) *The linear map $g \mapsto g \cdot \mu$ from $\mathcal{L}^1(\mu)$ to $\mathfrak{M}_\mu$ induces, by taking the quotient modulo μ, an isomorphism of the Riesz space $L^1(\mu)$ onto the Riesz space $\mathfrak{M}_\mu$.*

Proof. By Corollary (5.3), the linear map $g \mapsto g \cdot \mu$ from $\mathcal{L}^1(\mu)$ to $\mathfrak{M}_\mu$ induces, by taking the quotient modulo μ, a bijective linear map of $L^1(\mu)$ onto $\mathfrak{M}_\mu$ which transforms the order of $L^1(\mu)$ into the order of $\mathfrak{M}_\mu$ (that is to say, an isomorphism among these Riesz spaces). As a consequence, since $L^1(\mu)$ is a Riesz space, such is $\mathfrak{M}_\mu$ with the order inherited by $\mathfrak{M}$. Furthermore, thanks to Theorem (6.4), the space $\mathfrak{M}_\mu$ is solid in $\mathfrak{M}$. Thus, Proposition (2.4) yields that $\mathfrak{M}_\mu$ is a Riesz subspace of $\mathfrak{M}$, and, more precisely, an ideal of $\mathfrak{M}$. □

(8.2) Remark. From assertion (a) of the previous theorem, it follows, thanks to (2.6), that for every element ν in $\mathfrak{M}$, the following conditions are equivalent:

(a) ν is with base μ.

(b) $|\nu|$ is with base μ.

(c) ν^+ and ν^- are with base μ.

From assertion (b) of the same theorem, instead, it follows that, for every element g in $\mathcal{L}^1(\mu)$, the positive, negative and total variations of the measure $g \cdot \mu$ are

$$(g \cdot \mu)^+ = g^+ \cdot \mu, \qquad (g \cdot \mu)^- = g^- \cdot \mu, \qquad |g \cdot \mu| = |g| \cdot \mu,$$

(8.3) Theorem. *Let $(E, \mathcal{E})$ be a measurable space. Then:*

(a) *The Riesz space $\mathfrak{M}$ of all real measures on $(E, \mathcal{E})$ is Dedekind complete.*

(b) *Given a positive σ-finite measure μ on $(E, \mathcal{E})$, the subspace $\mathfrak{M}_\mu$ of $\mathfrak{M}$ of all the real measures with base μ is a band of $\mathfrak{M}$.*

Proof. In order to prove the assertion (a), thanks to (2.8), it suffices to prove that, an increasing directed family $(\mu_t)_{t \in T}$ of elements of $\mathfrak{M}_+$ with an upper bound ν_0 in $\mathfrak{M}$, has a least upper bound in $\mathfrak{M}$. To this end, it suffices to that this least upper bound it is the supremum of the family $(\mu_t)_{t \in T}$, that is to say, the positive set function ν defined on $\mathcal{E}$ by

$$\nu(A) = \sup_t \mu_t(A).$$

It is clear that this function is finite (because ν_0 is an upper bound) and additive (because it is pointwise limit of a directed family of additive functions). It remains to prove that ν is continuous on all increasing sequences of elements of $\mathcal{E}$. But this is not difficult to do. Indeed, if (A_n) is an increasing sequence of elements of $\mathcal{E}$, then, denoting by A the union of the A_n, one has

$$\begin{aligned}\nu(A) &= \sup_t \mu_t(A)\\ &= \sup_t \sup_n \mu_t(A_n)\\ &= \sup_n \sup_t \mu_t(A_n)\\ &= \sup_n \nu(A_n).\end{aligned}$$

Let us now prove assertion (b). To this end, thanks to (2.10), it suffices to prove that, if $(\mu_t)_{t \in T}$ is a directed family of increasing elements of $\mathfrak{M}_\mu$, with an upper bound in $\mathfrak{M}$, its supremum, calculated in $\mathfrak{M}$, also belongs to $\mathfrak{M}_\mu$. But this follows immediately from Radon-Nikodým Theorem. In fact, on each element of $\mathcal{E}$ over which the measure μ is zero, all the measures μ_t are also zero, hence their supremum must be zero as well. □

(8.4) Corollary. *For every σ-finite measure μ defined on the measurable space $(E, \mathcal{E})$, the quotient space $L^1(\mu)$ is a Dedekind complete Riesz space.*

Proof. Thanks to (8.1)(b), $L^1(\mu)$ is a Riesz space isomorphic to $\mathfrak{M}_\mu$. Since the latter is a band, whence Dedekind complete, such is the former. □

Chapter VI

The $\mathcal{L}^p$ Spaces

The key idea of functional analysis is to consider functions as 'points' in a space of functions. For example, take the space of all bounded, measurable functions on a finite measure space $(E, \mathcal{E}, \mu)$. This is indeed a vector space, as the sum of two such functions is bounded and measurable, and the product of such a function with a real constant is also bounded and measurable. Furthermore, for any two such functions, f and g, we know that the integral $\int fg \, \mathrm{d}\mu$ is finite. If we consider these functions as vectors, then this integral, that we can denote by $\langle g \mid f \rangle$, enjoys the same algebraic properties as an inner product: It is positive when $g = f$, it is symmetric, in the sense that $\langle g \mid f \rangle = \langle f \mid g \rangle$, and it is linear in f for fixed g. Using this inner product, it is possible to develop an analogue of Euclidean geometry in a space of functions, with a distance d defined by $d(f, g) = \langle f - g \mid f - g \rangle^{1/2}$, just as in a finite-dimensional vector space. In fact, if μ denotes the counting measure and if E is a finite set with n elements, $\langle g \mid f \rangle$ becomes the usual inner product in $\mathbb{R}^n$. But if μ is trace of the Lebesgue measure onto the interval $[0, 1]$, then, for the metric space of functions with distance d to be complete, we shall need to include some unbounded functions f such that $\int f^2 \, \mathrm{d}\mu < \infty$. Along the same lines, for any real number $p > 0$, given a measure μ on $(E, \mathcal{E})$, there is a collection of real functions f which are measurable and for which the integral $\int |f|^p \, \mathrm{d}\mu$ is finite. This collection is denoted by $\mathcal{L}^p(\mu)$ or $\mathcal{L}^p$. It is not immediately clear that this is a vector space, for it is not straightforward that, if f and g are in $\mathcal{L}^p$, then so is $f + g$: this is going to follow from an inequality that is going to be developed in §1. In §2 and §3, we shall give the formal definition of the spaces $\mathcal{L}^p$, and we shall discuss their main topological properties. Finally, throughout §4, we shall show what the topological dual of an $\mathcal{L}^p$ space looks like, and this will be very useful, in Chapter VIII, when we shall delve into all different types of convergence of sequences of functions.

DOI: 10.1201/9781003724766-VI

1. Convexity inequalities

In all the sequel, assume that a measure space $(E, \mathcal{E}, \mu)$ is assigned. This section is devoted to proving some important integral inequalities that generally go by the name of *convexity inequalities.* The first is named after the German mathematician Ludwig Otto Hölder (1859–1937).

(1.1) Proposition (Hölder inequality). *For every pair of measurable positive functions f, g and every pair a, b of strictly positive real numbers, with $a + b = 1$, the following inequality holds:*

$$\int f^a g^b \, \mathrm{d}\mu \leq \left(\int f \, \mathrm{d}\mu\right)^a \left(\int g \, \mathrm{d}\mu\right)^b.$$

Proof. If one of the two functions f and g is negligible, the inequality is trivial. Hence, we may assume that the two numbers $\int f \, \mathrm{d}\mu$ and $\int g \, \mathrm{d}\mu$ are both strictly positive. We may also assume that these two numbers are finite, because the inequality would once again be trivial. Moreover, dividing both sides by the right-hand side, and using the homogeneity of integrals, we may further assume that the two aforementioned integrals are both equal to 1. Thus, in the end, we only need to prove

$$\int f^a g^b \, \mathrm{d}\mu \leq 1.$$

Now, since the logarithmic function, $x \mapsto \log x$ is concave, for every pair of strictly positive real numbers x and y, the relation

$$\log x^a y^b = a \log x + b \log y \leq \log(ax + by)$$

holds. Since the exponential is an increasing function, we thus get

$$x^a y^b \leq ax + by.$$

This relation, which is true if x and y are (not necessarily finite) positive numbers, yields

$$f^a g^b \leq af + bg$$

which, in turn, entails

$$\int f^a g^b \, \mathrm{d}\mu \leq \int (af + bg) \, \mathrm{d}\mu = a \int f \, \mathrm{d}\mu + b \int g \, \mathrm{d}\mu = a + b = 1. \qquad \square$$

For every strictly positive real number p and every measurable function f, let us denote by $\|f\|_p$ the positive real number

$$\left(\int |f|^p \, \mathrm{d}\mu\right)^{1/p}.$$

Using this notation, Hölder's inequality can be written as follows.

(1.2) Proposition (Hölder-Schwarz's inequality). *For any pair f and g of measurable functions and any triple p, q, r of strictly positive real numbers, which satisfy the relation*

$$\frac{1}{p} + \frac{1}{q} = \frac{1}{r},$$

one has

$$\|fg\|_r \leq \|f\|_p \|g\|_q.$$

Proof. It suffices to apply Proposition (1.1) to the functions $|f|^p$ and $|g|^q$, with $a = r/p$ and $b = r/q$. □

A simple corollary of the previous inequality is the following.

(1.3) Corollary. *If μ is a probability measure (that is to say, if $\mu(E) = 1$), then the function $p \mapsto \|f\|_p$ is increasing in the interval $(0, +\infty)$.*

Proof. To see this, given two real numbers r and p, with $0 < r < p$, observe that $1/r > 1/p$, which means that it is possible to find a strictly positive real number q such that $1/r = 1/p + 1/q$. Hence,

$$\|f\|_r = \|f \cdot 1\|_r \leq \|f\|_p \|1\|_q = \|f\|_p.$$ □

The following proposition summarises the properties of the function $f \mapsto \|f\|_p$, and it is an immediate consequence of the properties of integrals.

(1.4) Proposition. *The function $f \mapsto \|f\|_p$, from the set of all measurable functions on $(E, \mathcal{E}, \mu)$ in the interval $[0, \infty]$, satisfies to the following properties:*

(a) *If $\|f\|_p = 0$, then f is μ-negligible.*

(b) *If $|f| \leq |g|$, then $\|f\|_p \leq \|g\|_p$.*

(c) *For every real number c, one has $\|cf\|_p = |c|\|f\|_p$.*

(d) *For every sequence (f_n) of measurable functions, such that the sequence $(|f_n|)$ is increasing, one has $\| \sup_n f_n\|_p = \sup_n \|f_n\|_p$.*

In the special case when $p \geq 1$, another fundamental convexity property holds, which is named after Hermann Minkowski (1864–1909), described variously as German, Polish, Lithuanian-German, or Russian.

(1.5) Proposition (Minkowski's inequality). *If $p \geq 1$, the inequality*

$$\|f + g\|_p \leq \|f\|_p + \|g\|_p$$

holds for every pair of measurable functions f and g.

Proof. With no loss of generality, we may assume that $p > 1$, because the case $p = 1$ corresponds to the well-known triangle inequality. Furthermore, we may also assume that $\|f + g\|_p$ is strictly positive, and that both the numbers $\|f\|_p$ and $\|g\|_p$ are finite. From the elementary inequality

$$|f + g|^p \leq 2^p\big(|f|^p + |g|^p\big)$$

it follows that the number $\|f+g\|_p$ is also finite. Dividing the two functions f and g by the pth power of this number, we may reduce ourselves to proving this inequality in the special case when $\|f+g\|_p = 1$, that is to say under the hypothesis that $\int |f + g|^p \,\mathrm{d}\mu = 1$. With this further hypothesis, it suffices to prove that $1 \leq \|f\|_p + \|g\|_p$. To this end, let us observe that

$$|f + g|^p = |f + g| \cdot |f + g|^{p-1} \leq |f| \cdot |f + g|^{p-1} + |g| \cdot |f + g|^{p-1}.$$

On the other hand, applying Hölder's inequality (see (1.1)) to the functions $|f|^p$ and $|f + g|^p$ (with $a = 1/p$ and $b = 1 - 1/p$), gives

$$\begin{aligned}\textstyle\int |f| \cdot |f + g|^{p-1} \,\mathrm{d}\mu &= \textstyle\int \big(|f|^p\big)^{1/p}\big(|f + g|^p\big)^{1-1/p} \,\mathrm{d}\mu \\ &\leq \big(\textstyle\int |f|^p \,\mathrm{d}\mu\big)^{1/p} \big(\textstyle\int |f + g|^p \,\mathrm{d}\mu\big)^{1-1/p} \\ &= \|f\|_p.\end{aligned}$$

Similarly, we can also prove that

$$\textstyle\int |g| \cdot |f + g|^{p-1} \,\mathrm{d}\mu \leq \|g\|_p.$$

As a result, we finally get to

$$1 = \textstyle\int |f + g|^p \,\mathrm{d}\mu \leq \|f\|_p + \|g\|_p. \qquad \square$$

(1.6) Remark. The Minkowski's inequality extends immediately, by induction, to the sum of n functions in $\mathcal{L}^p$. If (f_k) is a sequence of real functions in $\mathcal{L}^p$, then for every n,

$$\|f_1 + \cdots + f_n\|_p \leq \|f_1\|_p + \cdots + \|f_n\|_p.$$

Taking the limit as $n \to \infty$ and applying Beppo Levi's theorem to the increasing sequence of partial sums $(f_1 + \cdots + f_n)$, we obtain the countable convexity property:

$$\big\|\textstyle\sum_{n=1}^{\infty} f_n\big\|_p \leq \textstyle\sum_{n=1}^{\infty} \|f_n\|_p.$$

2. The $\mathcal{L}^p$ spaces

Throughout this section, let us continue to assume that we are assigned a measure space $(E, \mathcal{E}, \mu)$. Unless otherwise specified, we shall also assume that any measurable function is defined on $(E, \mathcal{E})$.

(2.1) Definition. Let p be a positive real number greater than 1. Let us denote by $\mathcal{L}^p(\mu)$ (or simply $\mathcal{L}^p$) the set of all functions f, measurable on $(E, \mathcal{E})$, such that the number $\|f\|_p$ is finite.

(2.2) Remark. From (1.4)(c) it follows that, if $f \in \mathcal{L}^p$ and c is a real number, then $cf \in \mathcal{L}^p$. In addition to that, by means of Minkowski's inequality, it follows that, if $f, g \in \mathcal{L}^p$, then $f + g \in \mathcal{L}^p$, hence $\mathcal{L}^p$ is a vector space. Since this vector space is stable under the operation $f \mapsto |f|$, it follows that $\mathcal{L}^p$ is, in fact, a Riesz space.

It is also worth noting that, when $p = 1$, the space $\mathcal{L}^1(\mu)$ is the vector space of all measurable functions that are integrable with respect to the measure μ, that we had already introduced in Chapter II §3.

(2.3) Remark. By means of (1.4)(a) and (c) and again by Minkowski's inequality, it is simple to recognise that the function $f \mapsto \|f\|_p$ is a semi-norm on $\mathcal{L}^p$. In the sequel, we shall always assume that $\mathcal{L}^p$ is always equipped with the topology induced by this semi-norm. Let us also recall that a fundamental system of neighbourhoods of an element f of $\mathcal{L}^p$ for this topology consists of sets of the form

$$\left\{g \in \mathcal{L}^p : \|f - g\|_p < \varepsilon\right\},$$

with ε a strictly positive integer. As a result, for a sequence (f_n) of elements of $\mathcal{L}^p$ to converge towards an element f of $\mathcal{L}^p$ according to this topology it is necessary and sufficient that the numeric sequence

$$\left(\|f_n - f\|_p\right)$$

converges towards 0, or equivalently, that the sequence

$$\int |f_n - f|^p \, d\mu$$

converges towards 0. In this case, we shall say that (f_n) **converges in mean of order** p (or **in** $\mathcal{L}^p$) to f, or simply **in mean** when $p = 1$.

In the rest of this section we are going to discuss in more detail the properties of the topology introduced on the space $\mathcal{L}^p$ by the semi-norm $f \mapsto \|f\|_p$, and in particular we aim to show that $\mathcal{L}^p$ is complete.

(2.4) Proposition. *For a sequence (f_n) of elements of $\mathcal{L}^p$ to converge to f in $\mathcal{L}^p$, it is necessary and sufficient that the sequences (f_n^+) and (f_n^-) converge in $\mathcal{L}^p$ to f^+ and f^-, respectively.*

Proof. The sufficient condition follows from the relations

$$f = f^+ - f^-, \qquad f_n = f_n^+ - f_n^-;$$

the necessary condition follows from the relations

$$|f_n^+ - f^+| \leq |f_n - f|, \qquad |f_n^- - f^-| \leq |f_n - f|. \qquad \square$$

(2.5) Proposition. *Let (f_n) be a sequence of elements of $\mathcal{L}^p$, and let f be an element of $\mathcal{L}^p$. For the sequence (f_n) to converge in $\mathcal{L}^p$ to f, it is sufficient that the sequence $(|f_n|^p)$ converges in $\mathcal{L}^1$ to $|f|^p$.*

Proof. It suffices to observe that $|f_n - f|^p \leq ||f_n|^p - |f|^p|$. □

Given a sequence (f_n) of elements of $\mathcal{L}^p$, one may ask under what conditions the sequence of partial sums $(f_1 + \cdots + f_n)$ admits a limit with respect to convergence in $\mathcal{L}^p$. The following theorem provides a sufficient condition. As we shall see immediately after proving this result, an important consequence of this theorem shall be that the space $\mathcal{L}^p$ is complete.

(2.6) Theorem. *Let (f_n) be a sequence of elements of $\mathcal{L}^p(\mu)$, and set*

$$f(x) = \begin{cases} \sum_n f_n(x) & \text{if } \sum_n |f_n(x)| < \infty, \\ 0 & \text{otherwise.} \end{cases}$$

Then, for every positive integer n,

$$\|f - \textstyle\sum_{k=1}^n f_k\|_p \leq \sum_{k \geq n+1} \|f_k\|_p. \tag{2.7}$$

If, in addition,

$$\textstyle\sum_n \|f_n\|_p < \infty, \tag{2.8}$$

then:

(a) *The function f belongs to $\mathcal{L}^p(\mu)$.*

(b) *The sequence $(f_1 + \cdots + f_n)$ converges towards f almost everywhere and in $\mathcal{L}^p$.*

Proof. From the definition of f, for every positive integer n,

$$|f - \textstyle\sum_{k=1}^n f_k| \leq \sum_{k=n+1}^{\infty} |f_k|.$$

Applying the countable convexity property (1.6), we obtain the relation (2.7).

Now suppose that (2.8) holds. The relation (2.7), with $n = 1$, shows that f belongs to $\mathcal{L}^p$. The same relation, letting n approach ∞, shows that the sequence of partial sums $(f_1 + \cdots + f_n)$ converges in $\mathcal{L}^p$ to f. Finally, the relation

$$\|\textstyle\sum_n |f_n|\|_p \leq \sum_n \|f_n\|_p < \infty$$

shows that the function $\sum_n |f_n|$ is finite almost everywhere. It follows, from the very definition of f, that the sequence $(f_1 + \cdots + f_n)$ converges almost surely to f. □

From this theorem, we deduce the following fundamental result.

(2.9) Theorem (Fischer-Riesz). *The semi-normed space $\mathcal{L}^p$ is complete. Moreover, every sequence that converges in $\mathcal{L}^p$ to an element f of $\mathcal{L}^p$ admits a subsequence converging to f almost everywhere.*

Proof. Let $(f_n)_{n\geq 0}$ be a Cauchy sequence in $\mathcal{L}^p$. This means that for every strictly positive real number ε, there exists a positive integer n_ε such that $\|f_n - f_m\|_p < \varepsilon$ for every pair n, m of integers greater than n_ε. We can thus define, by induction, a strictly increasing sequence (n_k) of positive integers such that $\|f_{n_k} - f_{n_{k-1}}\|_p < 2^{-k}$ for every positive index k. The partial sums of this sequence then, by the previous theorem, converge almost everywhere and in $\mathcal{L}^p$ to an element of $\mathcal{L}^p$. But since the sequence of these partial sums is simply the sequence $(f_{n_k} - f_{n_1})$, this means that (f_{n_k}) converges almost everywhere and in $\mathcal{L}^p$ to an element of $\mathcal{L}^p$. Since (f_n) is a Cauchy sequence in $\mathcal{L}^p$, it converges in $\mathcal{L}^p$ to the same element. □

(2.10) Remark. Let us denote by $L^p(\mu)$ (or simply L^p) the quotient of $\mathcal{L}^p(\mu)$ under the relation of equivalence modulo μ. Thus, the elements of L^p are the equivalence classes determined on $\mathcal{L}^p$ by the relation

$$\|f - g\|_p = 0,$$

that is to say,

$$f \equiv g \pmod{\mu}.$$

If we denote by $f \mapsto f^\bullet$ the canonical injection of $\mathcal{L}^p$ onto L^p, the norm of $f^\bullet$ in L^p is the number

$$\|f^\bullet\|_p = \|f\|_p$$

(which does not depend on the representative f chosen in the corresponding class). Since $\mathcal{L}^p$ is a complete semi-normed space, L^p is a complete normed space, that is to say, a Banach space.

Let us complete this section by proving that the simple integrable functions are dense in $\mathcal{L}^1$. More precisely, we shall now prove a slightly stronger result, that is to say every function in $\mathcal{L}^p$ may be approximated, in the norm $\mathcal{L}^p$, with arbitrary accuracy, by a sequence of simple functions in $\mathcal{L}^p$.

(2.11) Theorem. *The vector space $\mathcal{S}^p(\mu)$ of all simple functions in $\mathcal{L}^p$ is dense in the complete semi-normed space $\mathcal{L}^p$, for every $p \geq 1$.*

Proof. Let f be an element of $\mathcal{L}^p$. With no loss of generality, we may assume that f is positive. In fact, in the general case, it suffices to consider the decomposition $f = f^+ - f^-$ and apply the result to f^+ and f^- separately. In this further hypothesis, let (f_n) be an increasing sequence of positive simple functions which converges pointwise towards f. Since f is in $\mathcal{L}^p$ and $f_n \leq f$, it follows that $f_n \in \mathcal{L}^p$. By changing each of the f_n on a μ-negligible set, we may assume, with no loss of generality, that each of these functions is finite. Hence, for every index n, we have

$$|f_n - f| \leq |f_n| + |f| \leq 2|f|.$$

Taking the pth power of the latter inequality yields $|f_n - f|^p \leq 2^p|f|^p$, whence all the elements of the sequence $(|f_n - f|^p)$ are bounded from above, in absolute value, by an integrable function. Thus, applying Lebesgue's theorem on dominated convergence, one finally gets $\int |f_n - f|^p \, \mathrm{d}\mu \to 0$, and this completes the proof. □

3. The space $\mathcal{L}^\infty$

In the sequel, we shall continue to assume that a measure space $(E, \mathcal{E}, \mu)$ is assigned. If f is a measurable function defined on $(E, \mathcal{E})$, it is easy to check that the set I of all positive numbers c such that the relation $|f(x)| \leq c$ holds μ-almost everywhere, has a minimum element. In fact, if b denotes the infimum of I, and if (c_n) is a sequence of elements in I such that $\inf_n c_n = b$, observe that the set $\{x \in E : |f(x)| > b\}$ is bound to be negligible, because it is equal to the countable union of the negligible sets $\{x \in E : |f(x)| > c_n\}$.

This minimum shall be denoted by $\|f\|_\infty$ and shall be called the **essential minimum** of f. In other words, the essential minimum is defined as

$$\|f\|_\infty = \min\Big\{c \geq 0 : \mu\{|f| > c\} = 0\Big\}$$

This notation is justified by the following result.

(3.1) Proposition. *If f is a measurable function. Assume that either $\|f\|_p$ is finite for some strictly positive real number p, or that the measure μ is bounded, that is to say, that $\mu(E)$ is finite. Then,*

$$\lim_{p \to \infty} \|f\|_p = \|f\|_\infty$$

Proof. Let r be a strictly positive real number, such that $\|f\|_r$ is finite. In order to prove the inequality

$$\liminf_{p \to \infty} \|f\|_p \geq \|f\|_\infty,$$

let c be a real number with $0 < c < \|f\|_\infty$ and set $A = \{x \in E : |f(x)| > c\}$. By definition of $\|f\|_\infty$, A is not negligible, and $|f| \geq cI_A$. As a result of that, $\|f\|_p \geq c\|I_A\|_p = c\mu(A)^{1/p}$. Thus, taking $p = r$, it follows that the number $\mu(A)$ must be finite and that

$$\liminf_{p \to \infty} \|f\|_p \geq c.$$

In order to prove the inequality

$$\limsup_{p \to \infty} \|f\|_p \leq \|f\|_\infty,$$

assume that $c = \|f\|_\infty$ is a finite number, and observe that, for every real number p greater than r, one has

$$\|f\|_p \le \|f^{1-r/p}\|_\infty \|f^{r/p}\|_p = c^{1-r/p}\big(\|f\|_r\big)^{r/p}. \qquad \square$$

Let us observe that $\|f\|_\infty = 0$ if and only if the function f is negligible, and that $\|f\|_\infty$ is finite if and only if the function f is equivalent to a bounded function. In this case, we shall also say that the function f is **μ-essentially bounded**, or simply **essentially bounded**, if the reference to the specific measure may be inferred by the context.

It is easy to show that the results described in Proposition (1.4) keep on holding in the case where $p = \infty$, and so does the countable convexity inequality described in (1.6). The Hölder-Schwarz inequality (see (1.2)) keeps on holding for every triple p, q, r of elements of $(0, \infty]$, such that $1/p + 1/q = 1/r$ (setting $1/\infty = 0$). In other words, one has

$$\|fg\|_\infty \le \|f\|_\infty \|g\|_\infty, \qquad \|fg\|_r \le \|f\|_\infty \|g\|_r$$

for every pair f, g of measurable functions and every real number r.

(3.2) Definition. Let us denote by $\mathcal{L}^\infty(\mu)$ (or simply $\mathcal{L}^\infty$) the set of all functions f, measurable on $(E, \mathcal{E})$, such that $\|f\|_\infty$ is finite.

(3.3) Remark. Let us observe that, if $f \in \mathcal{L}^\infty$ and c is a real number, then $cf \in \mathcal{L}^\infty$. Furthermore, Minkowski's inequality, which holds for $p = \infty$ as well, yields that, if $f, g \in \mathcal{L}^\infty$, then $f + g \in \mathcal{L}^\infty$, hence $\mathcal{L}^\infty$ is a vector space. Since this vector space is stable under the operation $f \mapsto |f|$, it follows that $\mathcal{L}^\infty$ is, in fact, a Riesz space.

It is simple to show that the function $f \mapsto \|f\|_\infty$ is a semi-norm on $\mathcal{L}^\infty$. In the sequel, we shall always assume that $\mathcal{L}^\infty$ is equipped with the topology induced by this semi-norm. The following theorem shows that the semi-normed space $\mathcal{L}^\infty$ is complete.

(3.4) Theorem. *The semi-normed space $\mathcal{L}^\infty$ is complete. Moreover, every sequence that converges in $\mathcal{L}^\infty$ to an element f of $\mathcal{L}^\infty$ admits a subsequence converging to f almost everywhere.*

Proof. Let (f_n) be a Cauchy sequence in $\mathcal{L}^\infty$. By definition, for every pair of indices m and n, there exists a negligible set A_{mn} such that

$$|f_m(x) - f_n(x)| \le \|f_m - f_n\|_\infty \qquad \text{for every } x \in A_{mn}^c.$$

Observe that the union A of all sets A_{mn} for every pair of indices m and n is negligible, and, for every $x \in A^c$, the sequence $(f_n(x))$ converges towards some number $f(x)$. On the other hand, for every $x \in A$, set $f(x) = 0$. Then, by means of the measurability criterion by patchworking (see Theorem I(2.15)), the function f is measurable and, for any strictly positive real number ε, an for every x in the set A^c, one has

$$|f(x) - f_n(x)| \le \sup_{m \ge n} \|f_m - f_n\|_\infty \le \varepsilon,$$

for n sufficiently large. This proves that, on one side $\|f\|_\infty \le \|f_n\|_\infty + \varepsilon$, so $f \in \mathcal{L}^\infty$, and on the other that $\|f - f_n\|_\infty \to 0$ as $n \to \infty$, as desired. $\square$

(3.5) Remark. Let us denote by $L^\infty(\mu)$ (or simply L^∞) the quotient of $\mathcal{L}^\infty(\mu)$ under the relation of equivalence modulo μ. If we denote by $f^\bullet$ the equivalence class determined by the element f of $\mathcal{L}^\infty$, the norm of $f^\bullet$ in L^∞ is the number

$$\|f^\bullet\|_\infty = \|f\|_\infty$$

(which does not depend on the representative f chosen in the corresponding class). Since $\mathcal{L}^\infty$ is a complete semi-normed space, L^∞ is a complete normed space, that is to say, a Banach space.

(3.6) Remark. Unlike the spaces $\mathcal{L}^p$ with $p \in [1, \infty)$, the space $\mathcal{L}^\infty$ is generally *not* separable. To see this, let us consider an uncountable class $\mathcal{U}$ of measurable sets, such that, for every pair A, and B of elements of $\mathcal{U}$, the functions I_A and I_B are not equivalent modulo μ. In other words, we are requiring that $\|I_A - I_B\|_\infty = 1$ for every pair $A, B \in \mathcal{U}$. As a result, in $\mathcal{L}^\infty$, the open sets of the form

$$\left\{f \in \mathcal{L}^\infty : \|I_A - f\|_\infty < \tfrac{1}{2}\right\} \qquad \text{with } A \in \mathcal{U}$$

are pairwise disjoint.

4. Duality in the $\mathcal{L}^p$ spaces

Let us continue to assume that a measure space $(E, \mathcal{E}, \mu)$ is assigned. Let us also assume that the measure μ is finite. Given an element p of the interval $[1, \infty]$, let us recall that, for a linear form τ, defined on the vector space $\mathcal{L}^p$, to be continuous with respect to the topology induced by the semi-norm $f \mapsto \|f\|_p$, it is necessary and sufficient that there exists a positive real constant c such that

$$|\tau(f)| \le c\|f\|_p \qquad \text{for every element } f \text{ in } \mathcal{L}^p. \tag{4.1}$$

In other words, it is necessary and sufficient that the number $\|\tau\|$ defined by

$$\|\tau\| = \sup\Big\{|\tau(f)| : f \in \mathcal{L}^p,\ \|f\|_p \le 1\Big\} \tag{4.2}$$

is finite. When this happens, the number $\|\tau\|$ is the smallest positive real number c such that (4.1) holds. Furthermore, the set of all continuous linear forms on $\mathcal{L}^p$ is a vector space, which goes by the name of **topological dual** (or, simply, the **dual**) of $\mathcal{L}^p$, and the function $\tau \mapsto \|\tau\|$ is a norm on this vector space.

Let us denote by q the **conjugate exponent** of p, that is to say the element of $[1,\infty]$ defined by the condition

$$\frac{1}{p}+\frac{1}{q}=1,$$

with the usual arrangement to set $1/\infty = 0$. If g is an element of $\mathcal{L}^q$, then for every element f of $\mathcal{L}^p$ the Hölder-Schwarz inequality (see (1.2)) gives

$$\int |fg|\,\mathrm{d}\mu \le \|f\|_p\|g\|_q \tag{4.3}$$

and shows that the function fg is integrable. Thus, it is possible to consider, on $\mathcal{L}^p$, the linear form defined by

$$\theta_g(f)=\int fg\,\mathrm{d}\mu \qquad \text{for every element } f\in\mathcal{L}^p.$$

The aforementioned inequality (4.3) shows that θ_g is continuous and that its norm $\|\theta_g\|$ is bounded from above by $\|g\|_q$. Indeed,

$$\|\theta_g\| = \sup\Big\{\Big|\int fg\,\mathrm{d}\mu\Big| : f\in\mathcal{L}^p,\ \|f\|_p\le 1\Big\} \le \|g\|_q.$$

The aim for the rest of this section is to give a characterisation of the topological dual of the spaces $\mathcal{L}^p$, with $p\in[1,\infty]$. Specifically, the final goal of this section consists in showing that every element of the topological dual of $\mathcal{L}^p$, with $p\in[1,\infty)$, can be written in the form θ_g for some $g\in\mathcal{L}^q$, where q is the conjugate exponent of p. This is going to be done in three stages: First of all we shall prove the statement for the space $\mathcal{L}^1$, we shall do the proof in the general case for the space $\mathcal{L}^p$, with $p\in(1,\infty)$, and finally we shall prove a similar statement for the space $\mathcal{L}^\infty$, showing a counter-example that ascertains that not all linear forms on $\mathcal{L}^\infty$ are of the form θ_g.

(4.4) Theorem. *For every element g of $\mathcal{L}^\infty$, let us consider the continuous linear form θ_g, defined on $\mathcal{L}^1$ by the relationship*

$$\theta_g(f)=\int fg\,\mathrm{d}\mu \qquad \textit{for every } f \textit{ in } \mathcal{L}^1, \tag{4.5}$$

and set

$$\|\theta_g\| = \sup\Big\{\Big|\int fg\,\mathrm{d}\mu\Big| : f\in\mathcal{L}^1,\ \|f\|_1\le 1\Big\}.$$

Then, one has $\|\theta_g\| = \|g\|_\infty$. Furthermore, every continuous linear form on $\mathcal{L}^1$ can be written in the form θ_g for a suitable element g of $\mathcal{L}^\infty$.

Proof. Let τ be an arbitrary continuous linear form on $\mathcal{L}^1$, and let us set

$$\|\tau\| = \sup\Big\{|\tau(f)| : f \in \mathcal{L}^1,\ \|f\|_1 \le 1\Big\}.$$

Let us consider the set function ν defined on $\mathcal{E}$ by

$$\nu(A) = \tau(I_A) \qquad \text{for every element } A \text{ of } \mathcal{E}.$$

From the fact that τ is a linear form, it follows that ν is additive. Furthermore, for every element A of $\mathcal{E}$, one has

$$|\nu(A)| \le \|\tau\|\|I_A\|_1 = \|\tau\|\mu(A). \tag{4.6}$$

It is clear, therefore, that ν is a measure with base μ, that is to say, a measure of the form $\nu = g \cdot \mu$, where g is a suitable real function integrable with respect to μ. Furthermore, the relation (4.6) shows that, in the Riesz space of all measures with base μ, one has

$$|\nu| \le \|\tau\|\mu,$$

which can be translated, by making explicit the link with the density g of ν with respect to μ, into the form $|g| \cdot \mu \le \|\tau\|\mu$ or, equivalently, into the even more explicit form $\mu\{|g| \ge \|\tau\|\} = 0$. This ensures that the function g lies in $\mathcal{L}^\infty(\mu)$ and that $\|g\|_\infty \le \|\tau\|$. From the definition of ν, it follows that, on every element of $\mathcal{L}^1$ of the form I_A, the linear form τ coincides with the linear form θ_g defined by (4.5). By linearity, the coincidence between τ and θ_g extends to the class of all simple functions on $(E, \mathcal{E})$. However, this class is dense in $\mathcal{L}^1$, and this yields that the linear form τ coincides with θ_g. It remains to prove that $\|\theta_g\| = \|g\|_\infty$. To this end, let us start by observing that $\|\theta_g\| \le \|g\|_\infty$. From this inequality, and from the fact that, as we had already observed, $\|g\|_\infty \le \|\tau\| = \|\theta_g\|$, it follows the desired equality. □

The next theorem shows that a similar type of result holds in the more general case of any $\mathcal{L}^p$ space, with $p \in (1, \infty)$. Afterwards, we are also going to show that the case of $\mathcal{L}^\infty$ is slightly come complicated.

(4.7) Theorem. *Let p be a real number greater than 1, and let us denote by q the conjugate exponent of p. For every element g of $\mathcal{L}^q(\mu)$, let us consider the linear form θ_g defined on $\mathcal{L}^p$ by*

$$\theta_g(f) = \int fg\,\mathrm{d}\mu \qquad \textit{for every } f \textit{ in } \mathcal{L}^p, \tag{4.8}$$

and set

$$\|\theta_g\| = \sup\Big\{\big|\int fg\,\mathrm{d}\mu\big| : f \in \mathcal{L}^p,\ \|f\|_p \le 1\Big\}.$$

Then, one has $\|\theta_g\| = \|g\|_q$. Furthermore, every continuous linear form on $\mathcal{L}^p$ can be written in the form θ_g for a suitable element g of $\mathcal{L}^q$.

Proof. Let τ be a continuous linear form on $\mathcal{L}^p$. With no loss of generality we may assume that τ is not constantly zero. Let us set

$$\|\tau\| = \sup\Big\{|\tau(f)\| : f \in \mathcal{L}^p. \|f\|_p \leq 1\Big\}.$$

Let us consider the set function ν defined on $\mathcal{E}$ by

$$\nu(A) = \tau(I_A) \qquad \text{for every element } A \text{ of } \mathcal{E}.$$

From the fact that τ is a linear form, it follows that ν is additive. Furthermore, for every element A of $\mathcal{E}$, one has

$$|\nu(A)| \leq \|\tau\|\|I_A\|_p = \|\tau\|\big(\mu(A)\big)^{1/p}.$$

It is clear, therefore, that ν is a measure with base μ, that is to say, a measure of the form $\nu = g\cdot\mu$, where g is a suitable real function integrable with respect to μ. From the very definition of ν, it is clear that, for any simple function on $(E, \mathcal{E})$, one has

$$\tau(f) = \int f\,\mathrm{d}\nu = \int fg\,\mathrm{d}\mu. \tag{4.9}$$

It is also clear that the function g is non-negligible with respect to μ. Let us take a sequence (h_n) of simple functions on $(E, \mathcal{E})$ such that $0 \leq h_n \uparrow |g|$, and let us denote by γ the function that coincides with 1 on $\{g \geq 0\}$ and with -1 elsewhere, so that we may write $g = \gamma|g|$. Finally, let us set

$$f_n = h_n^{q/p}\gamma = h_n^{q-1}\gamma.$$

Thus, each of the functions f_n is simple, and we have

$$\int f_n g\,\mathrm{d}\mu = \tau(f_n) \leq \|\tau\|\|f_n\|_p.$$

On the other and, from the very definition of the function f_n, one has

$$f_n g = h_n^{q-1}\gamma\cdot\gamma|g| = h_n^{q-1}|g|, \qquad \|f_n\|_p = \Big(\int h_n^q\,\mathrm{d}\mu\Big)^{1/p}.$$

From these inequalities, we finally get

$$\int h_n^{q-1}|g|\,\mathrm{d}\mu \leq \|\tau\|\Big(\int h_n^q\,\mathrm{d}\mu\Big)^{1/p}.$$

Taking the limit as n approaches ∞, thanks to the fact that $0 \leq h_N \uparrow |g|$, one gets

$$\int |g|^p\,\mathrm{d}\mu \leq \|\tau\|\Big(\int |g|^q\,\mathrm{d}\mu\Big)^{1/p},$$

that is to say

$$\|g\|_q = \Big(\int |g|^q\,\mathrm{d}\mu\Big)^{1-1/p} \leq \|\tau\|.$$

This proves that the function g belongs to $\mathcal{L}^q$ and that $\|g\| \leq \|\tau\|$.

Finally, taking a linear form θ_g of the form (4.8), the relation (4.9) shows that $\tau(f) = \theta_g(f)$ for every simple function f on $(E, \mathcal{E})$. Since the space of all simple functions is dense in $\mathcal{L}^p$, it follows that $\tau = \theta_g$. It remains to prove that $\|\theta_g\| = \|g\|_q$. To this end, it suffices to observe that,

$$\|g\|_q \leq \|\tau\| = \|\theta_g\| \leq \|g\|_q,$$

and this completes the proof. □

(4.10) Remark. In the hypotheses of the previous theorem, it is easy to recognise that, for any pair g, h of elements of $\mathcal{L}^q$, and any pair a, b of real numbers, one has

$$\theta_{ag+bh} = a\theta_g + b\theta_h.$$

In other words, the map $g \mapsto \theta_g$ from $\mathcal{L}^q$ in the dual of $\mathcal{L}^p$ is linear. It is also evident that, if g is an element of $\mathcal{L}^q$ negligible with respect to μ (that is to say, which belongs to the zero of the quotient space L^q), the linear form θ_g is identically zero. Finally, it is clear that the linear form θ_g takes equal values on two elements of $\mathcal{L}^p$ that are equivalent modulo μ (that is to say, which belong to the same element of L^p) and can therefore be identified to a continuous linear form on L^p. Thus, we can see that the map $g \mapsto \theta_g$ induces, by projecting to the quotient, an injective linear map from L^q to the dual of L^p. Furthermore, the previous theorem shows that this map ranges over the whole dual space of L^p and transforms the norm of L^q into the norm of the dual of L^p. This is therefore a bijective isometry of the space L^q into the dual of the space L^p (enabling us to identify these two spaces).

A similar observation applies to Theorem (4.4): this theorem shows that it is possible to construct an isometric linear mapping from the space L^∞ in the dual of L^1. On the other hand, as we shall see momentarily, the next theorem shall show that it is only possible to construct an isometric linear mapping of the space L^1 into *a subspace* of the dual of L^∞: as we shall see in Example (4.13), this space may indeed be proper.

(4.11) Theorem. *For every element g of $\mathcal{L}^1$, let us consider the continuous linear form θ_g, defined on $\mathcal{L}^\infty$ by the relationship*

$$\theta_g(f) = \int fg \, \mathrm{d}\mu \qquad \text{for every } f \text{ in } \mathcal{L}^\infty, \tag{4.12}$$

and set

$$\|\theta_g\| = \sup\Big\{\big|\int fg \, \mathrm{d}\mu\big| : f \in \mathcal{L}^\infty, \ \|f\|_\infty \leq 1\Big\}.$$

Then, one has $\|\theta\| = \|g\|_1$. In general, however, not all continuous linear forms on $\mathcal{L}^\infty$ are of the form (4.12).

Proof. To prove the equality $\|\theta_g\| = \|g\|_1$, since one always has $\|\theta_g\| \leq \|g\|_1$ it suffices to prove the converse inequality. To this end, it suffices that, denoting by γ the function that coincides with 1 on $\{g \geq 0\}$ and with -1 elsewhere, one has

$$\|g\|_1 = \int |g| \,\mathrm{d}\mu = \int \gamma g \,\mathrm{d}\mu \leq \|\theta_g\|.$$

The final statement of the theorem may be proved displaying a suitable counter-example. □

(4.13) Example. In the hypotheses of the previous theorem, assume that μ is the trace of Borel-Lebesgue measure onto the interval $[0,1]$. Let us consider the subspace $\mathcal{C}([0,1])$ of $\mathcal{L}^\infty$ of all continuous function on $[0,1]$, and the linear form τ defined by $\tau(f) = f(1)$. Let us observe that, for every element f of $\mathcal{C}([0,1])$, one has

$$|\tau(f)| = |f(1)| \leq \max\big\{|f(x)| : 0 \leq x \leq 1\big\} = \|f\|_\infty.$$

Thanks to Hahn-Banach theorem, it is possible to extend τ to a continuous linear form on $\mathcal{L}^\infty$. Suppose, by contradiction, that this extension is of the form θ_g for a suitable element g of $\mathcal{L}^1$. If, for every integer n greater than 1, let us denote by f_n the restriction to $[0,1]$ of the function $x \mapsto x^n$, one gets

$$\tau(f_n) = \theta_g(f_n) = \int_0^1 x^n g(x) \,\mathrm{d}x.$$

This is indeed a contradiction. In fact, while the left-hand side equals 1, the right-hand side converges towards 0 as n approaches ∞ by virtue of Lebesgue's theorem on dominated convergence.

Chapter VII

Measures on a Topological Space

In the development of the theory of measures on a measurable space, we have assumed no underlying structure on the set E. It turns out, however, that, when the set E is equipped with some topological structure, the choice of the domain of definition of a measure becomes more restrictive, in order to account for the structure of the set E, and, as a result, the measures acquires a wealth of properties linked to this underlying topological structure. This chapter is therefore devoted to the study of measures defined on suitable topological spaces, and it will culminate in introducing a special type of measures, called 'Radon measures', that have very interesting properties, among which the ability to approximate the measure of a measurable set with the measure of more regular sets, and we shall delve, in particular, in the study of two fundamental properties of measures: 'regularity' and 'tightness', the first related, as we have said, to this ability to approximate sets with more regular ones, and the second ensuring that the measures behave well with compact sets.

In order to organically carry out this study, in §1, we shall introduce a special type of topological spaces, called 'Polish spaces', which turn out to be a rather natural environment to study measure theory. After studying the topological properties of these spaces, in §2, we shall discuss the two most significant σ-fields on a topological space and show the connection between them. In §3, we shall introduce the notions of 'regularity' and 'tightness' for a measure on a general topological space, and, in §4, we shall study the relation between these property and the topology for a metrizable space. It is within this section that we shall describe the fine properties of measures in Polish spaces. Having studied the case of metrizable spaces, §5 shall have the sole purpose of recalling the topological theory of locally compact Hausdorff spaces, and the most important properties of these spaces, while it is in §6 to §8 that we finally discuss the fine properties of Borel measures on a locally compact Hausdorff space and the connection between Borel and Baire measures.

DOI: 10.1201/9781003724766-VII

1. Polish spaces and their properties

In all the sequel, we shall assume that E is a non-empty set.

(1.1) Definition. A **Polish space** is a separable topological space such that there exists a metric, compatible with its topology, that makes it a complete metric space.

The definition of Polish spaces dates back to the French mathematician Roger Godement (1921–2016), member of the Bourbaki group. The origin of this peculiar terminology is brilliantly reconstructed in the following footnote, which appears in Godement's book [33].

> *Abréviation que, pour rire, j'ai suggérée à N. Bourbaki en 1949 après avoir appris le sujet dans la* Topologie *de Casimir Kuratowski et constaté la contribution des Polonais; j'aurais du rest aussi bien pu les appeler "polono-russes". Le plaisenterie, qui n'en était pas entièrement une, fut prise au sérieux et tous les experts ont depuis adopté cette étrange terminologie, généralement sans mentionner N. Bourbaki à l'exception de Kuratowski lui-même qui, dans un petit livre sur l'histoire des mathématiques polonaises, y voit à juste titre un hommage à celles-ci.*

This section is devoted to discuss a number of elementary properties of Polish spaces. To begin with, however, let us reflect for a moment on the definition. It is somewhat strange, at first glance, to define a Polish space as a topological space rather than a metric space. However, there are many topological spaces that are Polish, but have no complete metric that is particularly natural, or reasonable. Furthermore, many constructions and facts of interest in measure theory depend on the existence of a complete metric, but not on the choice of a particular metric. It is therefore more natural to deal with a Polish space rather than a complete separable metric space.

(1.2) Example. The space $\mathbb{R}^n$ with the usual topology, is a Polish space. More generally, each separable Banach space, with the topology induced by its norm is a Polish space.

(1.3) Proposition. *The product of a finite or infinite sequence of Polish spaces is Polish.*

Proof. Let (E_n) be a finite or infinite sequence of Polish space. With no loss of generality, we may assume that none of the E_n is empty. For each index n, let d_n be a metric on E_n, compatible with its topology, that makes E_n a complete metric space. After replacing d_n with $1 \wedge d_n$, if necessary, we may assume that $d_n(x,y) \leq 1$ for every pair x, y of elements of E_n. Then, for every pair $x = (x_n)$, $y = (y_n)$ in the product space $\prod_n E_n$, let us set

$$d(x,y) = \textstyle\sum_n 2^{-n} d_n(x_n, y_n).$$

It is easy to check that d is a metric on $\prod_n E_n$, that d is compatible with the product topology of $\prod_n E_n$, and that $\prod_n E_n$ is complete under d. □

(1.4) Proposition. *Every closed subset of a Polish space is Polish.*

Proof. Let E be a Polish space and let X be a closed subset of E. Since E is separable, such is X. It remains only to prove that the topology of X can be defined by a metric that makes it a complete metric space. Let d be a metric for E, compatible with its topology, which makes E a complete metric space. Then, the restriction of d to X is also a metric, and since X is closed, the topology induced by this restriction on X, which is compatible with the original topology, is complete. □

(1.5) Proposition. *Every open subset of a Polish space is Polish.*

Proof. Let us start by proving that, if E is a Polish space and U is an open subset of E, then U is homeomorphic to a closed subset of the product space $E \times \mathbb{R}$. To this end, let d be a metric on E which is compatible with the topology of E. Let us also assume that $U \neq E$. Consider the real function defined on E by

$$f(x) = d(x, U^c) \qquad \text{for } x \in E.$$

It is clear that $U = \{f > 0\}$. Let us consider the mapping g from U to $E \times \mathbb{R}$ defined by

$$g(x) = \big(x, 1/f(x)\big) \qquad \text{for } x \in U.$$

Thanks to the equality

$$g(U) = \Big\{(x,t) \in E \times \mathbb{R} : tf(x) = 1\Big\},$$

the set $g(U)$ is closed in the product space $E \times \mathbb{R}$. Furthermore, g is a homeomorphism from U onto $g(U)$: in fact, this is obviously an injective and continuous map from U to $g(U)$, whose inverse is the restriction to $g(U)$ of the canonical projection of $E \times \mathbb{R}$ onto the first term.

Now, thanks to (1.3), the product space $E \times \mathbb{R}$ is Polish, and, since $g(U)$ is a closed subset of $E \times \mathbb{R}$, it follows from (1.4) that $g(U)$ is also Polish, whence U is Polish as well. □

Let us recall that, if E is a topological space, a G_δ of E is any countable intersection of open sets in E. In other words, if $\mathcal{O}$ denotes the topology on E, a G_δ is an element of $\mathcal{O}_\delta$. The reason for this terminology is largely historical, for the topology used to be denoted, especially in the German tradition, by G, first letter of the German word *Gebiet*, open set. Let us just mention that the 'dual' notion is an F_σ of E, which is any countable union of closed sets in E. Unlike its dual notion, of Germanic heritage, this terminology originates from the French word *fermé*, closed set.

A very useful feature of Polish spaces is that every Polish subspace of a Polish space E is a G_δ of E. To this end, let us start by proving the following result.

(1.6) Theorem. *Let F be a Hausdorff space, and E a subspace of F, homeomorphic to a complete metric space. Then, E is the intersection of a G_δ of F and the closure of E in F.*

Proof. Let d be a distance on E, which is compatible with the topology on E, and such that the metric space (E, d) is complete. For every strictly positive integer n, let us denote by U_n the union of the open sets of F whose intersection with E has a diameter (under d), smaller than $1/n$. Let us also denote by $\overline{E}$ the closure of E in F. It suffices to prove that

$$E = \overline{E} \cap \bigcap_{n=1}^{\infty} U_n.$$

Now, it is clear that the left-hand side of this equality is included in the second. It suffices to prove the converse inclusion. To this end, let us consider an element x on the right-hand side, and let us denote by $\mathcal{U}$ the trace onto E of the filter of all neighbourhoods of x in F. Then, $\mathcal{U}$ is a filter base on F, which obviously converges towards x. On the other hand, $\mathcal{U}$ is also a filter on E. Furthermore, $\mathcal{U}$ is a Cauchy filter for the distance d, so it must converge in the subspace E. Since F is Hausdorff, the limit of a filter is unique, hence x must belong to E, which completes the proof. □

(1.7) Corollary. *In the same hypotheses as the previous theorem, let us assume that F is a Polish space. Then E is a G_δ of F.*

Proof. From the fact that F is metrizable, it follows that $\overline{E}$ is a G_δ of F. Indeed, if d is a metric on F which is compatible with its topology, then $\overline{E}$ coincides with the intersection of all the open sets

$$V_n = \left\{x \in F : d(x, E) < 2^{-n}\right\},$$

whence it is a G_δ of F. □

Another important feature of Polish space is that every Polish space is homeomorphic to a G_δ of the unitary hyper-cube $[0, 1]^{\mathbb{N}}$. To this end, let us start by proving this simple result due to U. Dini.

(1.8) Proposition (Dini's lemma). *Let E be a compact space. If (f_n) is a monotone sequence of continuous real functions defined on E, which converges pointwise towards a continuous real function f also defined on E, then the convergence $f_n \to f$ is uniform.*

Proof. With no loss of generality, we may assume that (f_n) is an increasing sequence, for if not it suffices to apply the statement to the sequence $(-f_n)$. Under this additional hypothesis, let us set $g_n = f_n - f$. For each strictly positive real number ε, it suffices to observe that the sequence of open sets $(\{g_n > \varepsilon\})$ is decreasing, and its intersection is empty. Hence, from the fact that E is a compact set, it follows that its terms are definitely empty. □

With the aid of Dini's lemma, we may now prove the following result.

(1.9) Theorem. *Let E be a compact metrizable space. Then, the set $\mathcal{C}(E)$ of all continuous real functions on E, equipped with the topology of the uniform convergence, is separable.*

Proof. Let d be a distance compatible with the topology of E, and set

$$B(x,r) = \Big\{y \in E : d(x,y) < r\Big\}$$

for every element x of E and every strictly positive real number r. For every strictly positive integer number k, the family $(B(x,1/k))_{x\in E}$ is an open covering of E. Thanks to the compactness of E, one can extract a finite subcovering of the form $(B(x,1/k))_{x\in D_k}$, where D_k is a finite part of E. This implies that the union of all D_k is a countable dense set in E. Thus, the topology of E admits a countable base. Choosing such a countable base $\mathcal{U}$, let us consider, for every strictly positive integer number n and every element U of $\mathcal{U}$, the continuous function $g_{n,U}$ defined on E by

$$g_{n,U}(x) = 1 \wedge nd(x, U^c).$$

Hence, $I_U = \sup_n g_{n,U}$. As a result, for every positive element f of $\mathcal{C}(E)$, one can write

$$\begin{aligned} f = \sup_{\alpha\in\mathbb{Q}_+} \alpha I_{\{f>\alpha\}} &= \sup_{\alpha\in\mathbb{Q}_+} \sup_{U\in\mathcal{U},\, U\subseteq\{f>\alpha\}} \alpha I_U \\ &= \sup_{\alpha\in\mathbb{Q}_+} \sup_{U\in\mathcal{U},\, U\subseteq\{f>\alpha\}} \sup_{n\in\mathbb{N}} \alpha g_{n,U}. \end{aligned}$$

In other words, if we denote by $\mathcal{G}$ the (countable) set of all functions of the form $\alpha g_{n,U}$, with $\alpha \in \mathbb{Q}_+$, $n \in \mathbb{N}$ and $U \in \mathcal{U}$, we can see that every positive element f of $\mathcal{C}(E)$ is the supremum of a part of $\mathcal{G}$. Let us denote by $\mathcal{H}$ the (again countable) set of all suprema of every finite family of elements of $\mathcal{G}$. Then f is a positive element of $\mathcal{C}(E)$, and it is the supremum of an *increasing sequence* of elements of $\mathcal{H}$. Thus, thanks to Dini's lemma, this increasing sequence uniformly converges towards f. Thus, thanks to the decomposition $f = f^+ - f^-$, every element of $\mathcal{C}(E)$ can be seen as the uniform limit of a sequence of elements of $\mathcal{H} - \mathcal{H}$. In other words, the countable set $\mathcal{H} - \mathcal{H}$ is dense in $\mathcal{C}(E)$, equipped with the topology of uniform convergence. □

(1.10) Proposition. *Let E be a separable metrizable space. Then E is homeomorphic to a subspace of the unit hyper-cube $[0,1]^{\mathbb{N}}$.*

Proof. Let us choose a metric d compatible with the topology of E. We may assume (after replacing d with $1 \wedge d$, if necessary), that d has values in $[0,1]$. Let us choose a sequence (a_n) of elements of E, dense in E, and let us denote by φ the mapping from E to $[0,1]^{\mathbb{N}}$ defined by

$$\varphi(x) = \big(\varphi_n(x)\big) \qquad \text{with } \varphi_n(x) = d(a_n, x).$$

It is clear that φ is continuous (because such are all its components). We wish to prove that φ is a homeomorphism from E onto $\varphi(E)$. Take an element x in E and a sequence (x_j) of elements of E, such that $(\varphi(x_j))$ converges towards $\varphi(x)$. It suffices to prove that (x_j) converges towards x. To this end, let us observe that, for every pair j, n of positive integers, on has

$$d(x_j, x) \le d(x_j, a_n) + d(a_n, x) = \varphi_n(x_j) + \varphi_n(x).$$

By hypothesis, as j tends towards ∞, $\varphi(x_j)$ tends towards $\varphi(x)$: in particular, the sequence $(\varphi_n(x_j))_j$ converges towards $\varphi_n(x)$. Thus,

$$\limsup_j d(x, x_j) \le 2\varphi_n(x).$$

As a result,

$$\limsup_j d(x, x_j) \le 2 \inf_n \varphi_n(x) = 0,$$

where the last equality is due to the fact that the sequence (a_n) is dense in E. This fully proves the proposition. □

(1.11) Corollary. *Every Polish space is homeomorphic to a* G_δ *of* $[0,1]^{\mathbb{N}}$.

Proof. Let E be a Polish space. Then E is homeomorphic to a Polish subspace of $[0,1]^{\mathbb{N}}$. Since the latter is also a Polish space (it is countable product of Polish spaces), this is a G_δ of $[0,1]^{\mathbb{N}}$ by virtue of (1.7). □

2. Borel and Baire measures

In the sequel, let us suppose that E is a topological space, and let us recall that $\mathcal{B}(E)$ denotes the Borel σ-field of E, that is to say, the smallest σ-field on E which contains all open sets of E. If E is a Hausdorff space, then all compact sets are, in particular, closed, hence they are Borel sets. For this reason, from now onwards, any topological space shall be assumed to be Hausdorff. Under this supplementary hypothesis, however, the σ-field generated by the compact sets may be strictly smaller than the Borel σ-field, as the following example shows.

(2.1) Example. Let E be a discrete topological space, that is to say, let us assume that the topology on E is the set $\mathcal{P}(E)$ of all subsets of E. Then, obviously $\mathcal{B}(E) = \mathcal{P}(E)$. On the other hand, the set of all compact sets consists just of the finite subsets of E. As a consequence, the σ-field generated by the compact sets coincides with the family $\mathcal{E}$ of all subsets A in E such that either A or A^c is countable. Thus, the σ-field $\mathcal{E}$ generated by the compact sets of E coincides with $\mathcal{B}(E)$ if and only if E is countable.

(2.2) Definition. Let E be a topological space. A measure defined on the Borel σ-field of E is called a **Borel measure** on E.

Observe immediately that, as a result of the fundamental criterion for the coincidence of two measures (see I(5.6)), a Borel measure is uniquely determined by its value on the open sets. Indeed, the following result holds by just recalling that the open sets are closed under finite intersection, whence they are a base for the Borel σ-field.

(2.3) Proposition. *If two Borel measures on a topological space E coincide on all open sets, then they coincide on all Borel sets.*

Let us recall from V(7.1) that a measure μ on a measurable space $(E, \mathcal{E})$ is concentrated on a measurable set C, or that a set C *carries* the measure μ, if $\mu(A) = \mu(A \cap C)$ for every A in $\mathcal{E}$. If the set E is a topological space, and the σ-field $\mathcal{E}$ coincides with the Borel σ-field, this notion can be further refined, connecting it with the topology of the space.

(2.4) Definition. Let E be a topological space. A Borel measure μ is said to be **separable** if it is carried by the closure of a countable subset of E. In this case, there exists the smallest closed set that carries μ, and it is called the **support** of μ, and it is denoted by $\mathrm{supp}(\mu)$. In particular, if the support of μ is compact, we shall say that μ has **compact support**.

Let us also introduce an important class of Borel measures which is going to be particularly useful, especially when dealing with a certain type of topological spaces.

(2.5) Definition. A Borel measure μ on a topological space E is **locally finite** if, for every element x in E, there exists a neighbourhood U of x such that the measure of U under μ is finite.

(2.6) Proposition. *If μ is a locally finite Borel measure on a Hausdorff topological space E, then μ is finite on any compact set. In other words, for every compact set K in E, $\mu(K)$ is finite.*

Proof. Let K be a compact set in E. Since E is Hausdorff, K is closed, hence measurable. For every element x in K, let U_x be an open neighbourhood of x such that $\mu(U_x)$ is finite. Then, the family $\{U_x : x \in K\}$ is an open covering of K. Since K is compact, there exist a finite sequence $x_1, \ldots, x_m$ of points in K such that $\{U_{x_1}, \ldots, U_{x_m}\}$ is an open covering of K. Thus,

$$\mu(K) \leq \mu\big(\textstyle\bigcup_{i=1}^m U_{x_i}\big) \leq \sum_{i=1}^m \mu(U_{x_i}) < \infty. \qquad \square$$

(2.7) Corollary. *If μ is a locally finite measure on a topological space E with compact support, then μ is finite.*

Proof. It suffices to observe that, if K is the compact support of μ, then, by virtue of Proposition (2.6), $\mu(E) = \mu(K) < \infty$. □

Besides the Borel σ-field, there is another σ-field on a topological space that is compatible with the topology.

(2.8) Definition. If E is a topological space, and if $\mathcal{C}(E)$ denotes the class consisting of all continuous real functions defined on E, the smallest σ-field on E that makes all these functions simultaneously measurable is called the **Baire σ-field** of E, and it shall be denoted by $\mathcal{B}_0(E)$. Every element of $\mathcal{B}_0(E)$ shall also be called a **Baire set** of E.

(2.9) Remark. Let us denote by $\mathcal{C}_b(E)$ the space of all bounded continuous real functions on E. For any real function f defined on E, the bounded function $x \mapsto \arctan f(x)$ is measurable if and only if f is measurable (regardless of which specific σ-field is assigned on E). Thus, $\mathcal{B}_0(E)$ is also the smallest σ-field for which all f in $\mathcal{C}_b(E)$ are measurable.

The following theorem shows the link between this and the Borel σ-field.

(2.10) Theorem. *If E is a topological space, the Baire σ-field of E is always included in the Borel σ-field of E. In other words,*

$$\mathcal{B}_0(E) \subseteq \mathcal{B}(E).$$

Furthermore, this inclusion reduces to an equality whenever the topological space E is metrizable.

Proof. The inclusion yields from the fact that every continuous function on E is in particular a Borel measurable function, that is to say a measurable function from $(E, \mathcal{B}(E))$ to $(\mathbb{R}, \mathcal{B}(\mathbb{R}))$. To complete the proof, let us now suppose that the topological space E be metrizable and let us denote by d any distance compatible with its topology. If A is a closed and non-empty subset of E, the identity $A = \{x \in E : d(x, A) = 0\}$ shows, thanks to the fact that the function $x \mapsto d(x, A)$ is continuous, that A is a Baire set. As a result, the σ-field $\mathcal{B}_0(E)$ contains the class of all closed subsets of E, hence it contains the Borel σ-field, and therefore these two σ-fields coincide. □

(2.11) Example. Let E be the product of uncountably many compact Hausdorff spaces, each having more than one point. A singleton is closed and hence a Borel set, but it is not a Baire set. To see this, let us start by observing that, by means of Tychonoff theorem, E is a compact Hausdorff space. Let us denote by $\mathcal{A}$ the sub-algebra of $\mathcal{C}(E)$ of all continuous real functions depending on only finitely many coordinates. It is clear that the set $\mathcal{A}$ *separates the points* of E, that is to say, if x and y are two distinct elements of E, there exists a function f in $\mathcal{A}$ such that $f(x) \neq f(y)$. By virtue of Stone-Weierstrass theorem, $\mathcal{A}$ is dense in $\mathcal{C}(E)$. Thus, for any f in $\mathcal{C}(E)$, there are f_n in $\mathcal{A}$ with $\sup_{x \in E} |f(x) - f_n(x)| < 1/n$. It follows that f depends at most on countably many coordinates, taking the union of the set of coordinates that the f_n depend on. The collection of Baire sets, then, depending only on countably many coordinates, includes a subset generating the Baire σ-field and is closed under complements and countable unions (again taking a countable union of sets of indices), so it is the entire Baire σ-field, which therefore does not contain singletons.

(2.12) Definition. Let E be a topological space. A measure defined on the Baire σ-field of E is called a **Baire measure** on E.

Since the Baire σ-field is a sub-σ-field of the Borel σ-field, an interesting question is whether, in those cases where this inclusion is proper, it is possible to extend a Baire measure to a Borel measure. We shall see in §8 that, if the topology is sufficiently well-behaved, it is indeed possible to extend every Baire measure to a unique Borel measure.

3. Regular measures, Radon measures

When the set E is equipped with a topology, and it is interpreted as a measurable space by equipping it with its Borel σ-field, it is often useful to be able to approximate general measurable sets by more tractable sets such as open sets, closed sets, or compact sets. In doing this, it may be good to recall once again that any closed subset of a compact set is compact, and in a Hausdorff space, any compact set is closed.

For this reason, regardless on whether it is explicitly stated or not, every topological space shall always be assumed to be Hausdorff.

(3.1) Definition. Let E be a Hausdorff topological space, and μ a Borel measure on E. With respect to μ, a Borel set A is said to be

(a) **outer regular** if

$$\mu(A) = \inf\Big\{\mu(O) : O \text{ open, } A \subseteq O\Big\};$$

(b) **inner weakly regular** if

$$\mu(A) = \sup\Big\{\mu(F) : F \text{ closed, } F \subseteq A\Big\};$$

(c) **inner regular** if

$$\mu(A) = \sup\Big\{\mu(K) : K \text{ compact, } K \subseteq A\Big\}.$$

The measure μ is said to be **outer regular** (resp. **inner** [**weakly**] **regular**) if, with respect to μ, every Borel set is outer regular (resp. inner [weakly] regular). Finally, the measure μ is said to be [**weakly**] **regular** if it is, at the same time, outer regular and inner [weakly] regular.

The notions of regularity introduced in the previous definition can be rephrased by saying that the measure of a Borel set can be approximated with open, closed or compact sets, respectively. These three notions of regularity are, in fact, extremely entangled, and mostly depend on the structure of the underlying topological space, which is quite unsurprising, because the amount of open, closed, and compact sets in a topological space would make more or less easy to use them for approximating the other Borel sets. Throughout this chapter, we shall see that, in different topological spaces, there are different connections between these properties of regularity, but often a subset of them is sufficient to imply them all. The next result shows an almost equivalent condition to weak regularity.

(3.2) Proposition. *Let E be a Hausdorff topological space, and μ a Borel measure on E. For the measure μ to be weakly regular it is sufficient that, for every strictly positive real number ε, there exist an open set O and a closed set F in E such that*

$$F \subseteq A \subseteq O, \qquad \text{and} \qquad \mu(O \setminus F) < \varepsilon.$$

Furthermore, if the measure μ is finite, this condition is also necessary.

Proof. Assume that μ is a weakly regular Borel measure on E and A a Borel set. Then, for every strictly positive real number ε, there exist an open set O and a closed set F such that $F \subseteq A \subseteq O$ and $\mu(O \setminus F) < \varepsilon$. In particular,

$$\mu(O) = \mu(A) + \mu(O \setminus A) < \mu(A) + \varepsilon,$$

which proves that μ is outer regular. Similarly,

$$\mu(A) = \mu(F) + \mu(A \setminus F) \leq \mu(F) + \mu(O \setminus F) < \mu(F) + \varepsilon$$

which proves that μ is inner weakly regular as well.

Conversely, let us now assume that μ is finite and both outer and inner weakly regular. Thus, for every strictly positive real number ε, there exist an open set O such that $A \subseteq O$ and $\mu(O) < \mu(A) + \frac{1}{2}\varepsilon$, and a closed set F such that $F \subseteq A$ and $\mu(A) < \mu(F) + \frac{1}{2}\varepsilon$. Thus,

$$\mu(O \setminus F) = \mu(O) - \mu(F) < \mu(A) + \tfrac{1}{2}\varepsilon - \mu(A) + \tfrac{1}{2}\varepsilon = \varepsilon,$$

which shows that μ is indeed weakly regular. □

The following simple result shows that for a finite Borel measure, the notion of regularity implies that of weak regularity. When the Borel measure is not finite, however, this implication may fail, unless there are some additional hypotheses on the topology of the underlying set.

(3.3) Proposition. *Let E be a Hausdorff topological space and μ a finite Borel measure. If μ is regular, then it is also weakly regular.*

Proof. Let A be a Borel set in E. Since E is a Hausdorff space, all compact sets are closed, hence the inclusion

$$\Big\{\mu(K) : K \text{ compact}, K \subseteq A\Big\} \subseteq \Big\{\mu(F) : F \text{ closed}, F \subseteq A\Big\}.$$

Since the supremum of the smaller set coincides with $\mu(A)$ by inner regularity, and the supremum of the bigger set is bounded above by $\mu(A)$, this shows that μ is inner weakly regular. Since μ is finite and both outer and inner weakly regular, μ is weakly regular. □

In the development of topological measure theory, that is to say of measure theory on (the Borel σ-field of) a topological space, Borel measures that are at the same time locally finite and inner regular turn out to have a distinguished role. Such measures are nowadays baptised after the Austrian mathematician Johann Radon (1887–1956). Precisely, let us give the following definition.

(3.4) Definition. On a topological space E, a **Radon measure** is a Borel measure that is at the same time locally finite and inner regular.

A Radon measure constitutes perhaps the most important class of measures for applications. As we shall see in the sequel, on many spaces (including Polish spaces), all Borel measures are Radon. This is not always true, however, as shown in the following example, due to the French mathematician Jean Dieudonné (1906–1992) who was, among other things, one of the founders of the Bourbaki group.

(3.5) Example. Let us denote by E the set of all ordinals which are smaller than or equal to the first uncountable ordinal ω_1. Then, E is a well-ordered uncountable set, with ω_1 as its maximal element. For every $\alpha < \omega_1$, the initial segment $[0, \alpha] = \{x \in E : x \leq \alpha\}$ is countable. Equip E with the 'order topology', that is to say, the one generated by the sub-base of open intervals of the form

$$\{x \in E : x < \beta\}, \qquad \{x \in E : \beta_1 < x < \beta_2\}, \qquad \{x \in E : x > \beta\},$$

where β, β_1, β_2 are ordinals. In this topology, the space E is compact.

Let us set $E_0 = E \setminus \{\omega_1\}$, and let $\mathcal{C}_0$ be the collection of all uncountable closed subsets of E_0 (with the topology induced from E).

We may now define a set function μ on the Borel σ-field $\mathcal{B}(E)$ as follows: For $A \in \mathcal{B}(E)$, let $\mu(A) = 1$ if A contains some element of $\mathcal{C}_0$, and $\mu(A) = 0$ otherwise. Let us prove that μ is a measure.

To this end, let us denote by $\mathcal{E}$ the class of all subsets A of E such that either A or A^c contains an element of $\mathcal{C}_0$. Then, $\mathcal{E}$ is a σ-field. Indeed, it is clear that the set E lies in $\mathcal{E}$, and that $\mathcal{E}$ is stable under complementation. To verify that $\mathcal{E}$ is stable under countable intersections, let us take a sequence (A_n) of elements of $\mathcal{C}_0$, and let us show that, then, one has $A = \bigcap_n A_n \in \mathcal{C}_0$. Indeed, if A were countable, there would exist an ordinal $\alpha < \omega_1$ such that $A \subseteq [0, \alpha]$. Since each A_n is uncountable and closed, for each n one can find $\alpha_n > \alpha$ such that A_n contains ordinals greater than α. Taking a subsequence, if necessary, we may assume that le sequence (α_n) is increasing and converges towards an ordinal $\beta > \alpha$, whence $\beta \in A$, for the sets A_n are closed, contradicting the fact that $A \subseteq [0, \alpha]$. Hence A is uncountable and closed, and thus A belongs to $\mathcal{C}_0$.

Let us now show that $\mathcal{B}(E) \subseteq \mathcal{E}$. Indeed, if $A \in \mathcal{B}(E)$ is closed and uncountable, then $A \cap E_0 \in \mathcal{C}_0$. On the other hand, if A is at most countable, then $A \subseteq [0, \alpha]$ for some $\alpha < \omega_1$, and hence A^c contains an element of $\mathcal{C}_0$. Thus, $A \in \mathcal{E}$ in either cases.

Finally, in order to prove that the set function μ is countably additive, let (A_n) be a sequence of pairwise disjoint Borel subsets of E. If, for some index n, the set A_n contains a set from $\mathcal{C}_0$, then $\mu(\bigcup_n A_n) = 1$. Otherwise, each A_n^c contains a set B_n from $\mathcal{C}_0$, and since the class $\mathcal{C}_0$ is stable under countable intersections, one has $\bigcap_n B_n \in \mathcal{C}_0$. On the other hand, the latter set is clearly included in the complement of $\bigcup_n A_n$, whence $\bigcup_n A_n$ does not intersect any element of $\mathcal{C}_0$, so $\mu(\bigcup_n A_n) = 0 = \sum_n \mu(A_n)$.

So, μ is a Borel measure on E. However, every element x of E_0 has a neighbourhood of measure zero, since singletons and countable sets do not intersect any member of $\mathcal{C}_0$. Thus, $\mu(K) = 0$ for every compact set K included in E_0. Moreover, $\mu(\{\omega_1\}) = 0$, and since $\mu(E_0) = 1$ and this shows that μ is not a Radon measure, because it is not inner regular.

Borel measures are not automatically locally finite, but in many cases this property follows from a weaker condition, and precisely the one of being finite on compact set, as shows the following result.

(3.6) Proposition. *On a first countable Hausdorff space, if a Borel measure is inner regular and finite on compacts, then it is locally finite, hence Radon.*

Proof. Let μ be an inner regular and finite on compacts Borel measure on E, and assume that E is first countable. By contradiction, assume that μ is not locally finite. This means that there exists a point x in E such that every open neighbourhood U of x has infinite measure. Let (V_n) be a countable base for the open neighbourhoods of x (which exists because E is first countable). By replacing the sequence (V_n) with $\big((V_1 \cap \cdots \cap V_n)\big)$, if necessary, we may assume that (V_n) is a decreasing sequence of open sets whose intersection is $\{x\}$. Since, for every index n, μ is inner regular, and $\mu(V_n) = \infty$, there exists a compact set $K_n \subseteq V_n$ such that $\mu(K_n) > n$.

Let us show that the set $K = \{x\} \cup \bigcup_n K_n$ is compact. In fact, if $\mathcal{U}$ is an open covering of K, then some U in $\mathcal{U}$ contains x and since (V_n) is a base for the open neighbourhoods of x, then $V_{n_0} \subseteq U$, for some index n_0. It thus follows that $K_n \subseteq V_n \subseteq V_{n_0} \subseteq U$ for every $n \geq n_0$. Since $K_1 \cup \cdots \cup K_{n_0}$ is a compact subset of K, it is covered by finitely many sets in $\mathcal{U}$. These, together with U, furnish the desired finite covering of K.

On the other hand, $\mu(K)$ must be finite, because μ is finite on compacts, but one also has

$$\mu(K) \geq \mu(K_n) \geq n \qquad \text{for every positive integer } n,$$

and this is the contradiction sought. □

4. Borel measures on a metrizable space

In general a Borel measure need not be regular. Indeed, the regularity of a Borel measure depends heavily on the topological structure of the space, and in particular on 'how many' open, closed, and compact sets there are. In rich spaces, measures tend to be regular more often, while, in very poor spaces, this may happen very rarely.

In this section, we shall focus on the case of metrizable spaces. As we are about to see, in such a space, all finite Borel measures are weakly regular, and, with some additional requirements, every measure is indeed regular.

(4.1) Theorem. *On a metrizable space E, any finite Borel measure μ is weakly regular.*

Proof. Let us denote by $\mathcal{R}$ the class of all Borel sets of E which are weakly regular with respect to μ. Then, obviously, $\mathcal{R} \subseteq \mathcal{B}(E)$. Since $\emptyset$ and E are at the same time open and closed sets, it readily follows that $\emptyset \in \mathcal{R}$ and $E \in \mathcal{R}$.

Let us prove that $\mathcal{R}$ is stable under the operation of complementation. To this end, if $A \in \mathcal{R}$ and ε is a strictly positive real number, then there exist an open set O and a closed set F such that $F \subseteq A \subseteq O$, and $\mu(O \setminus F) < \varepsilon$. Now, to conclude that $A^c \in \mathcal{R}$ it suffices to notice that F^c is an open set, O^c is a closed set, $O^c \subseteq A^c \subseteq F^c$ and $F^c \setminus O^c = O \setminus F$.

Let us also prove that $\mathcal{R}$ is stable under countable unions. To this end, let (A_n) be an infinite sequence of elements of $\mathcal{R}$, and set $A = \bigcup_n A_n$. We wish to prove that $A \in \mathcal{R}$. In fact, for every index n and every strictly positive real number ε, there exist an open set O_n, a closed set F_n such that

$$F_n \subseteq A_n \subseteq O_n \quad \text{and} \quad \mu(O_n \setminus F_n) < 3^{-n}\varepsilon.$$

Let us set $O = \bigcup_n O_n$ and $N = \bigcup_n F_n$. Since μ is a finite measure, by virtue of Proposition I(3.7), we may choose a positive integer k large enough that $\mu\big(N \setminus (F_1 \cup \cdots \cup F_k)\big) < \frac{1}{2}\varepsilon$. Let us set $F = F_1 \cup \cdots \cup F_k$. Then, O is an open set, F is a closed set in E, $F \subseteq A \subseteq O$ and

$$\begin{aligned} \mu(O \setminus F) &\leq \mu(O \setminus N) + \mu(N \setminus F) \\ &\leq \textstyle\sum_{n=1}^{\infty} \mu(O_n \setminus F_n) + \frac{1}{2}\varepsilon \\ &< \textstyle\sum_{n=1}^{\infty} 3^{-n}\varepsilon + \frac{1}{2}\varepsilon \\ &= \varepsilon. \end{aligned}$$

It is thus proved that $\mathcal{R}$ is a σ-field.

To prove that μ is weakly regular, it suffices to show that $\mathcal{R}$ contains all closed sets. To this end, let F be a closed set in E and let ε be a strictly positive real number. Since E is metrizable, C is a G_δ and therefore, there exists a sequence (O_n) of open sets in E such that $\bigcap_n O_n = C$. Since the class of all open sets is stable under finite intersection, with no loss of generality, the sequence (O_n) may be supposed to be decreasing, so that $\mu(O_n) \to \mu(F)$. Thus, we can always find a positive index k such that $\mu(O_k \backslash F) < \varepsilon$. Since F is closed, this proves that F is weakly regular with respect to μ, whence $F \in \mathcal{R}$. This completes the proof. □

It is interesting to wonder when weak regularity implies regularity for a Borel measure. This happens for a special class of Borel measures that have the property of being determined by their values on compact sets. Precisely, let us give the following definition.

(4.2) Definition. Let E be a Hausdorff topological space. A Borel measure μ is said to be **tight** if, for every strictly positive real number ε, there exists a compact set K such that $\mu(K^c) < \varepsilon$.

The condition expressed by the previous definition is equivalent to require that the measure μ is concentrated on a suitable countable union of compact sets. Thus, a tight measure is, in particular, separable.

(4.3) Remark. If μ is a tight measure on a Hausdorff topological space E, then any measure with base μ is also tight. In fact, if $\nu = f \cdot \mu$ is a measure with base μ, where f is a positive function in $\mathcal{L}^1(\mu)$, and if ε is a strictly positive real number, then, since μ is tight, there exists a compact set K such that $\mu(K^c) < (\int f \,\mathrm{d}\mu)^{-1}\varepsilon$. Thus, $\nu(K^c) = \int_{K^c} f \,\mathrm{d}\mu \leq \mu(K^c)(\int f \,\mathrm{d}\mu) < \varepsilon$.

(4.4) Theorem. *Let E be a metrizable space and μ a tight finite Borel measure on E. Then μ is regular.*

Proof. Let E be a metrizable space, and μ be a tight finite Borel measure on E. By virtue of Theorem (4.1), μ is weakly regular, hence, by means of Proposition (3.2), μ is outer regular and inner weakly regular. Thus, it suffices to show that μ is inner regular. To this end, let A be a Borel set of E, and let ε be a strictly positive real number. Since μ is tight, there exists a compact set K such that $\mu(K^c) < \frac{1}{2}\varepsilon$, and since μ is inner weakly regular, there also exists a closed set F such that $F \subseteq A$ and $\mu(A) < \mu(F) + \frac{1}{2}\varepsilon$, or, equivalently, $\mu(A \setminus F) < \frac{1}{2}\varepsilon$. Since E is Hausdorff, the set $C = F \cap K$ is compact, $C \subseteq F \subseteq A$, and

$$\begin{aligned}
\mu(A) &= \mu(C) + \mu(A \setminus C) \\
&\leq \mu(C) + \mu(A \setminus F) + \mu(A \setminus K) \\
&\leq \mu(C) + \mu(A \setminus F) + \mu(K^c) \\
&\leq \mu(C) + \tfrac{1}{2}\varepsilon + \tfrac{1}{2}\varepsilon \\
&= \mu(C) + \varepsilon,
\end{aligned}$$

which shows that μ is inner regular. □

The next result, which is due to the Polish-American mathematician, and member of the Manhattan Project, Stanisław Marcin Ulam (1909–1984), shows that all measures on a Polish space are tight.

(4.5) Theorem (Ulam). *Assume that E is a Polish space. Then every Borel measure on E is tight.*

To prove this theorem, we shall need the following result of general topology.

(4.6) Lemma. *Assume that E is a complete metric space, and let K be a closed set of E which is totally bounded, that is to say, for every strictly positive real number δ, there exists a finite partition of E, whose elements all have diameter smaller than δ. Then, K is compact.*

Proof. Let (x_i) be a sequence of elements of K and let us show that we can extract a convergent subsequence. To this end, let us denote by $\mathcal{G}$ the set of all subsets A of K such that the sequence is 'frequently' in A, that is to say such that set $\{i : x_i \in A\}$ is infinite. It is clear that K lies in $\mathcal{G}$, and that if the union of a finite sequence of subsets of K lies in $\mathcal{G}$, then at least one of the elements of this sequence must lie in $\mathcal{G}$. Thus, given an element A in $\mathcal{G}$, and a strictly positive real number δ, if we consider a finite partition of A whose elements all have diameter smaller than δ, one of the elements of this partition must lie in $\mathcal{G}$. One can therefore construct an increasing sequence (A_n) of elements of $\mathcal{G}$, with $\operatorname{diam}(A_k) < 1/k$ for every index k. Thus, there exists a strictly increasing sequence (i_k) of indices such that $x_{i_k} \in A_k$ for every index k. Since the sequence (x_{i_k}) thereby constructed is a Cauchy sequence, it must be convergent, because the space E is by hypothesis complete. □

We are now finally ready to prove Ulam's theorem.

Proof of Theorem (4.5). Let d be a metric compatible with the topology of E, and let μ be a Borel measure on E. Since E is separable, there exists, for every strictly positive integer n, a countable covering $\mathcal{R}_n$ of E of open sets with diameter smaller than $1/n$. For example, one can take all open balls with centre in a countable dense subset of E and radius $1/(2n)$. Thus, from the fact that $\mathcal{R}_n$ is a countable covering of E, for every strictly positive real number ε, we can always find a set U_n, which is the union of a finite number of elements of $\mathcal{R}_n$, such that $\mu(U_n^c) < 2^{-n}\varepsilon$. Thus, setting $B = \bigcap_n U_n$, one has $\mu(B^c) < \varepsilon$. Furthermore, the set B is totally bounded: In fact, for every n, it is included in U_n, which is the finite union of sets with diameter smaller than $1/n$. It follows that closure of B is also totally bounded. Thanks to Lemma (4.6), this closure is a compact set. Since its complement has measure, with respect to μ, strictly smaller than ε, the theorem is proved. □

(4.7) Corollary. *In a Polish space E, every finite Borel measure is regular.*

Proof. If μ is a finite Borel measure on E, by virtue of Theorem (4.1), it is a weakly regular measure and by means of Ulam's Theorem, it is also tight, hence regular, by means of Theorem (4.4). □

(4.8) Corollary. *On a Polish space E, every locally finite Borel measure is a σ-finite outer regular Radon measure.*

Proof. Since μ is, by hypothesis, locally finite, for every point x in E, there exists an open neighbourhood U_x of x with finite measure. The family of all these neighbourhoods U_x, as x varies in E, is an open covering of E. Since E is separable, this ensures that there exists a countable sub-covering. In other words, there exists a sequence (x_n) of points of E, such that (U_{x_n}) is an open covering of E, and such that $\mu(U_{x_n})$ is finite. For every index n, let us set $O_n = U_{x_1} \cup \cdots \cup U_{x_n}$. Then, (O_n) is an increasing sequence of open sets in E, with finite measure, that covers E. This proves that μ is σ-finite. Furthermore, let us denote by μ_n the trace of μ onto O_n, i.e. let us set

$$\mu_n(A) = \mu(A \cap O_n) \qquad \text{for every Borel set } A \text{ in } E.$$

Each of these measures is finite, hence regular. In particular, for every index n, μ_n is inner regular. It follows that, for every Borel set A in E

$$\begin{aligned}\mu(A) &= \sup_n \mu_n(A) \\ &= \sup_n \sup\Big\{\mu_n(K) : K \text{ compact, } K \subseteq A\Big\} \\ &= \sup\Big\{\sup_n \mu_n(K) : K \text{ compact, } K \subseteq A\Big\} \\ &= \sup\Big\{\mu(K) : K \text{ compact, } K \subseteq A\Big\},\end{aligned}$$

thereby proving that the measure μ is inner regular, hence Radon. The same argument shows that the measure μ is also outer regular. □

5. Locally compact Hausdorff spaces

Besides the case of metrizable spaces, there is another family of topological spaces, which quite frequently appears in applications, and which may not be metrizable, yet where measures turn out to be particularly well-behaved. Again, we shall always assume that every topological space is Hausdorff, to ensure that every compact set is closed.

Let us recall that a topological space E is **locally compact** if each of its point has a relatively compact neighbourhood, that is to say, every point has a neighbourhood whose closure is compact. Examples of locally compact Hausdorff spaces are $\mathbb{R}^n$, any space with a discrete topology (for example $\mathbb{Z}$), and a compact Hausdorff space. In general, however, a Banach space with its strong topology is not locally compact. Note that locally compact Hausdorff spaces need not be metrizable.

The following elementary results are going to be basic tools for what follows.

(5.1) Proposition. *Let E be a Hausdorff space, and let K and L be two disjoint compact subsets of E. Then, there are disjoint open subsets U and V of E such that $K \subseteq U$ and $L \subseteq V$.*

Proof. With no loss of generality, we may assume that K and L are both non-empty sets. Thus, let x be a given point in K, and observe that, since the space E is Hausdorff, for each point y in L, there exist two disjoint open sets $U(x, y)$ and $V(x, y)$ such that $x \in U(x, y)$ and $y \in V(x, y)$. Now observe that the collection of all such $V(x, y)$, as y varies in L, is an open covering of L, whence there exists a finite sub-covering $V(x, y_1), \ldots, V(x, y_s)$ of L. Set

$$U(x) = U(x, y_1) \cap \cdots \cap U(x, y_s), \quad \text{and} \quad V(x) = V(x, y_1) \cup \cdots \cup V(x, y_s).$$

Now, as x varies in K, the collection of all sets U_x, as x varies in K, is an open covering of K, whence there exist a finite sub-covering $U(x_1), \ldots, U(x_r)$ of K. It suffices then to take

$$U = U(x_1) \cup \cdots \cup U(x_r), \quad \text{and} \quad V = V(x_1) \cap \cdots \cap V(x_r)$$

and the theorem is whereby proved. □

The open sets U and V found in the previous proposition are said to **separate** the compact sets K and L. The next few results shall all utilise, more or less directly, Proposition (5.1).

(5.2) Proposition. *Let E be a locally compact Hausdorff space, let x be a point in E, and let U be an open neighbourhood of x. Then, x has an open neighbourhood whose closure is compact and included in U.*

Proof. Since E is locally compact, there is an open neighbourhood of x, W, whose closure is compact. By replacing W with $U \cap W$, if necessary, we may assume that W is included in U. By virtue of Proposition (5.1), there exist two open sets V and V' that separate the compact sets $\{x\}$ and $\overline{W} \setminus W$, respectively. The closure of $V \cap W$ is then compact and included in W, hence in U; thus $V \cap W$ is the required open neighbourhood of x. □

(5.3) Proposition. *Let E be a locally compact Hausdorff space, let K be a compact subset of E, and U an open subset of E that contains K. Then, there exists a relatively compact open subset V of E such that $K \subseteq V \subseteq \overline{V} \subseteq U$.*

Proof. Proposition (5.2) implies that each point in K has an relatively compact open neighbourhood included in U. Since K is compact and the collection of all these relatively compact open neighbourhoods forms an open covering for K, there exists a finite sub-covering. Let V be the union of this finite sub-covering; then V has the properties required. □

Let us also recall that a topological space is **σ-compact** if it is the union of a countable collection of compact sets. Obviously, every locally compact Hausdorff space which is second countable, that is to say which has a countable base for its topology, is also σ-compact, because this base can always be taken with relatively compact sets. For a σ-compact space, the following result holds.

(5.4) Proposition. *Let E be a locally compact σ-compact Hausdorff space. Then, there exists a sequence of relatively compact open sets (O_n) that covers E and, for every index n, the closure of O_n is included in O_{n+1}.*

Proof. Since E is σ-compact, there is a sequence of compact sets, (K_n), whose union is E. Starting from this sequence, let us construct the sequence (O_n) by induction. For $n = 1$, by virtue of Proposition (5.3), there exists a relatively compact open set O_1 containing K_1. Inductively, let us suppose that the open sets $O_1, \ldots, O_n$. Again, by virtue of Proposition (5.3), there exists a relatively compact open set O_{n+1} containing $\overline{O}_n \cup K_n$. The sequence (O_n) thereby created satisfies the conditions required. □

Let us also recall that a topological space is **normal** if it is Hausdorff, and each pair of disjoint closed sets can be separated by a pair of disjoint open sets. The following theorem, which is named after the Soviet mathematician Pavel Urysohn (Павел Самуилович Урысон, 1898–1924) is a milestone of topology and is an incredibly useful tool to work with locally compact spaces.

(5.5) Theorem (Urysohn's lemma). *Let X be a normal topological space, and let F and G be two disjoint closed subsets of E. Then, there exists a continuous function f, defined on E, with values in $[0,1]$, such that $f(x) = 0$ holds for every $x \in F$ and $f(x) = 1$ holds for every $x \in G$.*

The proof of this theorem is going to need the following result.

(5.6) Lemma. *In the same hypotheses as in (5.5), let Δ be the set of all dyadic rational numbers in $[0,1]$, that is to say, all numbers of the form $2^{-n}m$, where m and n are positive integers, and $0 \leq m \leq 2^n$. Then, there exists a countable family $\{U_r : r \in \Delta\}$ of open subsets of E such that $F \subseteq U_r \subseteq G^\circ$ for every r in Δ, and $\overline{U}_r \subseteq U_s$ for $r < s$.*

Proof. For every positive integer n, denote by Δ_n the set of all rational numbers of the form $2^{-n}m$, with $0 \leq m \leq 2^n$, and let us observe that

$$\Delta_n \subseteq \Delta_{n+1} \subseteq \Delta.$$

The proof proceeds by induction on n. For $n = 0$, then $\Delta_0 = \{0, 1\}$. Let us set $U_1 = G^c$. Since E is normal, there exist two disjoint open sets U and V, such that $F \subseteq U$ and $G \subseteq V$, that is to say

$$F \subseteq U \subseteq \overline{U} \subseteq V^c \subseteq G^c.$$

Set $U_0 = U$. Now, assume to have defined every U_r, when r belongs to Δ_n, for a certain positive integer n, and let us proceed to define it when r belongs to Δ_{n+1}. To this end, let $r = 2^{-(n+1)}m$, with $0 \leq m \leq 2^{n+1}$, and let us distinguish two cases.

- When $m = 2k$ is even, then $r = 2^{-(n+1)} \cdot (2k) = 2^{-n}k$, which means that r belongs to Δ_n, so the set U_r is already defined by inductive hypothesis.
- When $m = 2k + 1$ is odd, then r belongs to Δ_{n+1} but not to Δ_n. Let us observe that, in this case, one has $2^{-n}k < r < 2^{-n}(k+1)$. Since E is normal, there exist two disjoint open sets W and X, such that $\overline{U}_{2^{-n}k} \subseteq W$ and $\overline{U}_{2^{-n}(k+1)} \subseteq X$, that is to say

$$\overline{U}_{2^{-n}k} \subseteq W \subseteq \overline{W} \subseteq X^c \subseteq U_{2^{-n}(k+1)}.$$

Thus, we may set $U_r = W$.

The family $\{U_r : r \in \Delta\}$ clearly has the required properties. □

We are now ready to prove Urysohn's lemma.

Proof of Theorem (5.5). Let $\{U_r : r \in \Delta\}$ be the countable family defined in Lemma (5.6), and let us define the function f on E by setting

$$f(x) = \inf\Big\{r \in \Delta : x \in U_r\Big\} \qquad \text{for every } x \text{ in } E.$$

Since $F \subseteq U_r \subseteq G^c$ for every r in Δ, one clearly has $f(x) = 0$ for x in F, and $f(x) = 1$ for x in G. It only remains to prove that the function f is continuous. To this end, it suffices to show that the inverse image of each member of some sub-base for the topology of the set of values is open; let us observe that the sets of the form $(-\infty, s)$ and $(s, +\infty)$, for any real number t, is a sub-base of for the usual topology for the real line. Consequently, to show that f is continuous, it suffices to show that $\{f < s\}$ is open, and $\{f \leq s\}$ is closed, for every real number s. To this end, observe that $f(x) < s$ if and only if $x \in U_r$ for some $r < s$, that is to say

$$\{f < s\} = \bigcup_{r \in \Delta,\, r < s} U_r$$

which is an open set, because it is union of open sets. Conversely, one has $f(x) > s$ if and only if $x \notin U_r$ for some $r > s$, if and only if $x \in \overline{U}_t$ for every $t > s$ (because $\overline{U}_r \subseteq U_t$ for $r < t$), that is to say

$$\{f \leq s\} = \bigcap_{r \in \Delta,\, r > s} U_r = \bigcap_{r \in \Delta,\, r > s} \overline{U}_r,$$

which is a closed set, because it is intersection of closed sets. □

If E is a locally compact Hausdorff space, one of the most useful families of real functions on E, from the point of view of integration theory, is the class $\mathcal{C}_c(E)$, consisting of all continuous real functions that vanish outside a compact subset of E. If f is a real function on E, the **support** of f is the closure of the set $\{x : f(x) \neq 0\}$. Thus, $\mathcal{C}_c(E)$ may be described as the class of all continuous real functions on E with compact support.

The following fact about $\mathcal{C}_c(E)$, which is a simple consequence of Urysohn's lemma, is essential for the development of measure theory of locally compact Hausdorff spaces. Albeit obvious, it also is worth noticing that, if K is a compact subset of E, then any function in $\mathcal{C}_c(K)$ extends to a function on $\mathcal{C}_c(E)$, by simply setting it equal to zero on K^c.

(5.7) Proposition. *Let E be a locally compact Hausdorff space, let K be a compact subset of E, and let U be an open subset that includes K. Then, there exist a function f that belongs to $\mathcal{C}_c(E)$, satisfying $I_K \leq f \leq I_U$ and whose support is included in U.*

Proof. By means of Proposition (5.3), there exists a relatively compact open set V such that $K \subseteq V \subseteq \overline{V} \subseteq U$. Thanks to Urysohn's lemma, applied to the compact Hausdorff (whence normal) space $\overline{V}$, there exists a continuous function g, defined on $\overline{V}$, with values in $[0, 1]$, such that $g(x) = 1$ hold for each x in K, and $g(x) = 0$ holds for each x in $\overline{V} \setminus V$. Now, define the function f on E by requiring that f agree with g on $\overline{V}$ and vanish outside $\overline{V}$. The continuity follows because f is continuous on $\overline{V}$, and it is constant, hence continuous, on V^c. The support of f is included in $\overline{V}$, and so it is compact and included in U. □

In the rest of this section, we wish to study the conditions that make a locally compact space metrizable. To this end, let us start by proving the following.

(5.8) Proposition. *For a compact Hausdorff space to be a Polish space it is (necessary and) sufficient that the topology have a countable base.*

Proof. The condition is clearly *necessary*, for a compact metrizable space is separable and so has a countable base. Thus, let us prove that the condition is *sufficient*. To this end, suppose that E is a compact Hausdorff space and that it has a countable base $\mathcal{U}$, This implies that the space is separable, hence it is sufficient to show that E is also metrizable, for a compact metric space is always complete. Since E is, in particular, normal, for each pair U and V of elements of $\mathcal{U}$, with $\overline{U} \cap \overline{V} = \emptyset$, by Urysohn's lemma, there exists a continuous function f on E, that vanishes on $\overline{U}$, and has value 1 everywhere on $\overline{V}$. By choosing all possible pairs of elements of $\mathcal{U}$, it is thereby defined a sequence (f_n) of elements of $\mathcal{C}_c(E)$. Let us show that this sequence separates the points of E, and for this it suffices to show that, for each pair of distinct points x, y in E, there are two sets U and V in $\mathcal{U}$, with disjoint closures, the first containing x and the second containing y. To construct such sets, choose two disjoint neighbourhoods W_1 and W_2 of x and y, and, by virtue of Proposition (5.3) there exist two relatively compact open sets U and V such that $x \in U \subseteq \overline{U} \subseteq W_1$, and $y \in V \subseteq \overline{V} \subseteq W_2$. By making U and V smaller, if necessary, we may assume that they also belong to $\mathcal{U}$. Thus the required sets U and V exist, and the sequence (f_n) does separate the points in E.

Let us define the function d on $E \times E$ by setting

$$d(x, y) = \sum_n 2^{-n}|f_n(x) - f_n(y)|.$$

It is easy to use the fact that the functions f_n separate the points of E to check that d is indeed a metric on E and to use the fact that the function f_n are continuous (with respect to the topology on E), to check that the topology induced by d is weaker than the original topology.

Since the original topology makes E a compact space, while the topology induced by d is weaker and Hausdorff, the two topology must in fact be the same. Thus, the original topology on E is metrizable and, in fact, metrised by d. This is sufficient to complete the proof. □

Let us now generalise this result to the case of a locally compact Hausdorff space. To this end, we are going to need the **one-point compactification** of a locally compact Hausdorff space E, which is also sometimes referred to as the **Alexandrov compactification**, after the Soviet mathematician Pavel Alexandrov (Павел Сергеевич Александров, 1896–1982). The idea of the construction is in adjoining to the set E an additional point, called the **point at infinity**, ω. Let us denote by E^* the set $E \cup \{\omega\}$. To endow E^* with a topology, the open subsets of E^* shall be, by definition, all the open sets of E, together with the complements, with respect to E, of the compact subsets of E, which are the all the open sets containing ω. It is not difficult to check that this does define a topology on E, and that the trace of this topology on E coincides with the original topology on E.

Furthermore, this topology is Hausdorff. Indeed, if x and y are two distinct points in E^*, then, if they both belong to E, then they are separated by two open sets in E, hence in E^*. If one of them, say y, is the point at infinity, then we can always choose a relatively compact neighbourhood U of x, and observe that $V = E^* \setminus \overline{U}$ is, by definition a neighbourhood of ω, and U and V are disjoint sets that separate x and ω.

It remains to show that E^* is compact. To this end, let $\mathcal{U}$ be an open covering of E^*. Then, there exists V in $\mathcal{U}$ such that ω belongs to V. Thus, by definition $K = E^* \setminus V$ is a compact set in E and $\mathcal{U}$ is a covering of K, hence there exists a finite subcovering $\mathcal{U}_0$ for K. Then, the set $\mathcal{U}_0$, together with V, forms a finite covering for E^*. Thus E^* is indeed compact.

Let us now prove the following simple result.

(5.9) Lemma. *Let E be a locally compact Hausdorff space. If there is a countable base for the topology of E, then there is a countable base for the topology for the one-point compactification of E.*

Proof. Let us denote by E^* the one-point compatification of E, let $\mathcal{U}$ be a countable base for the topology of E, and let $\mathcal{U}_0$ be the collection of those elements U of $\mathcal{U}$, which are relatively compact. Arrange $\mathcal{U}_0$ into a sequence, say (U_n). For every index n, set $V_n = U_1 \cup \cdots \cup U_n$, and observe that (V_n) is now an increasing sequence of relatively compact open sets in E, whose union coincides with E. Thus, for every compact subset K of E, there is a positive integer n such that K is included in the union V_n. Thus, if U is an open neighbourhood in E^* of the point at infinity, and if $K = E^* \setminus U$, then there is a positive integer n such that $K \subseteq V_n$ and hence $E^* \setminus \overline{V}_n \subseteq U$. It follows that the set $\mathcal{U}$, together with the set of all $E^* \setminus \overline{V}_n$, with n positive integer, form a countable base for the topology of E^*. □

(5.10) Proposition. *Every locally compact Hausdorff space that has a countable base is Polish.*

Proof. Let E be a locally compact Hausdorff space whose topology has a countable base and let us denote by E^* the one-point compactification of E. By Lemma (5.9), E^* is a compact Hausdorff space with a countable base, hence, by Proposition (5.8), it is a Polish space. Since E is an open set of E^*, by Proposition (1.5), E is also Polish. □

6. Measures on a locally compact Hausdorff space

After so much general topology, it is time to move back to the realm of measure theory, and to reap the harvest of our topological labour. To this end, let us recall that, if E is a Hausdorff space, then every compact set is closed, hence a Borel set. In general, however, as we have already mentioned in Example (2.1), for the family of all compact sets it is not guaranteed to be a system of generator for the Borel σ-field. However, for this to happen, it suffices that E be σ-compact, as the next result shows.

(6.1) Proposition. *If E is a σ-compact Hausdorff space, then the family of all compact sets in E generates the Borel σ-field on E.*

Proof. Since every compact set is closed, hence a Borel set, it suffices to show that every element of a set of generators for the Borel σ-field of E lies in the σ-field generated by the compact sets. Hence, it suffices to show that every closed set in E is the countable union of compact sets. Indeed, since E is a σ-compact space, there exists a sequence (K_n) of compact sets, whose union is E. Thus, if F is a closed set in E, then $F = \bigcup_n F \cap K_n$. On the other hand, since $F \cap K_n$ is a closed subset of K_n, which is compact, it is compact as well. Thus, every closed set is the countable union of compact sets, and the result is proved. □

Something similar happens to the Baire σ-field.

(6.2) Proposition. *If E is a locally compact, σ-compact, Hausdorff space, then the family $\mathcal{C}_c(E)$ of all real functions on E with compact support generates the Baire σ-field on E.*

Proof. The Baire σ-field is, by definition, the smallest σ-field that makes measurable all bounded continuous real functions on E. As a result, the Baire σ-field is generated by the sets of the form $\{a < f < b\}$, where f is in $\mathcal{C}_b(E)$ and a and b are any two real numbers, with $a \leq b$. It suffices to show that this set lies in the σ-field generated by $\mathcal{C}_c(E)$. Since E is σ-compact, there exists a sequence (O_n) of relatively compact open sets in E whose union is E and such that, for every index n, the closure of O_n is included in O_{n+1}.

By Proposition (5.7), for every index n, there exists a function g_n in $\mathcal{C}_c(E)$ such that $I_{\overline{O}_n} \leq g_n \leq I_{O_{n+1}}$ and whose support is included in O_{n+1}. Hence, setting $f_n = g_n f$, it is clear that the function f_n has compact support and

$$\{a < f < b\} = \bigcup_n \{a < f_n < b\},$$

as required. □

The next result shows that, in a locally compact, σ-compact, Hausdorff space E, to ascertain whether two Borel measures are identical it suffices to check whether they coincide 'locally'. To this end, let us recall that, if μ is a Borel measure on E, and H is any Borel set in E, the measure μ_H defined on the Borel σ-field of E by

$$\mu_H(A) = \mu(A \cap H) \qquad \text{for every Borel set } A \text{ of } E$$

is called the trace of μ onto H (see I(3.2)).

(6.3) Definition. On a locally compact, σ-compact, Hausdorff space E, two Borel measures μ and ν **coincide locally** on a point x in E if there exists an open neighbourhood U of x such that $\mu_U = \nu_U$.

(6.4) Theorem (Localisation principle). *For two Borel measures μ and ν on a locally compact, σ-compact, Hausdorff space E to coincide it is (necessary and) sufficient that they coincide locally on each point of E.*

Proof. The condition is obviously *necessary*. Let us prove that the condition is also *sufficient*. By hypothesis, for every point x, there exists an open neighbourhood U_x of x such that $\mu_{U_x} = \nu_{U_x}$. Since E is σ-compact, there also exists a countable family of compact sets, (K_n) whose union is E. For every index n, the set $\{U_x : x \in K_n\}$ is an open covering of K_n, whence we may extract a finite sub-covering. Let us denote its union by U_n and observe that $\mu_{U_n} = \nu_{U_n}$. Inductively, let us set

$$V_1 = U_1, \qquad \text{and} \qquad V_{n+1} = U_{n+1} \setminus \bigcup_{i=1}^n V_i,$$

and observe that the sequence (V_n) is still a covering of E, whose elements are pairwise disjoint and $\mu_{V_n} = \nu_{V_n}$ for each index n. Thus, for every Borel set A, one has

$$\begin{aligned}\mu(A) &= \mu\big(\textstyle\bigcup_n A \cap V_n\big) = \sum_n \mu_{V_n}(A) \\ &= \textstyle\sum_n \nu_{V_n}(A) = \nu\big(\bigcup_n A \cap V_n\big) = \nu(A),\end{aligned}$$

which proves that μ and ν coincide. □

As regard to the regularity of measures in a locally compact Hausdorff space, the rich topology yields some discount involving the three notions of regularity, as shown in the following result.

(6.5) Proposition. *Let E be a locally compact, σ-compact, Hausdorff space, and μ a Borel measure on E finite on compacts. If μ is inner regular, then it is outer regular.*

Proof. Let us start by observing that, if μ is finite, then the set A is inner weakly regular if and only if A^c is outer regular, hence, if μ is finite, it is inner weakly regular if and only if it is outer regular. Furthermore, if μ is inner regular, it is also inner weakly regular, hence outer regular.

Let us suppose that μ is inner regular and not finite, and observe that, since E is σ-compact and μ is finite on compacts, then μ is necessarily σ-finite. More precisely, there exists a sequence (O_n) of relatively compact open sets in E, with finite measure, whose union is E, and such that the closure of O_n is included in O_{n+1}. Furthermore, let us denote by μ_n the trace of μ onto O_n, that is to say, let us set

$$\mu_n(A) = \mu(A \cap O_n) \qquad \text{for every Borel set } A \text{ in } E.$$

Each of these measures is finite and inner regular, then is also outer regular. It follows that, for every Borel set A in E,

$$\begin{aligned}\mu(A) &= \sup_n \mu_n(A) \\ &= \sup_n \sup\Big\{\mu_n(O) : O \text{ open}, O \subseteq A\Big\} \\ &= \sup\Big\{\sup_n \mu_n(O) : O \text{ open}, O \subseteq A\Big\} \\ &= \sup\Big\{\mu(O) : O \text{ open}, O \subseteq A\Big\},\end{aligned}$$

thereby proving that the measure μ is outer regular. □

As a consequence of the previous proposition, any Radon measure on a locally compact, σ-compact, Hausdorff space is always outer regular. This is, essentially, the main reason why, to be deemed a Radon measure, it is only required to be inner regular.

Let us complete this section discussing the relationship between integrability and continuity. In general, continuous functions on a topological space may be integrable or not, with respect to a measure. Interestingly, when the topological space is normal, the subset of all continuous integrable functions is always dense in the space $\mathcal{L}^1$, as the following theorem shows.

(6.6) Theorem. *If μ is a weakly regular measure on a normal topological space E. Then, the class of all of the continuous integrable function on E is dense in $\mathcal{L}^1$.*

Proof. Since, as we already know from VI(2.11), the class $\mathcal{S}^1(\mu)$ of integrable simple functions is dense in $\mathcal{L}^1$, it suffices to prove that an integrable measurable function coincides with the limit, in $\mathcal{L}^1$, of a sequence of elements of $\mathcal{C}(E)$. Actually, it suffices to prove less: Precisely, that The indicator function I_A of an arbitrary Borel set with finite measure is the limit, in $\mathcal{L}^1$, of a sequence of elements of $\mathcal{C}(E)$.

To this end, take a Borel measurable set A of E, and for every positive integer n, let O_n be an open set of E and F_n be a closed set of E, such that, $F_n \subseteq A \subseteq O_n$ and $\mu(O_n \setminus F_n) < 2^{-n}$. In particular, O_n shall have finite measure under μ. By means of Urysohn's lemma, there exists therefore a continuous function f_n defined on E with values in $[0, 1]$ which coincides with 0 on O_n^{c} and with 1 on F_n. Then,

$$\int f_n \, \mathrm{d}\mu = \int_{O_n} f_n \, \mathrm{d}\mu \leq \mu(O_n).$$

The thesis follows therefore observing that

$$\int |I_A - f_n| \, \mathrm{d}\mu = \int_{O_n \setminus F_n} |I_A - f_n| \, \mathrm{d}\mu \leq \mu(O_n \setminus F_n) \leq 2^{-n}$$

and taking the limit as n approaches infinity. □

7. Riesz-Markov's representation theorem

Historically, the famous Riesz-Markov's representation theorem states that, in a compact Hausdorff space, every monotone linear form on the Riesz space of all continuous real functions on E can be *represented* as the integral of a suitable Baire measure. In the rest of this section, we are going to show a little extension of this result in the case of a Hausdorff σ-locally compact, as a simple special case of Daniell's representation theorem (see III(4.1)). Let us start by proving the following simple topological fact.

(7.1) Theorem. *In a locally compact Hausdorff space E, the indicator function of a compact set is bounded from above by a continuous real function with compact support.*

Proof. Let K be a compact set. Since there is a finite covering of K made of open and relatively compact sets, K is contained in an open set U which is relatively compact, i.e. which is in turn contained in a compact set H. The set H, equipped with the topology induced by that of E, is therefore a compact space. Applying to this Urysohn's lemma, we can see that there exists a continuous function f from H to $[0,1]$ that coincides with the constant 1 on K and with 0 on $H \backslash U$. Since f is identically zero on the frontier of H, then the function that coincides with f on H and with 0 on $E \setminus H$ is a continuous function on E (with support included in H) and bounded from above by the indicator function of K. □

We can now state Riesz-Markov's representation theorem.

(7.2) Theorem (Riesz-Markov's representation theorem). *Assume that E is a locally compact Hausdorff space, and denote by $\mathcal{C}_c$ the Riesz space of all continuous real functions defined on E with compact support. For every monotone linear form L on $\mathcal{C}_c$, the following holds:*

(a) *L satisfies Daniell's condition.*

(b) *If the space E is σ-compact, that is to say, if it the space E is countable union of a family of compact sets, then there exists a Baire measure μ on E, such that*

$$L(f) = \int f \, \mathrm{d}\mu \qquad \text{for every element } f \text{ of } \mathcal{C}_c.$$

Proof. To prove that L satisfies Daniell's condition, let us consider a decreasing sequence (f_n) of functions in $\mathcal{C}_c$, such that $\inf_n f_n = 0$, and choose, by means of Theorem (7.1), an element g of $\mathcal{C}_c$ such that the indicator function of the support of f_1 is bounded from above by g. For every n, we have therefore $0 \le f_n \le \|f_n\| g$ (where $\|f_n\|$ denotes the uniform norm of f_n) and therefore

$$0 \le L(f_n) \le \|f_n\| L(g).$$

Since, thanks to Dini's lemma (Proposition (1.8), the sequence (f_n) converges uniformly towards 0, the last term of the previous relation converges towards zero. It is therefore proved statement (a). Now, statement (b) follows from Daniell's representation theorem: In fact, the hypothesis that the space is σ-compact ensures that the constant 1 belongs to $\mathcal{C}_{c\sigma}$ and that the σ-field generated by $\mathcal{C}_c$ coincides with the Baire σ-field of E. □

(7.3) Remark. If E is a σ-locally compact Hausdorff space, then, assertion (b) of Riesz-Markov's representation theorem may be re-stated by saying that the dual of the Riesz space $\mathcal{C}_c$ of all continuous real functions on E with compact support coincides with the Riesz space of all Baire measures on E.

The following theorem is an immediate consequence of Theorem III(4.4), by taking the topological hypotheses of the previous theorem and choosing the space $\mathcal{C}_c$ as the space $\mathcal{R}$. Let us also note that the class $\mathcal{C}_{c\sigma}$ (resp. $\mathcal{C}_{c\delta}$) consists of special lower (resp. upper) semi-continuous functions on the space E.

(7.4) Theorem. *Let μ be a Baire measure on the locally compact space E, σ-compact, and let us denote by $\mathcal{C}_c$ the Riesz space of all continuous function with compact support on E. Then, for each element f of $\mathcal{L}^1(\mu)$, one has*

$$\sup\left\{\int g\,\mathrm{d}\mu : g \in \mathcal{C}_{c\delta},\, g \leq f\right\} = \int f\,\mathrm{d}\mu = \inf\left\{\int h\,\mathrm{d}\mu : h \in \mathcal{C}_{c\sigma},\, f \leq h\right\}.$$

As a result, the Riesz space $\mathcal{C}_c$ is dense in $\mathcal{L}^1(\mu)$.

8. Extension of Baire measures

Having established that Baire measures furnish a representation of the dual of the space $\mathcal{C}_c(E)$, the next task consists in showing that all Baire measures can be extended to a Radon measure. This is the goal of the present very short section, achieved as a very simple application of Daniell's extension and representation theorems.

(8.1) Theorem. *Let E be a locally compact, σ-compact, Hausdorff space, and μ_0 a finite Baire measure on E. Then there exists a Radon measure μ on the Borel σ-field of E whose restriction to the Baire σ-field $\mathcal{B}_0(E)$ is μ_0.*

Proof. Let us set

$$L(f) = \int f\,\mathrm{d}\mu_0 \qquad \text{for every element } f \text{ of } \mathcal{C}_c(E).$$

By means of Riesz-Markov theorem, this is a Daniell's functional on the Riesz space $\mathcal{C}_c(E)$. Hence, by means of Daniell's extension theorem (see III(3.5)), this functional extends to a Daniell functional J defined on the monotone Riesz space $\mathcal{M}$ generated by $\mathcal{C}_c(E)$.

Let K be a compact set in E, and let us prove that the indicator function of K is an element of $\mathcal{M}$. By virtue of Theorem (7.1), we know that then I_K is bounded above by function in $\mathcal{C}_c(E)$. As a result, it suffices to show that I_K can be expressed as the supremum of an increasing sequence of functions in $\mathcal{C}_c(E)$. By virtue of Proposition (5.4), there exists a sequence (O_n) of relatively compacts open sets such that, for every index n, the closure of O_n is included in O_{n+1} and whose union is E.

Set $U_n = K \cap \overline{O}_n$. Then, (U_n) is a sequence of relatively compact open sets in K with the topology induced by E, whose union is K, such that, for every index n, the closure of U_n is included in U_{n+1}. By means of Proposition (5.7), there exists a function g_n in $\mathcal{C}_c(K)$ such that $I_{\overline{U}_n} \leq g_n \leq I_{U_{n+1}}$ and whose support is included in U_{n+1}, hence in K. Thus, the function g_n may be extended to a continuous function on E, with support included in K, hence an element of $\mathcal{C}_c(E)$. It is clear that $\sup_n g_n = I_K$, whence $I_K \in \mathcal{M}$. Since, by Proposition (6.1), the family of compact sets generates the Borel σ-field, this proves that the σ-field generated by $\mathcal{M}$ contains the Borel σ-field of E.

Thus, by virtue of Daniell's representation theorem, there exists a Borel measure μ such that $\mathcal{M} \subseteq \mathcal{L}^1(\mu)$ and

$$J(f) = \int f \, \mathrm{d}\mu \qquad \text{for every element } f \text{ of } \mathcal{M}.$$

Since $\mathcal{C}_c(E)$ is included in $\mathcal{M}$ and J coincides with L on this space, μ is an extension of μ_0.

The fact that μ is inner regular is an immediate consequence of III(4.4). It remains to prove that the measure μ is also locally finite. To this end, let x be a point in E. Since E is locally compact, there exists an open neighbourhood U of x included in a compact set K. By virtue of Theorem (7.1), there exists a real function g on E with compact support, such that $I_K \leq g$. Thus,

$$\mu(U) \leq \mu(K) \leq \int g \, \mathrm{d}\mu_0$$

and this proves that μ is also locally finite. □

Chapter VIII

Convergence and Uniform Integrability

The purpose of this chapter is to study the convergence of sequences of measurable functions on a given measure space, and also to begin the study of the convergence of sequence of measures on a measurable space. In §1, we shall discuss in detail the main modes of convergence for real functions on an abstract measure space $(E, \mathcal{E}, \mu)$, we shall delve into the links between these different modes of convergence, and we shall show some examples that ought to clarify when the different possible implications fail. In §2 and §3, we shall introduce the general definition of uniform integrability for a family of functions, we shall see the main properties of uniformly integrable families of functions, and we shall prove the famous Vitali's theorem, which shows how uniform integrability is a relaxation of the hypothesis of domination in Lebesgue's convergence theorem.

In the second part of the chapter, starting from §4 we shall start introducing, in the abstract setting, a mode of convergence for real measures on a measurable space, and we shall prove a result due to O. Nikodým that describes under which hypotheses the set-wise limit of a sequence of real measure is a real measure. Using this result, we shall then move to proving Vitali-Hahn-Saks theorem, which is an extension and completion of Nikodým's theorem. With the aid of Vitali-Hahn-Saks theorem, then, we can finally focus on Dunford-Pettis theorem, which links uniform integrability to compactness in the weak topology $\sigma(\mathcal{L}^1, \mathcal{L}^\infty)$, giving a proof that, unlike the traditional proofs, does not require results of weak compactness in topological vector spaces, but that remains entirely inside the realm of measure theory. The chapter concludes with the introduction of another type of convergence for measures on an abstract measurable space, which has the advantage of inducing on the space of real measures a structure of Banach space.

DOI: 10.1201/9781003724766-VIII

1. Convergence of functions

In the sequel, we shall assume that $(E, \mathcal{E}, \mu)$ is a given measure space, (f_n) a sequence of measurable functions defined on E, and f a measurable function also defined on E. Even if the set of indices is not explicitly mentioned, we shall always assume it to be the set of all strictly positive integers.

It is well-known that the sequence (f_n) converges **pointwise** towards the function f if $f_n(x) \to f(x)$, for every element x in E. When a measure is assigned, however, this definition may be relaxed slightly allowing the convergence to only happen 'almost everywhere', that is to say outside of a negligible set. This idea is formalised in the next definition.

(1.1) Definition. We shall say that (f_n) converges **almost everywhere** with respect to μ, or **μ-almost everywhere**, if there exists a measurable set A, negligible with respect to μ, that is to say such that $\mu(A) = 0$, such that $f_n(x) \to f(x)$ for every $x \in A^c$.

Beside convergence almost everywhere, the presence of the measure allows to define another type of convergence, which is called 'convergence in measure' or 'convergence in probability', when the measure is, in fact, a probability. Precisely, let us give the following definition.

(1.2) Definition. We shall say that (f_n) converges **in measure** towards f if, for every strictly positive real number ε, one has

$$\lim_n \mu\{|f_n - f| > \varepsilon\} = 0.$$

Let us immediately note that, in general, convergence in measure neither implies nor is implied by convergence almost everywhere. Nevertheless, there are some useful relations, listed by the following two propositions, between convergence in measure and convergence almost everywhere.

(1.3) Proposition. *If the measure μ is finite, and if the sequence (f_n) converges almost everywhere towards f, then (f_n) also converges towards f in measure.*

Proof. Assume that the sequence (f_n) converges almost everywhere towards f, and let us fix a strictly positive real number ε. For every index n, let us set $A_n = \{|f_n - f| > \varepsilon\}$. It is clear that the set $\limsup_n A_n$ is included in the set of all x such that the sequence $(f_n(x))$ does not converge towards $f(x)$ which, by assumption, has measure 0. Hence, $\mu(\limsup_n A_n) = 0$. Since, by virtue of Proposition I(3.9),

$$0 \leq \liminf_n \mu(A_n) \leq \limsup_n \mu(A_n) \leq \mu\big(\limsup_n A_n\big) = 0,$$

the sequence (f_n) converges in measure towards f. □

(1.4) Proposition. *If (f_n) converges in measure towards f, then there exists a subsequence of (f_n) that converges to f almost everywhere.*

Proof. Assume that the sequence (f_n) converges in measure towards f. This means that $\mu\{|f_n - f| > \varepsilon\} \to 0$ for every strictly positive real number ε. We shall use this fact to construct a strictly increasing sequence (n_k) of positive integers. This is done inductively. Precisely, for $k = 1$, let us choose a positive integer n_1 such that

$$\mu\{|f_{n_1} - f| > 1\} \leq 2^{-1}$$

and, then, assuming that the numbers $n_1, \ldots, n_{k-1}$ have been defined, let us choose $n_k > n_{k-1}$ such that

$$\mu\{|f_{n_k} - f| > k^{-1}\} \leq 2^{-k}.$$

Let us set $A_k = \{|f_{n_k} - f| > k^{-1}\}$, and observe that if x does not lie in the set $\limsup_k A_k$, then it must be $|f_{n_r}(x) - f(x)| < r^{-1}$ for every $r \geq k$, hence the sequence (f_{n_r}) converges towards f outside of $\limsup_k A_k$. Thus, to complete the proof, it suffices to show that this set is negligible. Indeed, for every index k, one has

$$\mu\left(\bigcup_{r\geq k} A_r\right) \leq \sum_{r\geq k} \mu(A_r) \leq \sum_{r\geq k} 2^{-r} = 2^{1-k}$$

thus, $\mu(\limsup_k A_k) \leq \lim_k 2^{1-k} = 0$, and the proof is complete. □

Besides pointwise convergence, there exists another well-known mode of convergence, called *uniform convergence.* Precisely, we shall say that the sequence (f_n) converges **uniformly** towards f if

$$\sup_{x\in E} |f_n(x) - f(x)| \to 0.$$

It is not difficult to recognise that, if (f_n) converges uniformly towards f, then it also converges pointwise towards f.

The next result shows how distant the notions of convergence almost everywhere and uniform convergence are, and it is customarily called Egorov's theorem, after Dmitri Egorov (Дмитрий Фёдорович Егоров, 1869–1931), a Soviet mathematician. However, it would be more accurate to refer to it as Severini-Egorov's theorem, to acknowledge the fact that the Italian mathematician Carlo Severini (1872–1951) and Egorov published independent proofs of this result in 1910 and 1911, respectively.

(1.5) Theorem (Severini-Egorov). *If the measure μ is finite, and (f_n) converges towards f almost everywhere, then for each strictly positive number ε, there exists a measurable subset A of E such that $\mu(A^c) < \varepsilon$ and the sequence (f_n) converges to f uniformly on A.*

Proof. Let ε be a strictly positive real number, and for each index n, let us set $g_n = \sup_{k \geq n} |f_k - f|$. It is easy to check that the functions g_n are all finite almost everywhere. Furthermore, the sequence (g_n) converges towards 0 almost everywhere, hence in measure (thanks to Proposition (1.3)). Thus, for each index k, we can choose a positive integer n_k such that

$$\mu\{g_{n_k} > k^{-1}\} < 2^{-k}\varepsilon.$$

Let us set $A_k = \{g_{n_k} \leq k^{-1}\}$ and $A = \bigcap_k A_k$. The set A satisfies

$$\mu(A^c) = \mu(\bigcup_k A_k^c) \leq \sum_k \mu(A_k^c) < \sum_k 2^{-k}\varepsilon = \varepsilon.$$

If δ is a strictly positive real number and if k is a strictly positive integer such that $k^{-1} < \delta$ then, since $A \subseteq A_k$, one has

$$|f_n(x) - f(x)| \leq g_{n_k}(x) \leq k^{-1} < \delta$$

holds for all x in A and all strictly positive integers n such that $n \geq n_k$; thus (f_n) converges to f uniformly on A. □

In Chapter VI, we have also introduced another type of convergence. Precisely, a sequence (f_n) of elements of $\mathcal{L}^1$ converges in mean (or in $\mathcal{L}^1$) towards a function f also in $\mathcal{L}^1$, if $\int |f_n - f| \, \mathrm{d}\mu \to 0$. The relationship between convergence in mean and convergence in measure is shown in the following.

(1.6) Proposition. *If a sequence (f_n) of integrable functions converges in mean towards f, then (f_n) also converges towards f in measure.*

Proof. The proof is a direct consequence of Markov's inequality (see II(2.4)). Indeed, if ε is a strictly positive real number, one has

$$\mu\{|f_n - f| > \varepsilon\} \leq \varepsilon^{-1} \int |f_n - f| \, \mathrm{d}\mu. \qquad \square$$

Convergence in mean, however, does not imply convergence almost everywhere. On the other hand, we already know from Fischer-Riesz theorem that if (f_n) converges in mean, there exists a subsequence of (f_n) which converges almost everywhere towards f. This is also implied by the fact that, if (f_n) converges in mean, then it converges in measure, and the same conclusion now follows from (1.4). Also, observe that neither convergence almost everywhere nor convergence in measure imply convergence in mean. To see this, consider the following example.

(1.7) Example. On the measure space $(\mathbb{R}, \mathcal{B}(\mathbb{R}), \lambda)$, where λ denotes the Borel-Lebesgue measure, let us define the sequence (f_n) by setting $f_n(x) = n$ on the interval $[0, 1/n]$ and 0 elsewhere. In other words, set

$$f_n = nI_{[0,1/n]}.$$

Then, (f_n) converges towards 0 almost everywhere, and in measure, but not in mean. For this, it suffices to observe that

$$\int |f_n| \, \mathrm{d}\mu = n \cdot \int_0^{1/n} \mathrm{d}\lambda = 1.$$

There are, however, supplementary hypotheses under which convergence almost everywhere or convergence in measure does indeed imply convergence in mean. The following proposition gives a simple example of such hypotheses, while in the next section we shall introduce a weaker condition that also entails convergence in mean. This implication shall be proved in §3, and in fact it is the content of Vitali's theorem.

(1.8) Proposition. *Let (f_n) be a sequence of integrable functions, and f an integrable function. If the sequence (f_n) converges towards f almost everywhere or in measure, and if there is a positive integrable function g such that, for every index n,*

$$|f_n| \leq g, \quad \text{and} \quad |f| \leq g$$

hold almost everywhere, then the sequence (f_n) converges to f in mean.

Proof. Let us start by assuming that the sequence (f_n) converges almost everywhere towards f, and hence the sequence $(|f_n - f|)$ converges almost everywhere towards 0. Then the hypothesis implies that

$$|f_n - f| \leq |f_n| + |f| \leq 2g$$

holds almost everywhere. Thus, we can use Lebesgue's theorem of dominated convergence to conclude that $\int |f_n - f| \, \mathrm{d}\mu \to 0$.

Let us now assume that (f_n) converges towards f in measure. Then, by means of Proposition (1.4), every subsequence of (f_n) has a subsequence that converges almost everywhere towards f, and so, from what we have just proved, in mean. Thus, if, by contradiction, the original sequence (f_n) did not converge towards f, there would exist a strictly positive real number ε and a subsequence (f_{n_r}) of (f_n) such that $\int |f_{n_r} - f| \, \mathrm{d}\mu \geq \varepsilon$ holds for every index r. Since this subsequence could have no subsequence that converges towards f in mean, this gives a contradiction. Thus, the sequence (f_n) must converge towards f in mean. □

More generally, if $p \in [1, \infty]$, we have shown in Chapter VI that $\mathcal{L}^p$ is a semi-normed space. A sequence (f_n) of elements of $\mathcal{L}^p$ is said to converge **strongly** to an element f of $\mathcal{L}^p$, if it converges in the topology induced by this semi-norm, that is to say if $\|f_n - f\|_p \to 0$, or, what is the same, if

$$\int |f_n - f|^p \, \mathrm{d}\mu \to 0.$$

In particular, the convergence in mean coincides with the strong convergence in $\mathcal{L}^1$ induced by the corresponding semi-norm.

On the other hand, when $p \in (1, \infty)$, and if q denotes the conjugate exponent of p, that is to say if $1/p + 1/q = 1$, we have also established that the bilinear form $(f, g) \mapsto \int fg \, \mathrm{d}\mu$ induces a *duality* between the space $\mathcal{L}^p$ and the space $\mathcal{L}^q$. By virtue of this duality, on the space $\mathcal{L}^p$ one can also consider the corresponding weak topology, $\sigma(\mathcal{L}^p, \mathcal{L}^q)$, which, in turn, induces a further mode of convergence. Precisely, let us introduce the following terminology.

(1.9) Definition. The **weak topology** $\sigma(\mathcal{L}^p, \mathcal{L}^q)$ is the smallest topology that makes continuous every functional of the form

$$f \mapsto \int fg \, d\mu \qquad \text{with } g \in \mathcal{L}^q.$$

We shall also say that a sequence (f_n) of elements of $\mathcal{L}^p$ converges **weakly** towards an element f of $\mathcal{L}^p$, if (f_n) converges towards f in the weak topology $\sigma(\mathcal{L}^p, \mathcal{L}^q)$, that is to say, if

$$\int f_n g \, d\mu \to \int fg \, d\mu \qquad \text{for every element } g \text{ in } \mathcal{L}^q.$$

Whilst the topology $\sigma(\mathcal{L}^p, \mathcal{L}^q)$ is rather simple, because the spaces $\mathcal{L}^p$ are all *reflexive* for $p \in (1, \infty)$, the topology $\sigma(\mathcal{L}^1, \mathcal{L}^\infty)$ possesses some more subtle properties that we are going discuss in more details in §7, when we will have developed all the necessary instruments. Nevertheless, in the meantime, we wish at least to show that the strong convergence in $\mathcal{L}^p$ implies the weak convergence in the topology $\sigma(\mathcal{L}^p, \mathcal{L}^q)$.

(1.10) Proposition. *Let $p \in [1, \infty]$, (f_n) a sequence of elements of $\mathcal{L}^p$, f an element of $\mathcal{L}^p$. If the sequence (f_n) converges strongly towards f, then the sequence (f_n) converges weakly towards f.*

Proof. Assume that (f_n) converges strongly towards f in $\mathcal{L}^p$, and let g be an element of $\mathcal{L}^q$. Then, by means of Hölder's inequality (see VI(1.2)),

$$\int (f_n - f) g \, d\mu \leq \|f_n - f\|_p \|g\|_q \to 0$$

which implies that $\int f_n g \, d\mu \to \int fg \, d\mu$. □

The following example shows the anti-climatic statement that the weak convergence does not imply the strong convergence. Hence why the terms 'strong' and 'weak' are used to describe these two types of convergence.

(1.11) Example. On the measure space $(\mathbb{R}, \mathcal{B}(\mathbb{R}), \lambda)$, where λ denotes the Borel-Lebesgue measure, let φ be a positive real function in $\mathcal{L}^p$. Assume that φ is not the constant 0, and that its support is included in a compact interval of the form $[-m, m]$, for some strictly positive real number m. For every index n, let us set $f_n(x) = \varphi(x - n)$. It is not difficult to recognise that the sequence (f_n) converges weakly towards 0. Indeed, for every $g \in \mathcal{L}^q$,

$$\int f_n(x) g(x) \, dx = \int \varphi(x - n) g(x) \, dx = \int \varphi(y) g(y + n) \, dy \to 0,$$

but it does not converge strongly towards zero, because, for every positive integer n, one has $\|f_n\|_p = \|\varphi\|_p \neq 0$.

2. Uniform integrability

Historically, the notion of uniform integrability was introduced by the Italian mathematician Giuseppe Vitali (1875–1932), at the beginning of the 20th century, as a relaxation of that condition of domination expressed in Lebesgue's theorem on dominated convergence. Subsequently, uniform integrability has been heavily utilised with success in probability theory, eventually earning its space amid the fundamental tools of probability.

The present section is devoted to present the definition of uniform integrability in the general setting of a (not necessarily finite) measure space, introduced by the American mathematician Gilbert A. Hunt (1916–2008), which is slightly stronger than the usual definition of uniform integrability, but reduces to it when the underlying measure is finite.

To this end, in all the sequel we shall assume that $(E, \mathcal{E}, \mu)$ is a given measure space. All functions, whether it is explicitly mentioned or not, are always assumed to be measurable on the measurable space $(E, \mathcal{E})$, and with the phrase 'integrable function' we shall mean, unless otherwise specified, an integrable function with respect to the measure μ.

Let us start by proving the following simple proposition.

(2.1) Proposition. *The following conditions are equivalent:*

(a) *f is integrable.*

(b) *For every strictly positive real number ε, there exists a strictly positive integrable function g such that $\int_{\{|f|>g\}} |f| \, \mathrm{d}\mu < \varepsilon$.*

Proof. (a) $\Rightarrow$ (b): Let us assume that the function f is integrable. In other words, assume that the measure $\nu = |f| \cdot \mu$ is finite. Let us choose an increasing sequence of integrable functions, (g_n), such that $g_n \uparrow |f|$, and observe that $\{|f| > g_n\} \downarrow \emptyset$; hence, by means of ν's continuity on decreasing sequences of sets (see I(3.7)), one has $\nu\{|f| > g_n\} \downarrow 0$. As a result of this, if ε is an assigned strictly positive real number, for n sufficiently large, one has

$$\int_{\{|f|>g_n\}} |f| \, \mathrm{d}\mu = \nu\{|f| > g_n\} < \varepsilon,$$

which proves condition (b).

(b) $\Rightarrow$ (a): Assume now that condition (b) holds. Then, choosing $\varepsilon = 1$, there exists a positive integrable function g such that $\int_{\{|f|>g\}} |f| \, \mathrm{d}\mu \leq 1$, whence

$$\int |f| \, \mathrm{d}\mu = \int_{\{|f|\leq g\}} |f| \, \mathrm{d}\mu + \int_{\{|f|>g\}} |f| \, \mathrm{d}\mu \leq \int g \, \mathrm{d}\mu + 1.$$

This shows that f is integrable. □

The previous result suggests the introduction of the following definition.

(2.2) Definition. Let $\mathcal{H}$ be an arbitrary family of integrable functions. We shall say that $\mathcal{H}$ is **uniformly integrable**, or, equivalently, that $\mathcal{H}$ is a set of uniformly integrable functions, if the condition (b) of the previous proposition holds *uniformly* for every element f of $\mathcal{H}$, that is to say, for every strictly positive real number ε, there exists a positive integrable function g such that $\sup_{f\in\mathcal{H}} \int_{\{|f|>g\}} |f|\,\mathrm{d}\mu < \varepsilon$. More concisely, we can also say that $\mathcal{H}$ is uniformly integrable if

$$\inf_{g\in\mathcal{L}^1_+(\mu)} \sup_{f\in\mathcal{H}} \int_{\{|f|>g\}} |f|\,\mathrm{d}\mu = 0.$$

Furthermore, if $(f_i)_{i\in I}$ is an arbitrary family of integrable functions, we shall say that it is uniformly integrable if the set of its terms is, or, what is the same, if

$$\inf_{g\in\mathcal{L}^1_+(\mu)} \sup_{i\in I} \int_{\{|f_i|>g\}} |f_i|\,\mathrm{d}\mu = 0,$$

In the latter case, we may also adopt a more informal language and allow ourselves to say that 'the f_i are uniformly integrable'.

Albeit completely obvious, it is yet perhaps worthwhile to mention that the two identical formulations of uniform integrability, the former non-parametrised, and the latter parametrised, are completely equivalent, the first being always reducible to the second by considering the family $(f)_{f\in\mathcal{H}}$, and the second to the first by taking the set $\{f_i : i \in I\}$. The choice of one or the other is, in large part, a matter of personal preference.

The following proposition shows that, when a family of functions is uniformly integrable, the functions belonging to the family are mostly concentrated onto a set with finite measure. Precisely, the following holds.

(2.3) Proposition. *Let $\mathcal{H}$ be an arbitrary family of uniformly integrable functions. Then, for every strictly positive real number ε, there exists a measurable set K, with $\mu(K) < \infty$, such that $\int_{K^c} |f|\,\mathrm{d}\mu < \varepsilon$.*

Proof. Since the family $\mathcal{H}$ is uniformly integrable, if ε is a fixed strictly positive real number, there exists a positive integrable function g such that

$$\int_{\{|f|>g\}} |f|\,\mathrm{d}\mu < \tfrac{1}{2}\varepsilon \qquad \text{for every } f \text{ in } \mathcal{H}.$$

Furthermore, since the measure $g\cdot\mu$ is, by assumption, finite, there exists a positive real number η such that $\int_{\{g\le\eta\}} g\,\mathrm{d}\mu < \frac{1}{2}\varepsilon$.

Thanks to Markov's inequality,

$$\mu\{g>\eta\} \le \eta^{-1} \int g\,\mathrm{d}\mu,$$

whence the measure of $\{g > \eta\}$ is finite. Furthermore, if f is any function in $\mathcal{H}$, one also has

$$\begin{aligned}\textstyle\int_{\{g\le\eta\}} |f|\,\mathrm{d}\mu &= \textstyle\int_{\{g\le\eta,\,g<|f|\}} |f|\,\mathrm{d}\mu + \int_{\{g\le\eta,\,g\ge|f|\}} |f|\,\mathrm{d}\mu \\ &\le \textstyle\int_{\{g<|f|\}} |f|\,\mathrm{d}\mu + \int_{\{g\le\eta\}} g\,\mathrm{d}\mu \\ &\le \tfrac{1}{2}\varepsilon + \tfrac{1}{2}\varepsilon \\ &= \varepsilon.\end{aligned}$$

Thus, the set $K = \{g > \eta\}$ satisfies the conditions required. □

(2.4) Corollary. *In the same hypotheses as in the previous proposition, there exists a measurable set E_0 such that the trace to E_0 of the measure μ is σ-finite, and such that all functions in $\mathcal{H}$ vanish almost everywhere with respect to μ outside E_0.*

Proof. By means of the previous proposition, for every positive integer n, there exists a measurable set K_n with $\mu(K_n) < \infty$, such that

$$\textstyle\int_{K_n^c} |f|\,\mathrm{d}\mu < \frac{1}{n} \qquad \text{for every function } f \text{ in } \mathcal{H}.$$

In addition, with no loss of generality, we may always assume that the sequence (K_n) is increasing, and let us set $E_0 = \bigcup_{n=1}^{\infty} K_n$. To complete the proof, it suffices to show that every f in $\mathcal{H}$ vanishes on E_0^c. To this end, observe that $E_0^c = \bigcap_{n=1}^{\infty} K_n^c$, and the conclusion follows from the fact that the equality $\int_{E_0^c} |f|\,\mathrm{d}\mu = 0$ holds for every function f in $\mathcal{H}$. □

The following result offers an equivalent condition to uniform integrability.

(2.5) Theorem. *A family $\mathcal{H}$ of integrable functions is uniformly integrable, if and only if the following two conditions hold:*

(a) *The family $\mathcal{H}$ is uniformly bounded in $\mathcal{L}^1(\mu)$, that is to say*

$$\textstyle\sup_{f\in\mathcal{H}} \int |f|\,\mathrm{d}\mu < \infty.$$

(b) *There exists an integrable function h such that, for every strictly positive real number ε, there exists a strictly positive real number δ such that, for every measurable set A,*

$$\textstyle\int_A h\,\mathrm{d}\mu < \delta \Rightarrow \sup_{f\in\mathcal{H}} \int_A |f|\,\mathrm{d}\mu < \varepsilon.$$

Proof. Let us start by proving that the conditions are *necessary*. To this end, let us suppose that $\mathcal{H}$ is uniformly integrable. Then, taking $\varepsilon = 1$, it follows from the definition that there exists a positive integrable function g such that $\sup_{f\in\mathcal{H}} \int_{\{|f|>g\}} |f|\,\mathrm{d}\mu < 1$, hence, for every function f in $\mathcal{H}$,

$$\textstyle\int |f|\,\mathrm{d}\mu = \int_{\{|f|\le g\}} |f|\,\mathrm{d}\mu + \int_{\{|f|>g\}} |f|\,\mathrm{d}\mu < \int g\,\mathrm{d}\mu + 1.$$

This shows that the family $\mathcal{H}$ is uniformly bounded in $\mathcal{L}^1(\mu)$, i.e. that condition (a) holds. In order to prove that condition (b) holds as well, let us fix a strictly positive real number ε, and let us observe that, thanks to the fact that $\mathcal{H}$ is uniformly integrable, there exists a positive integrable function g, such that

$$\sup_{f\in\mathcal{H}} \int_{\{|f|>g\}} |f| \, \mathrm{d}\mu < \tfrac{1}{2}\varepsilon.$$

Thus, for every function f in $\mathcal{H}$, and every measurable set A,

$$\int_A |f| \, \mathrm{d}\mu = \int_{A\cap\{|f|\leq g\}} |f| \, \mathrm{d}\mu + \int_{A\cap\{|f|>g\}} |f| \, \mathrm{d}\mu < \int_A g \, \mathrm{d}\mu + \tfrac{1}{2}\varepsilon.$$

It follows that, for every measurable set A such that $\int_A g \, \mathrm{d}\mu < \frac{1}{2}\varepsilon$, one has

$$\sup_{f\in\mathcal{H}} \int_A |f| \, \mathrm{d}\mu < \int_A g \, \mathrm{d}\mu + \tfrac{1}{2}\varepsilon < \varepsilon.$$

This shows that condition (b) is also satisfied.

Let us now prove that the two given conditions are *sufficient*. In other words, let us suppose that conditions (a) and (b) are satisfied. Let us fix a strictly positive real number ε, let h be a positive integrable function, and δ a strictly positive real number that satisfy condition (b). For any function f in $\mathcal{H}$ and any positive real number α, consider the obvious inequality

$$\int_{\{|f|>\alpha h\}} h \, \mathrm{d}\mu \leq \alpha^{-1} \int |f| \, \mathrm{d}\mu.$$

Thanks to (a), the integrals $\int |f| \, \mathrm{d}\mu$ are all bounded as f ranges over $\mathcal{H}$, hence there exists a sufficiently large positive real number α such that

$$\int_{\{|f|>\alpha h\}} h \, \mathrm{d}\mu < \delta \qquad \text{for every } f \text{ in } \mathcal{H}.$$

Taking $g = \alpha h$, and applying (b), this gives $\int_{\{|f|>g\}} |f| \, \mathrm{d}\mu < \varepsilon$ for every function f in $\mathcal{H}$, hence $\mathcal{H}$ is uniformly integrable. □

(2.6) Corollary. *If μ is a finite measure, a family $\mathcal{H}$ of integrable functions is uniformly integrable if and only if the following two conditions hold:*

(a) *The family $\mathcal{H}$ is uniformly bounded in $\mathcal{L}^1(\mu)$, that is to say*

$$\sup_{f\in\mathcal{H}} \int |f| \, \mathrm{d}\mu < \infty.$$

(b) *For every strictly positive real number ε, there exists a strictly positive real number δ such that, for every measurable set A,*

$$\mu(A) < \delta \Rightarrow \sup_{f\in\mathcal{H}} \int_A |f| \, \mathrm{d}\mu < \varepsilon.$$

Proof. It suffices to take $h = 1$ in Theorem (2.5). □

(2.7) Corollary. *Let us suppose that $\mathcal{H}$ is a set of uniformly integrable functions. Then, such is the convex hull of $\mathcal{H}$, that is to say the set* co$(\mathcal{H})$ *constituted by all the finite linear combinations of the form*

$$\sum_{k=1}^{n} a_k f_k, \qquad \textit{with} \quad f_k \in \mathcal{H}, \quad a_k \geq 0, \quad \sum_{k=1}^{n} a_k = 1.$$

Proof. Thanks to Theorem (2.5), the hypothesis that $\mathcal{H}$ be uniformly integrable means that one has $\sup_{f \in \mathcal{H}} \int |f| \, \mathrm{d}\mu < \infty$ and that there exists an integrable function h such that, for every strictly positive real number ε, there exists a positive real number δ such that, for every measurable set A, with $\int_A h \, \mathrm{d}\mu \leq \delta$, one has $\sup_{f \in \mathcal{H}} \int_A |f| \, \mathrm{d}\mu < \varepsilon$. It is not difficult to recognise that these conditions are also verified with the set co$(\mathcal{H})$ in lieu of $\mathcal{H}$. □

When the underlying measure is finite, the definition of uniform integrability can be further simplified, as shown in the next proposition.

(2.8) Proposition. *If the measure μ is finite, a family $\mathcal{H}$ of integrable functions is uniformly integrable if and only if*

$$\inf_{t \geq 0} \sup_{i \in I} \int_{\{|f| > t\}} |f| \, \mathrm{d}\mu = 0.$$

Proof. ($\Leftarrow$): This implication is completely obvious, for, if the measure μ is finite, constant functions are integrable.

($\Rightarrow$): Suppose that the family $\mathcal{H}$ is uniformly integrable, fix a strictly positive real number ε, and let g be a positive integrable function such that

$$\int_{\{|f| > g\}} |f| \, \mathrm{d}\mu < \tfrac{1}{2}\varepsilon \qquad \text{for every } f \text{ in } \mathcal{H}.$$

Since the measure $g \cdot \mu$ is, by assumption, finite, the set $\{g = \infty\}$ is negligible with respect to the measure μ, whence to the measure $g \cdot \mu$. Thus, by continuity of measures under decreasing sequences, it is possible to find a real number t such that $\int_{\{g > t\}} g \, \mathrm{d}\mu < \frac{1}{2}\varepsilon$. As a result, for every function f in $\mathcal{H}$,

$$\begin{aligned}
\int_{\{|f| > t\}} |f| \, \mathrm{d}\mu &= \int_{\{|f| > t, \, |f| > g\}} |f| \, \mathrm{d}\mu + \int_{\{|f| > t, \, |f| \leq g\}} |f| \, \mathrm{d}\mu \\
&\leq \int_{\{|f| > g\}} |f| \, \mathrm{d}\mu + \int_{\{g > t\}} g \, \mathrm{d}\mu \\
&= \tfrac{1}{2}\varepsilon + \tfrac{1}{2}\varepsilon \\
&= \varepsilon.
\end{aligned}$$

This completes the proof. □

A very elegant *sufficient* condition for uniform integrability is given by the following criterion, due to the Belgian mathematician Charles-Jean Étienne Gustave Nicolas, baron de la Vallée Poussin (1866–1962).

(2.9) Theorem (de la Vallée Poussin). *Let $\mathcal{H}$ be a family of integrable functions, and suppose that there exists a strictly positive real number δ such that the family $\mathcal{H}$ is uniformly bounded in $\mathcal{L}^{1+\delta}(\mu)$, that is to say,*

$$\sup_{f\in\mathcal{H}} \int |f|^{1+\delta}\,\mathrm{d}\mu < \infty. \tag{2.10}$$

Then, the family $\mathcal{H}$ is uniformly integrable.

Proof. Let ε be a fixed strictly positive real number, denote by c the supremum of all numbers of the form $\int |f|^{1+\delta}\,\mathrm{d}\mu$, as f varies over $\mathcal{H}$, and denote by h any positive integrable function. Let us observe that, for every positive integer n, one has $n\mu\{h > n\} = \int_{\{h>n\}} n\,\mathrm{d}\mu \le \int_{\{h>n\}} h\,\mathrm{d}\mu$. Since the function h is assumed to be integrable, the right-hand side of the previous inequality tends towards 0, as n approaches infinity, whence it is possible to find a positive integer n_0 such that

$$n\mu\{h > n\} < \tfrac{1}{4}\varepsilon \qquad \text{whenever } n \ge n_0.$$

Let n be a positive integer greater than n_0, let k be any positive integer number, consider the function $g = kI_{\{h\le n\}}$ which is integrable, because the measurable set $\{h \le n\}$ has finite measure under μ. Thus,

$$\begin{aligned}\int_{\{|f|>g\}} |f| \wedge n\,\mathrm{d}\mu &= \int_{\{|f|>k,\, h\le n\}} |f| \wedge n\,\mathrm{d}\mu + \int_{\{|f|>0,\, h>n\}} |f| \wedge n\,\mathrm{d}\mu\\ &\le k^{-\delta} \int_{\{|f|>k,\, h\le n\}} |f|^{1+\delta}\,\mathrm{d}\mu + n\mu\{h > n\}\\ &< k^{-\delta}c + \tfrac{1}{4}\varepsilon.\end{aligned}$$

As a result, for a sufficiently large value of k, the rightmost side of the previous chain of inequalities becomes strictly smaller than $\frac{1}{2}\varepsilon$, leading to the inequality $\int_{\{|f|>g\}} |f| \wedge n\,\mathrm{d}\mu < \frac{1}{2}\varepsilon$, for every $n \ge n_0$. Taking the limit as n approaches infinity, and applying Beppo Levi's monotone convergence theorem, this finally gives

$$\int_{\{|f|>g\}} |f|\,\mathrm{d}\mu = \lim_n \int_{\{|f|>g\}} |f| \wedge n\,\mathrm{d}\mu \le \tfrac{1}{2}\varepsilon < \varepsilon. \qquad \square$$

As we have already mentioned, uniform integrability was first introduced as a relaxation of the notion of dominance, for a family of sets. In line with this idea, the next proposition shows that a *sufficient* condition for uniform integrability is indeed the one of *domination*, which appears in Lebesgue's theorem on dominated convergence. We shall also see that this sufficient condition is in fact not necessary for uniform integrability.

(2.11) Proposition. *Let us suppose that the family $(f_i)_{i\in I}$ of measurable functions is dominated in $\mathcal{L}^1(\mu)$, that is to say, assume that there exists a function g in $\mathcal{L}^1(\mu)$ that dominates the family, which means that satisfies the relation $|f_i| \le g$ for every index i.*

Then, $(f_i)_{i\in I}$ is uniformly integrable.

Proof. Let us take a strictly positive real number ε. Then, thanks to (2.1), there exists a positive integrable function h such that $\int_{\{g>h\}} g \, \mathrm{d}\mu < \varepsilon$. Thus, for every index i, one has

$$\int_{\{|f_i|>h\}} |f_i| \, \mathrm{d}\mu \leq \int_{\{|f_i|>h\}} g \, \mathrm{d}\mu \leq \int_{\{g>h\}} g \, \mathrm{d}\mu < \varepsilon,$$

and this proves that the f_i are uniformly integrable. □

(2.12) Corollary. *Every finite family $(f_i)_{i\in I}$ of integrable functions is uniformly integrable.*

Proof. Indeed, the measurable function $g = \sup_{i\in I} |f_i|$ is integrable (as it is the supremum of a finite family of integrable functions), and evidently dominates the given family. Thus, the latter is uniformly integrable by virtue of the previous proposition. □

(2.13) Example. On the measurable space $(\mathbb{N}, \mathcal{P}(\mathbb{N}))$, let us consider the measure μ defined, for every positive integer n, by $\mu\{n\} = 2^{-n}$, and let us consider the family (f_n) defined by $f_n = 2^n n^{-1} I_{\{n\}}$. It is not difficult to see that this family is uniformly integrable. Indeed, since the constant 1 belongs to $\mathcal{L}^1(\mu)$, for $\int 1 \, \mathrm{d}\mu = \sum_n 2^{-n} = 1$, it suffices to observe that

$$\int_{\{f_n>1\}} f_n \, \mathrm{d}\mu = n^{-1} \qquad \text{for every integer } n \geq 1.$$

On the other hand, the smallest function that is an upper bound for (f_n) is the function g defined by $g(n) = 2^n n^{-1}$ which is not in $\mathcal{L}^1(\mu)$, for

$$\int g \, \mathrm{d}\mu = \sum_n n^{-1} = \infty.$$

3. Vitali's theorem

This section is devoted to state and prove a fundamental theorem, due to Giuseppe Vitali, which extends and clarifies, through the concept of uniform integrability, Lebesgue's theorem on dominated convergence. To this end, let us start by proving the following lemma, which is only a first, modest, extension of a special case of Lebesgue's theorem, consisting in replacing the hypothesis of pointwise (or almost everywhere) convergence with the, much weaker one, of convergence in measure.

(3.1) Lemma. *On a finite measure space $(E, \mathcal{E}, \mu)$, let (f_n) be a sequence of measurable functions, which converges in measure towards the constant 0, and which is dominated in $\mathcal{L}^1(\mu)$.*

Then, as n approaches infinity, one has $\int |f_n| \, \mathrm{d}\mu \to 0$.

Proof. It suffices to prove that the (bounded) real sequence $(\int |f_n| \, d\mu)$ does not have any cluster point different from zero. Let us denote by c one of these cluster points: This means that there exists a strictly increasing sequence of positive integers (m_k) such that, as k approaches infinity, one has $\int |f_{m_k}| \, d\mu \to c$. Thanks to (1.4), we shall suppose, with no loss of generality, that the subsequence (f_{m_k}) converges towards 0 almost everywhere. Hence, applying Lebesgue's theorem on dominated convergence gives $c = 0$, as required. □

In order to prove Vitali's theorem, we are going to need the following proposition as well, which ought not to be deprived of its own interest. Indeed, this special criterion of uniform integrability is distinct from the others that we have introduced thus far (and that concern arbitrary families of measurable functions), from the fact that it is only confined to the special case of *sequences* of measurable functions (to which one also further impose the preliminary condition of being *integrable*).

(3.2) Proposition. *Let (f_n) be a sequence of integrable functions on a measure space $(E, \mathcal{E}, \mu)$. Then, the following conditions are equivalent:*

(a) *The sequence (f_n) is uniformly integrable.*

(b) *One has $\inf_{g \in \mathcal{L}^1_+(\mu)} \limsup_n \int_{\{|f_n|>g\}} |f_n| \, d\mu = 0$.*

Proof. (a) ⇒ (b): Condition (a) is equivalent to the relation

$$\inf_{g \in \mathcal{L}^1(\mu)} \sup_n \int_{\{|f_n|>g\}} |f_n| \, d\mu = 0,$$

which implies condition (b) thanks to the obvious inequality

$$\limsup_n \int_{\{|f_n|>g\}} |f_n| \, d\mu \leq \sup_n \int_{\{|f_n|>g\}} |f_n| \, d\mu.$$

(b) ⇒ (a): Assume that condition (b) holds. Then, given any strictly positive real number ε, there exists a positive integrable function g_0, such that

$$\limsup_n \int_{\{|f_n|>g_0\}} |f_n| \, d\mu < \varepsilon.$$

Correspondingly, there exists an integer $m \geq 1$ such that

$$\int_{\{|f_n|>g_0\}} |f_n| \, d\mu \leq \varepsilon \qquad \text{for } n > m.$$

On the other hand, since the m-tuple $(f_1, \ldots, f_m)$ is uniformly integrable (see (2.12)), there exists a positive integrable function g_1 such that

$$\int_{\{|f_n|>g_1\}} |f_n| \, d\mu \leq \varepsilon \qquad \text{for } n \leq m.$$

Thus, taking $g = g_1 \vee g_2$, the supremum of g_1 and g_2, we can see that the relation

$$\int_{\{|f_n|>g\}} |f_n| \, d\mu \leq \varepsilon$$

occurs for every integer n, and this proves that the sequence (f_n) is uniformly integrable. □

We are now finally ready to state and prove Vitali's theorem.

(3.3) Theorem (Vitali). *On a measure space $(E, \mathcal{E}, \mu)$, let (f_n) be a sequence of integrable functions and let f be a measurable function. Then, the following conditions are equivalent:*

(a) *f is integrable, and the sequence (f_n) converges towards f in $\mathcal{L}^1(\mu)$.*

(b) *The sequence (f_n) is uniformly integrable and converges in measure to f.*

Proof. (a) $\Rightarrow$ (b): Assume that condition (a) holds. Then, let us first of all observe that the sequence (f_n) converges in measure towards f. In fact, thanks to Markov's inequality, for every strictly positive real number ε,

$$\mu\{|f_n - f| > \varepsilon\} \leq \varepsilon^{-1} \int |f_n - f| \, \mathrm{d}\mu \to 0.$$

Setting $g_n = |f_n - f|$, we can see that (g_n) is a sequence of integrable functions, for which condition (b) of the previous proposition holds: In fact, for every positive integrable function h, one has $\limsup_n \int_{\{g_n > h\}} g_n \, \mathrm{d}\mu = 0$.

Hence, the sequence (g_n) is uniformly integrable, and therefore such is the collection of the g_n and f, whence also the sequence (f_n) (see (2.7)), thanks to the inequality $\frac{1}{2}|f_n| \leq \frac{1}{2}g_n + \frac{1}{2}|f|$.

(b) $\Rightarrow$ (a): Assume that condition (b) holds. If (m_k) is a strictly increasing sequence of positive integers, such that (f_{m_k}) converges almost everywhere towards f, one has, thanks to Fatou's lemma,

$$\int |f| \, \mathrm{d}\mu \leq \liminf_k \int |f_{m_k}| \, \mathrm{d}\mu < \infty$$

(where the last inequality follows from the fact that the sequence (f_n), being uniformly integrable, is bounded in $\mathcal{L}^1(\mu)$: see condition (a) of (2.5)). It is thus proved that the function f is integrable. It remains to prove that the sequence (f_n) converges in $\mathcal{L}^1(\mu)$. Setting $g_n = |f_n - f|$, this is now equivalent to demonstrate that, as n approaches infinity, one has $\int g_n \, \mathrm{d}\mu \to 0$.

To this end, let us observe that, since the sequence (f_n) is, by hypothesis, uniformly integrable, such is the set of the f_n and f, and, thanks to the inequality $\frac{1}{2}g_n \leq \frac{1}{2}|f_n| + \frac{1}{2}|f|$, the sequence (g_n) is uniformly integrable as well (see (2.7)). By virtue of Proposition (2.4), we may well assume, with no loss of generality, that the measure μ is σ-finite, whence that there exists an increasing sequence (E_r) of measurable sets with finite measure, such that $\bigcup_r E_r = E$. Let us denote by μ^r the trace of μ onto E_r, and let us fix a positive index r. Taking any strictly positive real number ε, there exists a positive integrable function h such that $\sup_n \int_{\{g_n > h\}} g_n \, \mathrm{d}\mu^r \leq \frac{1}{2}\varepsilon$. This yields, for each n,

$$\begin{aligned} \int g_n \, \mathrm{d}\mu^r &= \int_{\{g_n \leq h\}} g_n \, \mathrm{d}\mu^r + \int_{\{g_n > h\}} g_n \, \mathrm{d}\mu^r \\ &\leq \int_{\{g_n \leq h\}} g_n \wedge h \, \mathrm{d}\mu^r + \tfrac{1}{2}\varepsilon \\ &\leq \int g_n \wedge h \, \mathrm{d}\mu^r + \tfrac{1}{2}\varepsilon. \end{aligned}$$

Furthermore, applying Lemma (3.1) to the sequence $(g_n \wedge h)$, which is dominated by the integrable function h, one can see that, as n tends towards infinity, the limit $\int g_n \wedge h \, d\mu^r \to 0$ holds. From the previous inequality, it follows therefore that, for n sufficiently large, $\int g_n \, d\mu^r \leq \varepsilon$ holds. Taking now the supremum with respect to r, and observing that the left-hand side of the previous inequality then tends towards $\int g_n \, d\mu$, the proof is complete. □

4. Set-wise topology on the space of real measures

Let us now turn our attention to the convergence of a sequence of real measures over a given measurable space. This shall be done in the 'abstract' setting, that is to say without assuming any topological structure on the underlying set, and we shall defer the treatment of convergence of measures on a topological space in the next chapter.

So, focusing on the abstract setting, let us start by assuming that a measurable space $(E, \mathcal{E})$ is assigned. Throughout this section, with the phrase 'a sequence of real measures' we shall mean a sequence of real measures all defined on this measurable space, that is to say, a sequence of elements of the Riesz space $\mathfrak{M}(E, \mathcal{E})$.

(4.1) Definition. Assume that the Riesz space $\mathfrak{M}(E, \mathcal{E})$ is equipped with the **set-wise topology**, which means the least fine of all topologies on this space which make the linear functional $\mu \mapsto \mu(A)$ continuous, for every set A in $\mathcal{E}$. If (μ_n) is a sequence of real measures on $(E, \mathcal{E})$, and μ is a real measure on $(E, \mathcal{E})$, we shall say that (μ_n) converges **set-wise** towards μ if it converges with respect to this topology.

Let us observe immediately that $\mathfrak{M}(E, \mathcal{E})$ with this topology is a separated space, thus the uniqueness of limit holds. Furthermore, it is easy to check that the topology of set-wise convergence is compatible with the vector structure of $\mathfrak{M}(E, \mathcal{E})$, making this space a *topological vector space*. In particular, $\mathfrak{M}(E, \mathcal{E})$ is *regular*, as one can easily see by noting that, for any element μ_0 of $\mathfrak{M}(E, \mathcal{E})$, a fundamental system of neighbourhoods for μ_0 is the finite intersection of closed set of the form

$$\left\{\mu \in \mathfrak{M}(E, \mathcal{E}) : \left|\mu(A) - \mu_0(A)\right| \leq \varepsilon\right\},$$

with A an element of $\mathcal{E}$, and ε a strictly positive real number.

Let us also observe that, if λ is an assigned positive real measure on $(E, \mathcal{E})$, the set of all real measures with base λ is a closed set in $\mathfrak{M}(E, \mathcal{E})$: in fact, this set is identical to the intersection of the sets of the form

$$\left\{\mu \in \mathfrak{M}(E, \mathcal{E}) : \mu(A) = 0\right\}$$

for every element A of $\mathcal{E}$, which is negligible with respect to λ, that is to say, such that $\lambda(A) = 0$.

(4.2) Proposition. *On a measurable space $(E, \mathcal{E})$, assume that the σ-field $\mathcal{E}$ is countably generated. Then, every non-empty compact subset of $\mathfrak{M}(E, \mathcal{E})$ is metrizable.*

Proof. Let K be a non-empty compact subspace of $\mathfrak{M}(E, \mathcal{E})$, and let us denote by $\mathcal{O}_K$ its topology, that is to say the topology induced on K by the topology of $\mathfrak{M}(E, \mathcal{E})$. Let (A_n) be a sequence of elements of $\mathcal{E}$ such that the set $\{A_n\}$ is a base for the σ-field $\mathcal{E}$ (see I(5.5)). For every index n, let us denote by c_n the maximum value taken on K by the continuous real map $\mu \mapsto |\mu(A_n)|$. Finally, let us set, for every pair μ, ν of elements of K,

$$d(\mu, \nu) = \sum_{n \geq 1} 2^{-n}(1 + c_n)^{-1}|\mu(A_n) - \nu(A_n)|.$$

This defines a metric on K. (Observe that two elements μ and ν in K such that $d(\mu, \nu) = 0$ are identical, for they coincide on each of the sets A_n which form a base for $\mathcal{E}$.) Let us denote by $\mathcal{O}_d$ the topology associated to the distance d. According to this topology, for a sequence of elements of K to converge towards an element μ of K, it is necessary and sufficient that this sequence converge towards μ on each element A_n of the base. In other words, $\mathcal{O}_d$ is the least fine of all topologies on K that makes all the functions of the form $\mu \mapsto \mu(A_n)$ continuous, for $n \geq 1$. Thus, $\mathcal{O}_d$ is less fine than $\mathcal{O}_K$. Since $\mathcal{O}_K$ is compact, and $\mathcal{O}_d$ is separated, it follows that $\mathcal{O}_K$ coincides with $\mathcal{O}_d$. This proves that the subspace K of $\mathfrak{M}(E, \mathcal{E})$ is metrizable. □

5. Nikodým's theorem

This section is devoted to proving the an important result due to O. Nikodým. To this end, we are first going to prove the theorem in the special case where the real measures are all defined on the σ-field $\mathcal{P}(\mathbb{N})$ of all subsets of $\mathbb{N}$, and then we shall extend it to the most general case. Indeed, amongst all real measures on this special σ-field, there is, in particular, the *counting measure*, which shall be denoted by κ. The corresponding Banach space $L^1(\kappa)$ shall be denoted, as it is customary, by ℓ^1. The norm of an element f of ℓ^1 is

$$\|f\| = \int |f| \, \mathrm{d}\kappa = \sum_{j \in \mathbb{N}} |f(j)|.$$

A very useful characteristic of the counting measure is that every other measure μ on $\mathcal{P}(\mathbb{N})$ can always be written in the form $f \cdot \kappa$, where f is an element of ℓ^1. Thus, $f \mapsto f \cdot \kappa$ is a linear mapping from the vector space ℓ^1 to the vector space $\mathfrak{M}(\mathbb{N}, \mathcal{P}(\mathbb{N}))$ of all real measures on $\mathcal{P}(\mathbb{N})$. This map carries the Banach structure from ℓ^1 onto $\mathfrak{M}(\mathbb{N}, \mathcal{P}(\mathbb{N}))$, which is therefore a Banach space itself. The norm of an element $\mu = f \cdot \kappa$ of $\mathfrak{M}(\mathbb{N}, \mathcal{P}(\mathbb{N}))$ is

$$\|\mu\| = \|f\| = |\mu|(\mathbb{N}).$$

In this special case, Nikodým's theorem can be stated as follows.

(5.1) Theorem. *Let (f_n) be a sequence of elements of ℓ^1, and set $\mu_n = f_n \cdot \kappa$. Then, the following conditions are equivalent:*

(a) $\|\mu_n\| \to 0$, *or, what is the same,* $\|f_n\| \to 0$.

(b) $\mu_n(A) \to 0$ *for every subset A of $\mathbb{N}$.*

To prove this theorem we shall need the following simple result.

(5.2) Proposition. *Let $(E, \mathcal{E})$ be a measurable space and μ a real measure on the σ-field $\mathcal{E}$. Then, for every measurable set A, there exists a measurable set B, included in A, such that $|\mu|(A) \leq 2|\mu(B)|$.*

Proof. Let H and K be two disjoint measurable sets, such that μ^+ is concentrated on H and μ^- is concentrated on K, and let A be any measurable set. From the relation $|\mu|(A) = \mu^+(A) + \mu^-(A)$, it follows that, in the case when $\mu^+(A) \geq \mu^-(A)$, one has

$$|\mu|(A) \leq 2\mu^+(A) = 2\mu(A \cap H) = 2|\mu(A \cap H)|,$$

whilst, in the other case, one has

$$|\mu|(A) \leq 2\mu^-(A) = 2|\mu(A \cap K)|. \qquad \square$$

We are now ready to prove the special case of Nikodým's theorem.

Proof of Theorem (5.1). The implication (a) $\Rightarrow$ (b) is obvious. Thus, it suffices to prove the implication (b) $\Rightarrow$ (a). To this end, assume that condition (b) holds, and, by contradiction, assume that condition (a) does not. Then, taking a subsequence, if necessary, and multiplying all the μ_n by the same real number, we may assume that $\|\mu_n\| > 1$ for every index n. Note that, thanks to condition (b), the sequence $(|\mu_n|)$ converges towards zero on every singleton, hence on every finite part of $\mathbb{N}$. Hence, taking any pair p and m of strictly positive integers, it is always possible to find an integer n, greater than m, such that

$$|\mu_n|\big([0, p)\big) < \tfrac{1}{8}.$$

Furthermore, since $|\mu_n|$ has finite total mass, it is possible to find an integer q, greater than p, such that one has

$$|\mu_n|\big([q, \infty)\big) < \tfrac{1}{8},$$

in such a way that one has $|\mu_n|\big([p, q)^c\big) < \frac{1}{4}$. From this simple remark, we can now construct, by induction, two strictly increasing sequences of positive integers, (n_k) and (p_k), such that, for every index k, one has

$$|\mu_{n_k}|\big([p_k, p_{k+1})^c\big) < \tfrac{1}{4}, \quad \text{hence} \quad |\mu_{n_k}|\big([p_k, p_{k+1})\big) > \tfrac{3}{4}.$$

Using Proposition (5.2), let us now choose, for every index k, a subset A_k of $[p_k, p_{k+1})$ such that one has

$$|\mu_n(A_k)| \geq \tfrac{1}{2}\mu_{n_k}\big([p_k, p_{k+1})\big).$$

Let us set $A = \bigcup_k A_k$. Then, for every index k, since $A \setminus A_k \subseteq [p_k, p_{k+1})^c$,

$$|\mu_{n_k}(A) - \mu_{n_k}(A_k)| \leq |\mu_{n_k}|\big([p_k, p_{k+1})^c\big) < \tfrac{1}{4},$$

whence

$$\begin{aligned} |\mu_{n_k}(A)| &\geq |\mu_{n_k}(A_k)| - \tfrac{1}{4} \\ &\geq \tfrac{1}{2}|\mu_{n_k}|\big([p_k, p_{k+1})\big) - \tfrac{1}{4} \\ &\geq \tfrac{3}{4} - \tfrac{1}{4} = \tfrac{1}{2}, \end{aligned}$$

which yields $\mu_{n_k}(A) \not\to 0$, which, in turn, violates (b). □

(5.3) Theorem. *Let (μ_n) be a sequence of real measures on $\mathcal{P}(\mathbb{N})$, such that, for every subset A of $\mathbb{N}$, the sequence $\big(\mu_n(A)\big)$ has a finite limit. Then the real function μ, defined on $\mathcal{P}(\mathbb{N})$ by setting*

$$\mu(A) = \lim_n \mu_n(A)$$

is a real measure.

Proof. For every index n, let us write μ_n in the form $\mu_n = f_n \cdot \kappa$, where f_n is an element of ℓ^1. Thus, it suffices to show that the sequence (f_n) is Cauchy in ℓ^1. Indeed, in this case, then, denoted by f the limit of this sequence, then (μ_n) converges towards $f \cdot \kappa$ in every subset A of $\mathbb{N}$, hence, by uniqueness of the limit, the function μ would coincide with the real measure $f \cdot \kappa$.

By contradiction, suppose that the sequence (f_n) is not Cauchy in ℓ^1. Hence, there exists a strictly increasing sequence (n_k) of positive integers, such that the numeric sequence $(\|f_{n_k} - f_{n_{k+1}}\|)$ does not converge towards 0. On the other hand, thanks to Theorem (5.1), this violates the fact that the sequence of real measures $(\mu_{n_k} - \mu_{n_{k+1}})$ converges towards 0 on any subset of $\mathbb{N}$, and this gives the contradiction sought. □

We are now ready to prove Nikodým's theorem in its full generality.

(5.4) Theorem (Nikodým). *Let $(E, \mathcal{E})$ be a measurable space, and (μ_n) a sequence of real measures, defined on $\mathcal{E}$, such that, for every element A of $\mathcal{E}$, the sequence $\big(\mu_n(A)\big)$ has a finite limit. Then, the real function μ, defined on $\mathcal{E}$ by setting*

$$\mu(A) = \lim_n \mu_n(A) \qquad \textit{for every element } A \textit{ of } \mathcal{E}$$

is a real measure.

Proof. Let (B_j) be a sequence of pairwise disjoint elements of $\mathcal{E}$, and let us set $B = \bigcup_j B_j$. It suffices to prove that the two relations

$$\sum_{j\in\mathbb{N}} |\mu(B_j)| < \infty, \qquad \sum_{j\in\mathbb{N}} \mu(B_j) = \mu(B) \tag{5.5}$$

hold. To this end, for every index n, consider the real measure ν_n defined by

$$\nu_n(A) = \mu_n\big(\textstyle\bigcup_{j\in A} B_j\big) = \sum_{j\in A} \mu_n(B_j) \qquad \text{for every } A \text{ in } \mathcal{P}(\mathbb{N}),$$

that is to say, the measure $\nu_n = f_n \cdot \kappa$, where $f_n(j) = \mu_n(B_j)$. Then, for every subset A of $\mathbb{N}$, one has $\nu_n(A) \to \mu(\bigcup_{j\in A} B_j)$. From Theorem (5.3) it follows therefore that the sequence (f_n) converges in ℓ^1 towards a certain element f of ℓ^1. Since this convergence implies that the sequence (f_n) converges pointwise towards f, one has, for every element j of $\mathbb{N}$,

$$f(j) = \lim_n f_n(j) = \mu(B_j).$$

This shows that the first of the relations in (5.5) holds, and yields

$$\begin{aligned} \big|\mu_n(B) - \textstyle\sum_{j\in\mathbb{N}} \mu(B_j)\big| &= \big|\textstyle\sum_{j\in\mathbb{N}} \mu_n(B_j) - \sum_{j\in\mathbb{N}} \mu(B_j)\big| \\ &= \textstyle\sum_{j\in\mathbb{N}} \big|\mu_n(B_j) - \mu(B_j)\big| \\ &= \textstyle\sum_{j\in\mathbb{N}} \big|f_n(j) - f(j)\big| \to 0. \end{aligned}$$

Thus, being $\mu(B) = \lim_n \mu_n(B)$, the second of the relations in (5.5) holds. This completes the proof. □

(5.6) Remark. By virtue of Nikodým's theorem, let us observe that, for a sequence (μ_n) of real measures on a measurable space $(E, \mathcal{E})$ to converge on each element of $\mathcal{E}$, it is necessary and sufficient that, for every measurable function g with values in $[0, 1]$, the real sequence $(\int g \, d\mu_n)$ converges. In fact, the thesis is evident when g is the indicator function of an element of $\mathcal{E}$, hence it follows for any simple function, hence for any positive measurable function with values in $[0, 1]$, hence for any (integrable) function with values in $[0, 1]$.

6. Vitali-Hahn-Saks theorem

This section is devoted to proving an interesting result, introduced, in different forms, and with independent proofs, by the Italian Giuseppe Vitali, the German Hans Hahn, and Polish Stanisław Saks (1897–1942), in 1907, 1922, and 1933, respectively. Vitali-Hahn-Saks theorem enhances and completes Nikodým's theorem, offering an important complement to it when measures are based on a fixed positive measure.

Before proving this theorem, let us prove the following simple result that shows how a positive real measure on a measurable space induces a pseudo-metric structure on the σ-field which, in fact, makes it complete.

(6.1) Proposition. *Given a measurable space $(E, \mathcal{E})$ and a positive real measure λ on $\mathcal{E}$, let us denote by d the function that, to each pair (A, B) of elements of $\mathcal{E}$, associates the number $d(A, B)$ defined as*

$$d(A,B) = \int |I_A - I_B| \, \mathrm{d}\lambda = \lambda\{I_A \neq I_B\}.$$

Then:

(a) *d is a pseudo-metric on $\mathcal{E}$.*

(b) *The pseudo-metric space obtained equipping E with d is complete.*

(c) *For every element f of $\mathcal{L}^1(\lambda)$, the map $A \mapsto \int_A f \, \mathrm{d}\lambda$, from $\mathcal{E}$ to $\mathbb{R}$ is continuous according to the pseudo-metric d.*

Proof. Assertion (a) is obvious. Let us now prove assertion (b). To this end, consider a sequence (A_n) of element of $\mathcal{E}$ which we assume to be Cauchy with respect to the pseudo-metric d. By definition of d, this means that the sequence (I_{A_n}) is Cauchy in $\mathcal{L}^1(\lambda)$. Since this space is complete, the latter sequence converges, in $\mathcal{L}^1(\lambda)$, towards an element f of $\mathcal{L}^1(\lambda)$. As a result, a subsequence extracted from this sequence must converge towards f almost everywhere. Thus, f coincides almost everywhere with the indicator function of an element A of $\mathcal{E}$. This element is also the limit of the sequence (A_n) according to the pseudo-metric d.

Let us finally prove assertion (c). To this end, let f be an element of $\mathcal{L}^1(\lambda)$, and (A_n) a sequence of elements of $\mathcal{E}$ which converges, according to the pseudo-metric d, towards an element A of $\mathcal{E}$. Since the sequence (fI_{A_n}) is dominated in $\mathcal{L}^1(\lambda)$, whence it is uniformly integrable, and it converges in measure towards fI_A, it also converges towards fI_A in $\mathcal{L}^1(\mu)$, by virtue of Vitali's theorem. This completes the proof. □

We shall also utilise this simple criterion to establish when an additive function is, in fact, a measure absolutely continuous with respect to another.

(6.2) Proposition. *Let $(E, \mathcal{E})$ be a measurable space, λ a positive real measure on $\mathcal{E}$, and μ an additive function defined on $\mathcal{E}$. Then, the following conditions are equivalent:*

(a) *μ is a real measure, absolutely continuous with respect to λ.*

(b) *μ converges towards zero, under the filter base constituted of all the sets of the form $\{A \in \mathcal{E} : \lambda(A) < \delta\}$, with $\delta > 0$.*

Proof. (a) $\Rightarrow$ (b): Assume that μ is absolutely continuous with respect to λ, and assume by contradiction that condition (b) does not hold. This means that, for every strictly positive real number ε and for every positive integer n, there exists an element A_n in $\mathcal{E}$, with $\lambda(A_n) < 1/n$ and $\mu(A_n) \geq \varepsilon$. By replacing (A_n) with $(A_1 \cap \cdots \cap A_n)$, we may also assume, with no loss of generality, that this sequence is decreasing, and, setting $A = \bigcap_n A_n$, one also has $\lambda(A) = 0$ and $\mu(A) \geq \varepsilon$, which contradicts the fact that μ is, by assumption, absolutely continuous with respect to λ.

(b) $\Rightarrow$ (a): Assume that condition (b) holds. This shows, in particular, that the set function μ is continuous on decreasing sequences, hence a real measure. In addition, the condition $\lambda(A) = 0$ implies that A is a cluster point for the filter base of all the sets of the form $\{A \in \mathcal{E} : \lambda(A) < \delta\}$, whence $\mu(A) = 0$, by virtue of condition (b), and the proposition is fully proved. □

The final ingredient needed to prove Vitali-Hahn-Saks theorem is a result from general topology, usually called the *Baire's category theorem*, named after the French mathematician René-Louis Baire (1874–1932), who first proved this statement in a special case, while, in its most general form, it was actually proved by the German mathematician Felix Hausdorff (1868–1942), who may be considered the father of modern topology.

(6.3) Theorem (Baire). *In a complete pseudo-metric space, the intersection of a countable family of open dense subsets of E is itself dense in E.*

Proof. Let (O_n) be a countable family of dense open sets, let U be any open set, let x_1 be a point in $O_1 \cap U$, and let B_1 be a ball (of a suitable radius r_1) centred at x_1 and contained in $O_1 \cap U$. Since O_2 is dense, there must exist a point x_2 in $O_2 \cap B_1$, and since O_2 is open, there also exists a strictly positive real number r_2 such that the open ball $B_2 = B(x_2, r_2)$ is included in O_2. With no loss of generality, we may also assume that r_2 is at the same time smaller than both $\frac{1}{2}r_1$ and $r_1 - d(x_1, x_2)$. Then, $\overline{B}_2 \subseteq B_1$. Inductively, this process guarantees the existence of a sequence (B_n) of open balls with centres (x_n) and radii (r_n), respectively, such that

$$B_n \subseteq O_n, \qquad \overline{B}_n \subseteq B_{n+1}, \qquad r_n \downarrow 0.$$

Let us fix any positive integer m. Then, for every pair n, k of integers greater than m, one always has $x_n \in B_m$, $x_k \in B_m$, whence $d(x_n, x_k) < 2r_m$, which means that (x_n) is a Cauchy sequence. Since the pseudo-metric space E is assumed to be complete, this sequence converges towards an element x of E, and since $x_n \in B_m$ for every $n > m$, one has $x_n \in \overline{B}_m \subseteq O_m$. This yields that $x \in \bigcap_m O_m$, and therefore the set $\bigcap_m O_m$ is dense in E. □

(6.4) Corollary. *In a complete pseudo-metric space, the countable union of closed sets which contain no non-empty open sets contains no non-empty open sets.*

Proof. Let (F_n) be a countable family of closed sets each of which does not contain any non-empty open set, and let us set $O_n = F_n^c$. Then (O_n) is a countable family of open dense sets, hence, by Baire's theorem, the intersection $\bigcap_n O_n$ is dense. Thus, if U is any non-empty open set, there must exist a point $x \in \bigcap_n O_n$ such that $x \in U$, whence $x \notin O^c = \bigcup_n F_n$, and this guarantees that U cannot be included in $\bigcup_n F_n$. □

With all these prerequisites, we can now focus on the goal of this section: Vitali-Hahn-Saks theorem, which we can finally state and prove.

(6.5) Theorem (Vitali-Hahn-Saks). *Let $(E, \mathcal{E})$ be a measurable space, and (μ_n) a sequence of real measures, defined on $\mathcal{E}$, such that, for every element A of $\mathcal{E}$, the sequence $\big(\mu_n(A)\big)$ has a finite limit. In addition, let us suppose that there is a normalised positive measure λ on $(E, \mathcal{E})$ such that, for every index n, the measure μ_n is with base λ, that is to say of, μ_n is of the form $f_n \cdot \lambda$, where f_n is an element of $\mathcal{L}^1(\lambda)$.*

Then, the f_n are uniformly integrable on $(E, \mathcal{E}, \mu)$.

Proof. Let ε be a strictly positive real number, and, for every positive integer m, let us set

$$\mathcal{F}_m = \Big\{A \in \mathcal{E} : \sup_{n \geq m} \big|\textstyle\int_A (f_n - f_m)\,\mathrm{d}\lambda\big| \leq \frac{1}{2}\varepsilon\Big\}.$$

By means of (6.1)(c), each of these sets is closed under the topology induced by the pseudo-metric d defined in (6.1). Furthermore, for every element A of $\mathcal{E}$, the sequence $(\int_A f_n \,\mathrm{d}\lambda)$ is Cauchy in $\mathbb{R}$, which yields that the union of the sets $\mathcal{F}_m$ coincides with $\mathcal{E}$. By means of Baire's theorem (see (6.4)), at least one of the sets $\mathcal{F}_m$ must contain a non-empty open set. In other words, there exists an element B of $\mathcal{F}_m$ and a strictly positive real number δ, such that the set $\{A \in \mathcal{E} : d(A, B) < \delta\}$ is included in $\mathcal{F}_m$. For every element A of $\mathcal{E}$, the relations

$$B \cap A^c \subseteq B \subseteq B \cup A, \qquad (B \cup A) \setminus (B \cap A^c) = B \cap A$$

show that, as soon as $\lambda(A) < \delta$, the set A is the difference of two elements of $\mathcal{E}$ that, having, under d, a distance from B strictly smaller than δ, both belong to $\mathcal{F}_m$; that is to say

$$\lambda(A) < \delta \Rightarrow \sup_{n \geq m} \big|\textstyle\int_A (f_n - f_m)\,\mathrm{d}\lambda\big| < \varepsilon.$$

As a result, for every $n \geq m$, and every element A of $\mathcal{E}$, with $\lambda(A) < \delta$,

$$\big|\mu_n(A) - \mu_m(A)\big| = \big|\textstyle\int_A (f_n - f_m)\,\mathrm{d}\lambda\big| < \varepsilon.$$

In conclusion, this shows that it is possible to find an integer m and a strictly positive real number δ such that the following implication holds for every element A of $\mathcal{E}$:

$$\text{(6.6)} \qquad \lambda(A) < \delta \Rightarrow \sup_{n > m} |\mu_n(A) - \mu_m(A)| < \varepsilon.$$

Since, each of the m positive measures $|\mu_j|$ with $1 \leq j \leq m$ is with base λ, we can always assume (after substituting, if necessary, δ with an even smaller strictly positive real number) that, for every element A of $\mathcal{E}$, the following implication also holds:

$$\lambda(A) < \delta \Rightarrow \sup_{1 \leq j \leq m} |\mu_j|(A) < \varepsilon.$$

Thanks to the triangular inequality, for every integer n and every element A of $\mathcal{E}$, with $\lambda(A) < \delta$, one has $\mu_n^+(A) < 2\varepsilon$ and therefore also

$$\mu_n^+(A) = \int_A f_n^+ \,\mathrm{d}\lambda = \int_{A \cap \{f_n > 0\}} f_n \,\mathrm{d}\lambda = \mu_n\big(A \cap \{f_n > 0\}\big) < 2\varepsilon.$$

Similarly, one can also see that, for every integer n and every element A of $\mathcal{E}$, with $\lambda(A) < \delta$, one has $\mu_n^-(A) < 2\varepsilon$, hence $|\mu_n|(A) < 4\varepsilon$. Thus, for every element A of $\mathcal{E}$, one finally has

$$\lambda(A) < \delta \Rightarrow \sup_n \int_A |f_n| \,\mathrm{d}\lambda < 4\varepsilon. \tag{6.7}$$

Thanks to Theorem (2.5), the aim is now confined to proving that the sequence (f_n) is uniformly bounded in $\mathcal{L}^1(\lambda)$. To this end, by Corollary I(4.8), consider a finite partition $\mathcal{P}$ of E, composed of measurable sets each of which be, under λ, either an atom or a set with measure smaller than δ. It suffices to show that, for every element A of $\mathcal{P}$, one has

$$\sup_n \int_A |f_n| \,\mathrm{d}\lambda < \infty.$$

Now, when $\lambda(A) < \delta$, this follows from (6.7). It thus suffices to show this when A is an atom of λ. In this case, for every index n, the function f_n coincides almost everywhere on A with a constant. Denote by a_n this constant, the relation $\mu_n(A) \to \mu(A)$ becomes $a_n\lambda(A) \to \mu(A)$, whence $\sup_n |a_n| < \infty$, and therefore

$$\sup_n \int_A |f_n| \,\mathrm{d}\lambda = \sup_n |a_n|\lambda(A) < \infty.$$

This completes the proof. □

As a final remark let us observe that the request that there exists a measure λ such that all the real measures μ_n are absolutely continuous with respect to it, in the statement of Vitali-Hahn-Saks theorem, is not a limitation, because, for any sequence of real measures on a measurable space, there always exists a measure with respect to which all the measures in the sequence are absolutely continuous, as shown in the following very simple result due to H. Hahn.

(6.8) Proposition (Hahn). *Given a sequence (μ_n) of real measures on the measurable space $(E, \mathcal{E})$, there always exists a positive real measure λ on $(E, \mathcal{E})$ such that each of the μ_n is with base λ.*

Proof. It suffices to set $\lambda = \sum_n 2^{-n}\big[1 + |\mu_n|(E)\big]^{-1}|\mu_n|$. □

7. Dunford-Pettis theorem

In the rest of this section, we shall assume to be given a finite measure λ on a measurable space $(E, \mathcal{E})$, which may be assumed, with no loss of generality, to be normalised, that is to say, with $\lambda(E) = 1$. Let us also consider, between the two spaces $\mathcal{L}^1$ and $\mathcal{L}^\infty$, the duality induced by the bilinear form defined by $(f,g) \mapsto \int fg\,\mathrm{d}\lambda$. Thus, on the space $\mathcal{L}^1$, alongside the usual topology, we may also consider, as we had already mentioned towards the end of §1, the *weak topology*, that is to say $\sigma(\mathcal{L}^1, \mathcal{L}^\infty)$. This is, by definition, the least fine of all the topologies on $\mathcal{L}^1$ that make all the linear forms $f \mapsto fg\,\mathrm{d}\lambda$ continuous, when g is an element of $\mathcal{L}^\infty$.

If $\mathcal{H}$ is a part of $\mathcal{L}^1$, and if $\mathcal{F}$ is a filter on $\mathcal{H}$, for an element h to be the limit, in the topology $\sigma(\mathcal{L}^1, \mathcal{L}^\infty)$, of the family $(f)_{f\in\mathcal{H}}$, under the filter $\mathcal{F}$, it is necessary and sufficient that, for every element g of $\mathcal{L}^\infty$, the real number $\int hg\,\mathrm{d}\lambda$ is the limit, under $\mathcal{F}$, of the family of real numbers $(\int fg\,\mathrm{d}\lambda)_{f\in\mathcal{H}}$. If, in addition, the set $\mathcal{H}$ is bounded in $\mathcal{L}^1$, that is to say if the relation

$$\sup_{f\in\mathcal{H}} \int |f|\,\mathrm{d}\lambda < \infty$$

holds, then, for h to be the limit of the family $(f)_{f\in\mathcal{H}}$ under the filter $\mathcal{F}$, it is necessary and sufficient that, for ever measurable set A, the real number $\int_A h\,\mathrm{d}\lambda$ is the limit, under $\mathcal{F}$, of the family of real numbers $(\int_A f\,\mathrm{d}\lambda)_{f\in\mathcal{H}}$. This follows from the fact that the vector space generated by the functions of the form I_A with $A \in \mathcal{E}$ (that is to say the vector space of all simple functions) is dense in $\mathcal{L}^\infty$.

Roughly speaking, Dunford-Pettis theorem gives a characterisation of weak compactness in $\mathcal{L}^1$ according to the topology $\sigma(\mathcal{L}^1, \mathcal{L}^\infty)$. In order to prove this result, however, we are going to need some prerequisites that are borrowed from probability theory. In fact, using these tools, and a clever idea that appears in the beautiful *Probabilités at Potentiel*, by the French mathematicians Claude Dellacherie and Paul-André Meyer (1934–2003), one may avoid more tedious and obscure compactness results from the theory of topological vector spaces, and provide a proof that is fully included into the realm of measure theory.

(7.1) Proposition. *Let $\mathcal{U}$ be a sub-σ-field of $\mathcal{E}$, and let us denote by λ^0 the restriction of λ to $\mathcal{U}$. Then, for every measurable function g, defined on $(E,\mathcal{E})$, with values in $[0,1]$, there exists a function g_0, measurable under $(E,\mathcal{U})$ such that*

$$\int fg\,\mathrm{d}\lambda = \int fg_0\,\mathrm{d}\lambda \qquad \text{for every } f \in \mathcal{L}^1(\lambda^0)$$

Proof. Let us set $\mu = g\cdot\lambda$, and let us denote by μ^0 the restriction of μ to $\mathcal{U}$. Since $\mu^0 \leq \lambda^0$, there exists a function g_0, measurable on $(E,\mathcal{U})$, with values in $[0,1]$, such that $\mu^0 = g_0\cdot\lambda^0$. Thus, for every element f of $\mathcal{L}^1(\lambda^0)$, one has

$$\int gf\,\mathrm{d}\lambda = \int f\,\mathrm{d}\mu = \int f\,\mathrm{d}\mu^0 = \int fg_0\,\mathrm{d}\lambda^0 = \int fg_0\,\mathrm{d}\lambda. \qquad \square$$

(7.2) Proposition. *Let (μ_n) be a sequence of real measures on $(E,\mathcal{E})$. Then, there exists a sub-σ-field $\mathcal{U}$ of $\mathcal{E}$, generated by a countable base, such that the set-wise convergence of every subsequence of $(\mu^0_{n_k})$, extracted from (μ^0_n), where μ^0_n denotes the restriction of μ_n to $\mathcal{U}$, implies the set-wise convergence of the subsequence (μ_{n_k}).*

Proof. Let π be a probability measure such that all the measures μ_n are absolutely continuous with respect to π. Such a measure does exist by virtue of Hahn's result (see (6.8)). For every index n, let us choose an element f_n of $\mathcal{L}^1(\pi)$, such that $\mu_n = f_n \cdot \pi$. Let us also define $\mathcal{U}$ to be the σ-field generated by the set $\{f_n : n \geq 1\}$. For every measurable function f, defined on $(E, \mathcal{E})$, with values in $[0, 1]$, let us denote by g_0 the corresponding function described in (7.1). Thus, for every index n, one has

$$\int g \, \mathrm{d}\mu_n = \int g f_n \, \mathrm{d}\pi = \int g_0 f_n \, \mathrm{d}\pi = \int g_0 \, \mathrm{d}\mu_n = \int g_0 \, \mathrm{d}\mu_n^0.$$

Thus, for every strictly increasing infinite sequence of indices (n_k) such that the sequence $(\mu_{n_k}^0)$ converges on every element of $\mathcal{E}$, thanks to (5.6), it follows that both the sequences $(\int g_0 \, \mathrm{d}\mu_{n_k}^0)$ and $(\int g \, \mathrm{d}\mu_{n_k})$ converge. Once again, thanks to (5.6), this is sufficient to conclude that the σ-field $\mathcal{U}$ possesses the required properties. □

(7.3) Proposition. *Let $\mathfrak{K}$ be a non-empty compact subset of $\mathfrak{M}(E, \mathcal{E})$. Then, from every sequence of elements of $\mathfrak{K}$ one can extract a subsequence that converges in the topology of $\mathfrak{M}(E, \mathcal{E})$.*

Proof. Let (μ_n) be a sequence of elements of $\mathfrak{K}$, consider the sub-σ-field $\mathcal{U}$ given by Proposition (7.2), denote by $\mathfrak{M}$ the Riesz space $\mathfrak{M}(E, \mathcal{E})$ equipped with the topology of set-wise convergence, and by $\mathfrak{M}_0$ the space $\mathfrak{M}(E, \mathcal{U})$ of all real measures on $(E, \mathcal{U})$ equipped with the topology of set-wise convergence. Then, the map from $\mathfrak{M}$ to $\mathfrak{M}_0$ that associates to each real measure μ its restriction μ^0 to the σ-field $\mathcal{U}$, is clearly continuous. Thus, the image, under this map, of the compact set $\mathfrak{K}$, that is to say the set $\mathfrak{K}^0$ of all restriction of the elements of $\mathfrak{K}$ to the σ-field $\mathcal{U}$, is a compact set in $\mathfrak{M}_0$. Furthermore, thanks to (4.2), the topology induced on $\mathfrak{K}^0$ and $\mathfrak{M}_0$ are metrizable. As a result, we may extract from (μ_n^0) a subsequence $(\mu_{n_k}^0)$ that converges in the topology of $\mathfrak{M}_0$, whereby the sequence (μ_{n_k}) converges in the topology of $\mathfrak{M}$. □

We are finally ready to prove the main theorem of this section, which is named after the two American mathematicians Nelson J. Dunford (1906–1986) and Billy J. Pettis (1913–1979).

(7.4) Theorem (Dunford-Pettis). *Let $\mathcal{H}$ be a part of $\mathcal{L}^1$. The following conditions are equivalent:*

(a) *$\mathcal{H}$ is uniformly integrable.*

(b) *In the space $\mathcal{L}^1$, equipped with the topology $\sigma(\mathcal{L}^1, \mathcal{L}^\infty)$, the set $\mathcal{H}$ is relatively compact.*

(c) *From any sequence of elements of $\mathcal{H}$, it is always possible to extract a subsequence that is convergent according to the topology $\sigma(\mathcal{L}^1, \mathcal{L}^\infty)$.*

Proof. (a) $\Rightarrow$ (b): Assume that $\mathcal{H}$ is uniformly integrable, and observe that, then, such is the closure of $\mathcal{H}$ in the topology $\sigma(\mathcal{L}^1, \mathcal{L}^\infty)$. Hence, with no loss of generality, we may assume that $\mathcal{H}$ is closed in this topology. As a result, it suffices to prove that, with respect to an arbitrary ultrafilter $\mathcal{U}$ assigned on $\mathcal{H}$, the filtered family $(f)_{f\in\mathcal{H}}$ has a limit in the topology $\sigma(\mathcal{L}^1, \mathcal{L}^\infty)$. To this end, let us observe that condition (a) implies, in particular, that $\mathcal{H}$ is uniformly bounded in $\mathcal{L}^1$. Therefore, for each measurable set A, the filtered family of real numbers $(\int_A f\,\mathrm{d}\lambda)_{f\in\mathcal{H}}$ admits, with respect to the ultrafilter $\mathcal{U}$, a finite limit $\nu(A)$. It is clear that the set function ν thereby defined on $\mathcal{E}$ is bounded and additive. On the other hand, condition (a) also entails that, for every strictly positive real number ε, there exists a strictly positive real number δ, such that, for every measurable set A, with $\lambda(A) < \delta$, one has $\sup_{f\in\mathcal{H}} \int_A |f|\,\mathrm{d}\lambda < \varepsilon$, and therefore $|\nu(A)| < \varepsilon$. This proves that the set function ν is, in fact, a real measure with base λ, whence there exists an element h of $\mathcal{L}^1$ such that $\nu = h\cdot\lambda$. The function h is, with respect to the ultrafilter $\mathcal{U}$, the limit of the family $(f)_{f\in\mathcal{H}}$ in the topology $\sigma(\mathcal{L}^1, \mathcal{L}^\infty)$.

(b) $\Rightarrow$ (c): This is an immediate consequence of (7.3), by considering $\mathfrak{K}$ to be the closure in the space $\mathfrak{M}(E, \mathcal{E})$ of the band of all real measures with base λ and density function in $\mathcal{H}$, or, what is the same, of the form $f\cdot\lambda$ with $f \in \mathcal{H}$.

(c) $\Rightarrow$ (a): Assume that condition (c) holds, and suppose, by contradiction, that the set $\mathcal{H}$ were not uniformly integrable. It is easy to check, then, that there exists an infinite sequence (f_n) of elements of $\mathcal{H}$ such that no subsequence extracted therefrom be uniformly integrable. Nevertheless, thanks to Vitali-Hahn-Saks theorem, it follows that no subsequence extracted by $\mathcal{H}$ may converge in the topology $\sigma(\mathcal{L}^1, \mathcal{L}^\infty)$, in violation of condition (c). □

8. Uniform convergence of real measures

In this short section, we shall introduce another mode of convergence for real measures in the abstract case, beside the set-wise convergence already introduced. To this end, let $(E, \mathcal{E})$ be an assigned measurable space and let us denote by $\mathfrak{M}$ the Riesz space of all real measures on $(E, \mathcal{E})$.

Let us briefly recall that, if μ is such a real measure, one always has the decomposition $\mu = \mu^+ - \mu^-$, where μ^+ and μ^- are two positive and mutually singular real measures on $(E, \mathcal{E})$. Using this decomposition, the *total variation* of μ is defined as $|\mu| = \mu^+ + \mu^-$. Let us give the following definition.

(8.1) Definition. The function $\mu \mapsto \|\mu\|$ defined on $\mathfrak{M}$ by

$$\|\mu\| = |\mu|(E) = \mu^+(E) + \mu^-(E)$$

is called **variation norm**. The corresponding notion of convergence is called the **convergence in variation**.

(8.2) Remark. It is easy to check that the function $\mu \mapsto \|\mu\|$ is indeed a norm, and it is also rather obvious that, for a sequence (μ_n) of elements of

$\mathfrak{M}$, the convergence in variation towards zero is equivalent to the **uniform convergence** of (μ_n) towards zero, that is to say the relation

$$\sup_{A\in\mathcal{E}} |\mu_n(A)| \to 0.$$

To this end, it suffices to observe that the previous relation is equivalent of the two relations

$$\sup_{A\in\mathcal{E}}[\mu_n(A)] \to 0, \qquad \sup_{A\in\mathcal{E}}[-\mu_n(A)] \to 0$$

holding simultaneously, which can be, more concisely, written as

$$\mu_n^+(E) \to 0, \qquad \mu_n^-(E) \to 0.$$

As a result of this, equipping the space $\mathfrak{M}$ with the variation norm makes it a Banach space. In fact, a Cauchy sequence in this norm converges uniformly, hence set-wise, and its set-wise limit is, by virtue of Nikodým's theorem, a real measure.

It remains an open question to wonder whether, for an infinite sequence of elements of $\mathfrak{M}$, convergence in variation (that is to say uniform convergence) is actually stricter than set-wise convergence. To answer this question, let us recall that, by virtue of Hahn's result (see (6.8)), given an element μ of $\mathfrak{M}$ and an infinite sequence (μ_n) of elements of $\mathfrak{M}$, it is always possible to find a finite positive real measure λ such that

$$\mu = f\cdot\lambda \quad \text{and} \quad \mu_n = f_n\cdot\lambda \quad \text{for every index } n$$

with f, f_n elements of $\mathcal{L}^1(\lambda)$. Furthermore, with no loss of generality, λ may always be taken with total mass $\lambda(E) = 1$. Hence, let us observe that, thanks to the relation

$$\|\mu_n - \mu\| = |\mu_n - \mu|(E) = \int |f_n - f|\,\mathrm{d}\lambda,$$

the convergence of (μ_n) towards μ in variation is equivalent to the convergence of (f_n) towards f in $\mathcal{L}^1(\lambda)$, whence the relation

$$\int |f_n - f|\,\mathrm{d}\lambda \to 0.$$

Conversely, the set-wise convergence of (μ_n) towards μ, that is to say the relation

$$\int_A f_n\,\mathrm{d}\lambda \to \int_A f\,\mathrm{d}\lambda \qquad \text{for every element } A \text{ of } \mathcal{E},$$

which, by Vitali-Hahn-Saks theorem, implies the uniform integrability of the f_n, is equivalent to the convergence of the infinite sequence (f_n) towards f in the weak topology $\sigma(\mathcal{L}^1(\lambda), \mathcal{L}^\infty(\lambda))$.

Thus, it is simple to construct an example of a sequence (f_n) which converges in the weak topology $\sigma(\mathcal{L}^1(\lambda), \mathcal{L}^\infty(\lambda))$ but not in the strong topology of $\mathcal{L}^1(\lambda)$. To this end, however, it is convenient to introduce a criterion, for the convergence of a sequence in the weak topology $\sigma(\mathcal{L}^1(\lambda), \mathcal{L}^\infty(\lambda))$ which is based on the notion of *fundamental class.*

(8.3) Definition. We shall call a **fundamental class** for the real measures defined on the measurable space $(E, \mathcal{E})$, a class $\mathcal{D}$ of bounded measurable functions on $(E, \mathcal{E})$, such that the only real measure μ on $(E, \mathcal{E})$ satisfying the relation

$$\int g \, \mathrm{d}\mu = 0 \qquad \text{for every element } g \text{ of } \mathcal{D}$$

is the measure identically zero.

(8.4) Example. If $\mathcal{I}$ is a base for the σ-field $\mathcal{E}$, in the sense clarified in Definition I(5.5), then, the set $\{I_A : A \in \mathcal{I}\}$ is a fundamental class for the measures on $(E, \mathcal{E})$. In fact, if μ is a real measure on $(E, \mathcal{E})$ which verifies, for every element A of $\mathcal{I}$, the relation $\mu(A) = 0$, that is $\mu^+(A) = \mu^-(A)$, then the two positive real measures μ^+ and μ^-, being identical on $\mathcal{I}$, must be identical, by virtue of the fundamental criterion for the coincidence of two measures (see I(5.6)).

Herewith the announced criterion of convergence.

(8.5) Proposition. *On the measurable space $(E, \mathcal{E})$, let λ be a positive real measure. Let also $\mathcal{D}$ be a fundamental class. Let (f_n) be a sequence of elements of $\mathcal{L}^1(\lambda)$ and f an element of $\mathcal{L}^1(\lambda)$. Then, the following conditions are equivalent:*

(a) *The sequence (f_n) converges to f in the topology $\sigma(\mathcal{L}^1(\lambda), \mathcal{L}^\infty(\lambda))$.*

(b) *The sequence (f_n) is uniformly integrable and, for every element g of the class $\mathcal{D}$, the real sequence $(\int g f_n \, \mathrm{d}\lambda)$ converges towards $\int g f \, \mathrm{d}\lambda$.*

Proof. Since $\mathcal{L}^1(\lambda)$ is a topological vector space, with no loss of generality, we may assume that the function f is identically zero, by substituting, if necessary, the sequence (f_n) with $(f_n - f)$.

(a) $\Rightarrow$ (b): Assume that condition (a) holds. The first part of condition (b) is thus satisfied by virtue of Vitali-Hahn-Saks theorem. The second part of condition (b), instead, is obvious.

(b) $\Rightarrow$ (a): Assume that condition (b) holds, and, by contradiction, assume that there is a function h in $\mathcal{L}^\infty(\lambda)$ such that the sequence $(\int h f_n \, \mathrm{d}\mu)$ does not converge towards zero. This means that it is possible to extract from the latter sequence a subsequence that converges towards a real number b which is not zero. On the other hand, by virtue of Dunford-Pettis theorem, the first part of condition (b) guarantees that the closure of $\{f_n : n \geq 1\}$ is a compact set in the topology $\sigma(\mathcal{L}^1(\lambda), \mathcal{L}^\infty(\lambda))$, whence it is possible to further extract from the aforementioned subsequence a subsequence that weakly converges towards h. Let us denote by (f_{n_k}) this latter subsequence and observe that, for every element g of $\mathcal{D}$, the second part of condition (b) yields

$$\int g h \, \mathrm{d}\lambda = \lim_k \int g f_{n_k} \, \mathrm{d}\lambda = \lim_n \int g f_n \, \mathrm{d}\lambda = 0.$$

Hence, the function h is negligible with respect to λ, but this is impossible because $\int h \, \mathrm{d}\lambda = \lim_k \int f_{n_k} \, \mathrm{d}\lambda \neq 0$. □

To complete this section, let us now exhibit an example in which a sequence converges weakly but not strongly.

(8.6) Example. Let E be the interval $[0,1]$, $\mathcal{E}$ the Borel σ-field of the interval $[0,1]$, and λ the restriction of the Borel-Lebesgue to this σ-field. For every strictly positive real number n, let us set

$$f_n = \sum_{1 \leq j \leq 2^n} (-1)^j I_{[2^{-n}(j-1), 2^{-n}j]}.$$

It is clear that $|f_n| = 1$ for every index n, thus the sequence (f_n) is uniformly integrable. For every strictly positive integer n, let us also set

$$\mathcal{I}_n = \Big\{ [0, 2^{-n}j] : 1 \leq j \leq 2^n \Big\}.$$

Finally, let us set $\mathcal{I} = \bigcup_n \mathcal{I}_n$. It is clear that $\mathcal{I}$ is a base for the Borel σ-field of $(0,1]$. Indeed, $\mathcal{I}$ generates this σ-field, it is stable under the operation of binary intersection and contains the whole interval $[0,1]$ as an element. Furthermore, for every index n, and every element A in $\mathcal{I}_n$, one has

$$\int_A f_{n+h} \, \mathrm{d}\lambda = 0 \qquad \text{for } h \geq 1.$$

Thanks to (8.5), this suffices to infer that the sequence (f_n) converges towards zero in the topology $\sigma(\mathcal{L}^1(\lambda), \mathcal{L}^\infty(\lambda))$. However, it is clear that the convergence does not occur in the topology defined by the semi-norm of $\mathcal{L}^1(\lambda)$.

Chapter IX

Weak Convergence of Probability Measures

This very short chapter collects some important details the notion of weak convergence of a family of probability measures on a metrizable space. In particular, after having described the general definition of weak convergence for real measures, we shall confine ourselves on the study of the weak topology of the space of probability measures, starting by proving a nice criterion, which some authors call 'The *Portmanteau* Theorem', and then moving onto the proper study of the topology of the space of probability measures, showing that this space inherits some topological properties from the underlying space. In particular, we all see that, if the underlying space is compact, then such is the space of all probability measures, and the same happens when the underlying space is Polish. Hence, in these cases, it is always possible to ensure weak convergence by only checking it on a countable family of functions. The chapter terminates with the proof of Prokhorov's theorem, which shows a condition, that generalises that of a tight measure, and that ensures weak sequential compactness of a set of measures.

1. Weak convergence on a metrizable space

In the rest of this chapter, we shall assume that E is a metrizable space. We shall also assume that E is a measurable space with the Borel σ-field induced by its topology, and, whether or not it be explicitly stated, we shall always assume that any measure is defined on the Borel σ-field of E, whence it is a Borel measure on E.

To simplify the notations, we shall write $\mathfrak{M}(E)$ to denote the Riesz space of all Borel real measures, that is to say the set of all real measures defined on $\mathcal{B}(E)$, and, similarly, we shall write $\mathfrak{M}_+(E)$ to denote the cone of all positive Borel

DOI: 10.1201/9781003724766-IX

real measures on E. Let us also introduce a new notation, and denote by $\mathfrak{P}(E)$ the convex set of all Borel probability measures on E.

In the sequel, we shall focus on the convergence of a 'filtered family' of Borel probability measures. By this term, we shall mean a family $(\mu_t)_{t\in T}$ of elements of $\mathfrak{P}(E)$ in whose set of indices T a filter has been assigned, and the convergence shall be intended, unless otherwise specified, to be with respect to this filter. First of all, however, let us give the definition of convergence in the general setting of a Borel real measure

(1.1) Definition. The **weak topology** on $\mathfrak{M}(E)$ is the smallest topology on $\mathfrak{M}(E)$ that makes continuous every functional of the form

$$\mu \mapsto \int f\,\mathrm{d}\mu \qquad \text{with } f \in \mathcal{C}_b(E).$$

The notion of convergence associated to this topology shall be referred to as **weak convergence**. Clearly, with respect to this topology, a filtered family $(\mu_t)_{t\in T}$ of elements of $\mathfrak{M}(E)$ converges towards an element μ of $\mathfrak{M}(E)$, if and only if

$$\int f\,\mathrm{d}\mu_t \to \int f\,\mathrm{d}\mu \qquad \text{for every element } f \text{ of } \mathcal{C}_b(E).$$

When this happens, we shall say that (μ_t) **converges weakly** towards μ.

Since a real measure μ may be identified by the linear form $f \mapsto \int f\,\mathrm{d}\mu$, defined on the space $\mathcal{C}_b(E)$ of all bounded continuous functions on E, it is clear that the weak topology is Hausdorff.

Furthermore, the space $\mathfrak{M}_+(E)$ of all positive real measures on E is clearly a closed subspace of $\mathfrak{M}(E)$, and since $\mathfrak{P}(E)$ may be written in the form

$$\mathfrak{P}(E) = \{\mu \in \mathfrak{M}_+(E) : \mu(E) = 1\},$$

this is clearly a closed convex set of $\mathfrak{M}(E)$, and may be considered as a topological space with the weak topology induced by the topology of $\mathfrak{M}(E)$.

From now onwards, we shall solely focus on the convex space $\mathfrak{P}(E)$ endowed with the topologies (strong and weak) induced by $\mathfrak{M}(E)$.

(1.2) Example. On an arbitrary metric space (E, d), and let (x_n) be a sequence of points of E, which converges towards the point x in E. Then, the sequence (ε_{x_n}) of probability measures on E weakly converges towards the measure ε_x. In fact, if f is a bounded continuous function on E, one has

$$\int f\,\mathrm{d}\varepsilon_{x_n} = f(x_n) \to f(x) = \int f\,\mathrm{d}\varepsilon_x.$$

On the other hand, if (x_n) does not converge towards the point x, then there exists a positive real number c such that the relation $d(x_n, x) > c$ frequently holds. This implies that the limit $f(x_n) \to f(x)$ fails, whence (ε_{x_n}) does not converge weakly towards ε_x. Therefore, (ε_{x_n}) converges weakly towards ε_x if and only if (x_n) converges towards x.

The following proposition gives a few criteria for weak convergence, that some authors refer to as 'The *Portmanteau* Theorem', apparently because it represents a 'suitcase' or 'container' of various equivalent definitions of weak convergence of measures.

(1.3) Theorem. *Given a filtered family $(\mu_t)_{t\in T}$ of elements of $\mathfrak{P}(E)$, and an element μ of $\mathfrak{P}(E)$, the following conditions are equivalent:*

(a) *The family $(\mu_t)_{t\in T}$ weakly converges towards μ.*

(b) $\int f\, d\mu = \lim_t \int f\, \mathrm{d}\mu_t$ *for every numeric Lipschitz continuous function f, bounded on E.*

(c) $\int f\, \mathrm{d}\mu \leq \liminf_t \int f\, \mathrm{d}\mu_t$ *for every lower semi-continuous function f, bounded from below on E.*

(d) $\int f\, \mathrm{d}\mu \geq \limsup_t \int f\, \mathrm{d}\mu_t$ *for every upper semi-continuous function f, bounded from above on E.*

(e) $\int f\, \mathrm{d}\mu = \lim_t \int f\, \mathrm{d}\mu_t$ *for every numeric Borel measurable function f, bounded on E, for which the set of discontinuity points is negligible with respect to μ.*

Proof. The implication (a) $\Rightarrow$ (b) is obvious. Let us prove the implication (b) $\Rightarrow$ (c). To this end, assume that condition (b) holds, and let d be a bounded metric compatible with the topology of E. Then, condition (c) is satisfied thanks to the fact that, every lower semi-continuous function f, bounded from below in E, is the supremum of the increasing sequence of Lipschitz continuous function f_n defined for every index n by:

$$f_n(x) = \inf_{y\in E}\big[f(y) + nd(x,y)\big].$$

The equivalence between (c) and (d) follows from the fact that the functions intervening in (d) are exactly the opposite of those intervening in (c).

The implication [(c) and (d)] $\Rightarrow$ (e) is easy to prove, using the fact that, for any Borel measurable function f, the set of its discontinuity points is the Borel set $\{f_* < f^*\}$, where f_* and f^* are the lower and upper semi-continuous regularisation of f, respectively.

Finally, the implication (e) $\Rightarrow$ (a) is obvious. □

(1.4) Theorem. *Let μ be a probability measure and $(\mu_t)_{t\in T}$ a filtered family of probability measures on E. Then, the following conditions are equivalent.*

(a) *The family $(\mu_t)_{t\in T}$ weakly converges towards μ.*

(b) $\mu(E) = \lim_t \mu_t(E)$, *and* $\mu(U) \leq \liminf_t \mu_t(U)$, *for every open set U.*

(c) $\mu(E) = \lim_t \mu_t(E)$, *and* $\mu(F) \geq \limsup_t \mu_t(F)$, *for every closed set F.*

(d) $\mu(B) = \lim_t \mu_t(B)$, *for every Borel measurable subset B of E, whose boundary ∂B is negligible under μ.*

Proof. The implication (a) $\Rightarrow$ (b) follows from the previous theorem, thanks to the fact that the indicator function of an open set is lower semi-continuous. The equivalence between (b) and (c) follows from the fact that closed sets are the complement of open sets. The implication [(b) and (c)] $\Rightarrow$ (d) follows from the fact that, for every set, the boundary is the difference between the closure and the internal part of that set. Finally, the implication (d) $\Rightarrow$ (a) can be proved using the fact that every continuous and bounded numeric function f is the uniform limit of a linear combination of indicator functions of sets of the form $\{a < g \leq b\}$, where $\mu\{g = a\} = \mu\{g = b\} = 0$. □

(1.5) Corollary. *In the same hypotheses as the previous proposition, let D be a Borel subset of E (which is going to be considered as a subspace of E). Let us suppose that the probability measures μ and μ_t are concentrated on D, and let us denote by λ and λ_t their restriction to the Borel σ-field of D. Then, the following conditions are equivalent:*

(a) *The family $(\lambda_t)_{t \in T}$ weakly converges towards λ.*

(b) *The family $(\mu_t)_{t \in T}$ weakly converges towards μ.*

Proof. The implication (a) $\Rightarrow$ (b) is straightforward: In fact, if condition (a) is satisfied, and if f is an element of $\mathcal{C}_b(E)$, then the restriction of f to D is an element of $\mathcal{C}_b(D)$. Thus,

$$\int f \, \mathrm{d}\mu_t = \int_D f \, \mathrm{d}\lambda_t \to \int_D f \, \mathrm{d}\lambda = \int f \, \mathrm{d}\mu.$$

Let us now prove the implication (b) $\Rightarrow$ (a). To this end, assuming that condition (b) is satisfied, it suffices to show, thanks to Proposition (1.3), that, if f is a lower semi-continuous function, bounded from below on D, one has $\int f \, \mathrm{d}\lambda \leq \liminf_t \int f \, \mathrm{d}\lambda_t$. To this end, let us consider on E the function g that coincides with f on D and with ∞ otherwise. In each point of D, this function is lower semi-continuous (that is to say, it coincides with its lower semi-continuous regularisation g_*). Thus, taking into account hypothesis (b), one has

$$\begin{aligned} \int f \, \mathrm{d}\lambda = \int g \, \mathrm{d}\mu = \int g_* \, \mathrm{d}\mu &\leq \liminf_t \int g_* \, \mathrm{d}\mu_t \\ &= \liminf_t \int g \, \mathrm{d}\mu_t \\ &= \liminf_t \int f \, \mathrm{d}\lambda_t. \end{aligned}$$ □

The previous corollary can be rephrased, in an equivalent fashion, as follows.

(1.6) Corollary. *Let D be a Borel subset of E (which is going to be considered as a subspace of E), and let us denote by $\mathfrak{P}(E, D)$ the set of all probability measures on E which are concentrated on D. Then, the map ϱ that to such a probability measure associates its restriction to D is a homeomorphism of the space $\mathfrak{P}(E, D)$, subspace of $\mathfrak{P}(E)$, to the space $\mathfrak{P}(D)$.*

Proof. The map ϱ is bijective: Its inverse is the map that to each probability measure λ on D associates its image, under the canonical injection of D on E, that is to say the probability measure $A \mapsto \lambda(A \cap D)$. The fact that ϱ is also a homeomorphism follows from the previous corollary. $\square$

(1.7) Definition. We shall call a **determinant class** for the weak convergence in E any subset $\mathcal{H}$ of $\mathcal{C}_b(E)$ such that, for every probability measure μ and any filtered family $(\mu_t)_{t\in T}$ of probability measures on E, the weak convergence of $(\mu_t)_{t\in T}$ towards μ is equivalent to the following relation

$$\int f \,\mathrm{d}\mu_t \to \int f \,\mathrm{d}\mu \qquad \text{for every element } f \text{ of } \mathcal{H}.$$

(1.8) Remark. If f is lower semi-continuous, bounded from below on E, then, since f is the supremum of an increasing sequence (f_n) of elements of $\mathcal{C}_b(E)$, the functional $\mu \mapsto \int f \,\mathrm{d}\mu$ is the supremum of the increasing sequence of continuous functionals $\mu \mapsto \int f_n \,\mathrm{d}\mu$: as such, this is a lower semi-continuous function. Clearly, an analogous result holds for upper semi-continuous functions, bounded from above.

2. Topology of the space of probability measures

In the rest of this section, we shall continue to focus on the study of the topological properties of the closed convex space $\mathfrak{P}(E)$. Again, we shall assume that E is a metrizable space. Initially, we shall also assume that E is compact. In this case, in fact, the following special case of Riesz-Markov's theorem holds.

(2.1) Theorem. *Let E be a compact metrizable space. Then, for every monotone linear form L on $\mathcal{C}_b(E)$, with $L(1) = 1$, there exists one (and only one) probability measure μ on E, such that*

$$L(f) = \int f \,\mathrm{d}\mu \qquad \textit{for every element } f \textit{ of } \mathcal{C}_b(E).$$

From this theorem, the following immediately holds.

(2.2) Theorem. *Let E be a compact metrizable space, and let $\mathcal{H}$ be a subset of $\mathcal{C}_b(E)$, dense in $\mathcal{C}_b(E)$ for the uniform (strong) topology. Then, for a filtered family $(\mu_t)_{t\in T}$ of probability measures on E, the following conditions are equivalent:*

(a) *There exists a probability measure μ on E such that $(\mu_t)_{t\in T}$ weakly converges towards μ.*

(b) *For every element f of $\mathcal{H}$, the family of real numbers $(\int f \,\mathrm{d}\mu_t)_{t\in T}$ is a Cauchy filtered family.*

(2.3) Corollary. *In the same hypotheses as in the previous theorem, $\mathcal{H}$ is a determinant class for the weak convergence in E (as intended in (1.7)).*

We are now ready to show that, when E is a compact metrizable space, such is $\mathfrak{P}(E)$. In other words, the space of probability measure on E retains the compactness from the space E itself.

(2.4) Theorem. *Let E be a compact metrizable space. Then, such is $\mathfrak{P}(E)$.*

Proof. Thanks to (2.1), it is possible to identify $\mathfrak{P}(E)$ with the set of all monotone linear forms L on $\mathcal{C}_b(E)$, such that $L(1) = 1$, with the topology of pointwise convergence. This space, being a closed subspace of the product space $\mathbb{R}^{\mathcal{C}_b(E)}$, is compact. Thus, such is the space $\mathfrak{P}(E)$.

To prove that $\mathfrak{P}(E)$ is also metrizable, let us consider a countable subset $\mathcal{H}$ of $\mathcal{C}_b(E)$ which is dense in $\mathcal{C}_b(E)$ for the uniform topology. We may assume that $\mathcal{H}$ is described in the form $\mathcal{H} = \{f_k : k \geq 1\}$. To every pair μ, ν of probability measures on E, one can thus set

$$d(\mu, \nu) = \textstyle\sum_{k\geq 1} 2^{-k} \big[1 \wedge | \int f_k \, \mathrm{d}\mu - \int f_k \, \mathrm{d}\nu|\big].$$

It is thereby defined a metric d on the space $\mathfrak{P}(E)$. Thanks to (2.3), the topology defined by this distance is nothing else but the weak topology on $\mathfrak{P}(E)$: in fact, given an element μ on $\mathfrak{P}(E)$ and a filtered family $(\mu_t)_{t\in T}$ of elements of $\mathfrak{P}(E)$, the relation $d(\mu_t, \mu) \to 0$ holds if and only if one has $\int f_k \, \mathrm{d}\mu \to \int f_k \, \mathrm{d}\mu$ for evert k. □

Let us now relax the hypotheses on the underlying space and assume that it is a Polish space. In this case, the following theorem holds.

(2.5) Theorem. *Let D be a Polish space. Then,*

(a) *The space $\mathfrak{P}(D)$ is Polish.*

(b) *For the weak convergence of probability measures on D, there exists a determinant class which is countable.*

Proof. With no loss of generality, thanks to VII(1.11), it is possible to assume that D coincides with the countable intersection of open subsets of a suitable compact metrizable space E, which, if one will, may be taken equal to the product $[0, 1]^{\mathbb{N}}$. Thanks to (1.6), the space $\mathfrak{P}(D)$ is thus homeomorphic to the space $\mathfrak{P}(E, D)$ of all probability measures on E concentrated on D. From (2.4) it follows that the space $\mathfrak{P}(E)$ is compact and metrizable, hence Polish. Thus, it suffices to show that $\mathfrak{P}(E, D)$ can be expressed as a countable intersection of open sets in $\mathfrak{P}(E)$. To this end, let (G_n) be a sequence of open sets in E, such that $D = \bigcap_n G_n$, and let us observe that the space $\mathfrak{P}(E, D)$ can be written as

$$\begin{aligned}
\mathfrak{P}(E, D) &= \big\{\mu \in \mathfrak{P}(E) : \mu(D) = 1\big\} \\
&= \textstyle\bigcap_n \big\{\mu \in \mathfrak{P}(E) : \mu(G_n) = 1\big\} \\
&= \textstyle\bigcap_n \bigcap_k \big\{\mu \in \mathfrak{P}(E) : \mu(G_n) > 1 - (1/k)\big\}.
\end{aligned}$$

The final term of this chain of inequalities is a countable intersection of open sets: In fact, thanks to (1.3), each of the functionals of the form $\mu \mapsto \mu(G_n)$ is lower semi-continuous on $\mathfrak{P}(E)$. Assertion (a) is thus proved.

To prove assertion (b), it suffices to observe that, if $\mathcal{H}$ is a countable determinant class for the weak convergence in E, then the class constituted of all restrictions to D of the functions in $\mathcal{H}$ is, thanks to (1.5), determinant for the weak convergence in D. □

(2.6) Remark. As Corollary (2.3) shows, in the special case when the space E is compact and metrizable, the existence of a countable class that is determinant for the weak convergence of probability measures on E can be deduced simply from the existence of a dense countable class in the space $\mathcal{C}_b(E)$, equipped with the topology of uniform convergence. This property, however, does not extend to the case of a Polish space. In fact, this statement already fails in the special case of the real line: Indeed, the space $\mathcal{C}_b(\mathbb{R})$, with the topology of uniform convergence, is not separable. To prove the latter claim, choose a continuous function f on $\mathbb{R}$ with support in $[0,1]$, and consider, for every element ε in $\{0,1\}^{\mathbb{Z}}$, the function g_ε defined on $\mathbb{R}$ by

$$g_\varepsilon(x) = \textstyle\sum_{n\in\mathbb{Z}} \varepsilon(n) f(x-n).$$

Then, the family $(g_\varepsilon)_{\varepsilon\in\{0,1\}^{\mathbb{Z}}}$ is an uncountable family of elements of $\mathcal{C}_b(\mathbb{R})$ such that, for every pair $\varepsilon, \varepsilon'$ of distinct indices, one has $\|g_\varepsilon - g_{\varepsilon'}\| = \|f\| > 0$.

3. Prokhorov's theorem

In this section, E continues to be a metrizable space and all measures are supposed to be Borel positive real measures defined on E.

(3.1) Proposition. *Let K be a compact subset of E, let c be a strictly positive real number. Then, the set of all real measures concentrated on K with total mass bounded by c is a compact set in the weak topology.*

Proof. The set under examination is the one of all measures μ such that

$$\mu(K^c) \le 0, \qquad \mu(E) \le c.$$

Thus, this set is closed, thanks to the fact that the function $\mu \mapsto \mu(K^c)$ is lower semi-continuous and that the function $\mu \mapsto \mu(E)$ is continuous. To prove that this set is, in fact, compact, let us consider a filtered family $(\mu_t)_{t\in T}$, assume that on the set of indices T we are given an ultrafilter, and let us prove that, with respect to this ultrafilter, the assigned family weakly converges (towards a suitable measure). To this end, let us consider the linear form L defined on the space of all continuous and bounded functions on E by

$$L(f) = \lim_t \textstyle\int f \,\mathrm{d}\mu_t = \lim_t \int_K f \,\mathrm{d}\mu_t.$$

Since K is compact, Dini's lemma (see Proposition VII(1.8)) yields that L is a Daniell's functional. Hence, by virtue of Daniell's representation theorem, there exists a unique measure λ such that $L(f) = \int f \, \mathrm{d}\lambda$, for every function f in $\mathcal{C}_b(E)$. It is clear that this measure is the limit of (μ_t) with respect to the weak topology, and this completes the proof. □

Let us recall that, a Borel measure μ on E is 'tight' if, for every strictly positive real number ε, there exists a compact set K such that $\mu(K^c) < \varepsilon$. This definition may be generalised as follows.

(3.2) Definition. A set $\mathcal{H}$ of measure on E is **uniformly tight** if the total masses of the measures that compose this set are all bounded from above by the same real constant and if, for every strictly positive real number ε, there exists a compact set K such that

$$\sup_{\mu \in \mathcal{H}} \mu(K^c) < \varepsilon.$$

It is clear that, if $\mathcal{H}$ is uniformly tight, then, each of the elements of $\mathcal{H}$ is tight. Furthermore, if $\mathcal{H}$ is composed of only one measure, this definition becomes the familiar definition of tight measure.

We may now prove the following compactness criterion due to the Soviet mathematician Yuri Prokhorov (Юрий Васильевич Прохоров, 1929–2013).

(3.3) Theorem (Prokhorov). *If $\mathcal{H}$ is a set of uniformly tight measures, then it is relatively compact in the weak topology.*

Proof. With no loss of generality, we may assume that the total masses of the measures in $\mathcal{H}$ are all bounded from above by the constant 1. Furthermore, let us suppose that an increasing sequence (K_n) of compact sets in E is assigned, such that, for each index n, one has

$$\sup_{\mu \in \mathcal{H}} \mu(K_n^c) < 1/n.$$

Let us consider a filtered family $(\mu_t)_{t \in T}$ of elements of $\mathcal{H}$, such that the filter assigned on T is an ultrafilter, and let us prove that, with respect to this ultrafilter, the given family weakly converges (towards a suitable measure). To this end, observe that, for every index n, thanks to Proposition (3.1), the family $(I_{K_n} \cdot \mu_t)_{t \in T}$ weakly converges towards a measure λ_n with total mass smaller than 1. Since the sequence (λ_n) thereby obtained is clearly increasing, it converges set-wise (hence weakly) towards a measure λ. It is easy to check that λ is the limit of $(\mu_t)_{t \in T}$ in the weak topology. Indeed, it suffices to observe that, for a continuous function f on E, with $|f| \leq 1$,

$$\left|\int f \, \mathrm{d}\lambda - \int f \, \mathrm{d}\mu_t\right| \leq \left|\int f \, \mathrm{d}\lambda - \int f \, \mathrm{d}\lambda_n\right| + \left|\int f \, \mathrm{d}\lambda_n - \int_{K_n} f \, \mathrm{d}\mu_t\right| + 1/n,$$

whence

$$\limsup_t \left|\int f \, \mathrm{d}\lambda - \int f \, \mathrm{d}\mu_t\right| \leq \left|\int f \, \mathrm{d}\lambda - \int f \, \mathrm{d}\lambda_n\right| + 1/n.$$ □

Chapter X

Disintegration of Measures

The notions of 'conditional probability' and 'disintegration' are extremely important concepts in probability theory, but these concept are also very interesting in the wider context of measure theory, finding applications in several other branches of mathematical analysis, such as differential equations, and optimal transportation.

This chapter is devoted to introducing, in the abstract setting of measure theory, the notion of disintegration of a measure with respect to another, and the related notion of conditional measure with respect to a σ-field. In particular, in §1 and §2, we shall give the definition of disintegration and conditional measure, and we shall delve into the study of their main properties, establishing the link between these two concepts. In §3, instead, we shall focus on the conditions that ensure the existence of a disintegration, studying in particular the spaces, called *regular* (for the concept of disintegration), which always ensure the existence of a disintegration. With the aid of Ionescu-Tulcea's theorem, we shall prove that every Polish space is regular, that is to say, that every probability measure on a Polish space admits a disintegration with respect to any probability measure. Having introduced the case where the disintegration exists, in §4 we are going to show an example where the disintegration does not exist. The chapter finally terminates with a general result, due to M. Jiřina, that guarantees the existence of a disintegration in a space that does not necessarily have a topological structure, but where a notion similar to that of compactness is still available.

For the purpose of this chapter, any measure considered is assumed to be a probability measure, that is to say a positive real measure with total mass 1. Let us observe that, with some very minimal changes, the results would also extend to any positive real measure, but the study of disintegrations, being borrowed from probability theory, and being generally applied to probability measures, looks more natural with this extra assumption.

DOI: 10.1201/9781003724766-X

1. Disintegration of a measure on a product space

(1.1) Definition. Let $(E, \mathcal{E})$ and $(F, \mathcal{F})$ be two measurable space and let us denote by ξ_E and ξ_F the projections from $E \times F$ onto E and F, respectively. Let λ be a probability measure on the product space $(E \times F, \mathcal{E} \otimes \mathcal{F})$. The measures $\mu = \xi_E(\lambda)$ and $\nu = \xi_F(\lambda)$, images of λ under the maps ξ_E and ξ_F, respectively, are called the **marginals** of λ **relative** to the first and the second component, respectively.

In the sequel, we shall denote by $\Pi(\mu, \nu)$ the set of all probability measures on the product space $(E \times F, \mathcal{E} \otimes \mathcal{F})$ whose marginals relative to the first and second components are the measures μ and ν, respectively. In symbols:

$$\Pi(\mu, \nu) = \Big\{\lambda \in \mathfrak{P}(E \times F, \mathcal{E} \otimes \mathcal{F}) : \xi_E(\lambda) = \mu,\ \xi_F(\lambda) = \nu\Big\}.$$

Let us start by observing that, in the same hypotheses and with the same notations as in the previous definition, the two marginals, μ and ν, are defined by the identities

$$\mu(A) = \lambda(A \times F), \qquad \nu(B) = \lambda(E \times B),$$

for every A in $\mathcal{E}$ and every B in $\mathcal{F}$. In particular, the marginals are both normalised, hence probability measures.

The first, interesting, question is whether the measure λ coincides or not with the product of the two measures μ and ν. In other words, one may wonder whether it is true that $\lambda = \mu \otimes \nu$. We have already shown in Chapter IV, §2 that this is not the case but it is useful to show here another simple example.

(1.2) Example. On the measurable space $(\mathbb{R}^2, \mathcal{B}(\mathbb{R}^2))$, let us consider the measure μ, image of the trace λ of the Borel-Lebesgue measure onto the interval $[0,1]$, under the map $t \mapsto (t,t)$. The marginals of μ obviously coincide with λ, but it easy to check that $\mu \neq \lambda \otimes \lambda$. To this end, it suffices to observe that the measure μ is concentrated on the set

$$D = \big\{(x, y) : x = y\big\}$$

and $\mu(D) = 1$, whilst this set is negligible with respect to $\lambda \otimes \lambda$.

Thus, probability measures on the product space $(E \times F, \mathcal{E} \otimes \mathcal{F})$ may not be of the form $\mu \otimes \nu$, with μ a measure on $(E, \mathcal{E})$ and ν a measure on $(F, \mathcal{F})$. In other words, while it is certainly true that $\mu \otimes \nu$ lies in $\Pi(\mu, \nu)$, or, what is the same, while it is true that the marginals of $\mu \otimes \nu$ are the measures μ and ν, it is also true that this is not the only probability measure on the product space $(E \times F, \mathcal{E} \otimes \mathcal{F})$ which possesses this very property.

In Chapter IV, we have seen another way of constructing a measure on the product space $(E \times F, \mathcal{E} \otimes \mathcal{F})$ consisting in taking the product of a measure on $(E, \mathcal{E})$ and a kernel N from $(E, \mathcal{E})$ to $(F, \mathcal{F})$. Hence, it would be natural to wonder whether, given an element λ of $\Pi(\mu, \nu)$, it is possible to find a Markov kernel $N = (N(x, \cdot\,))_{x \in E}$, from $(E, \mathcal{E})$ to $(F, \mathcal{F})$, such that

$$\lambda = \mu \otimes N.$$

For this equality to hold, of course, it is necessary that $\nu = \mu N$, that is to say, it is necessary that

$$\nu(B) = \textstyle\int N(x, B)\, \mu(\mathrm{d}x) \qquad \text{for every } B \text{ in } \mathcal{F}$$

but this condition, alone, is generally not sufficient to ensure that $\lambda = \mu \otimes N$.

(1.3) Example. Let us take the measurable spaces $E = F = \{0, 1\}$ with their corresponding discrete σ-fields, and let $\mu = \nu = \frac{1}{2}\varepsilon_0 + \frac{1}{2}\varepsilon_1$. Let us also define the kernel N by setting $N(x, \cdot\,) = \varepsilon_x$ with $x = 0, 1$. Then, it is simple to check that

$$\mu N = \textstyle\int N(x, \cdot\,)\, \mu(\mathrm{d}x) = \frac{1}{2}\varepsilon_0 + \frac{1}{2}\varepsilon_1 = \nu.$$

On the product space $\{0, 1\}^2$, with its discrete σ-field, consider the measure

$$\lambda = \tfrac{1}{2}\varepsilon_{(0,1)} + \tfrac{1}{2}\varepsilon_{(1,0)}.$$

It is not difficult to recognise that μ and ν are the marginals of this measure. On the other hand, by virtue of IV(1.17),

$$\begin{aligned} \mu \otimes N &= \textstyle\int \big(\varepsilon_x \otimes N(x, \cdot\,)\big)\mu(\mathrm{d}x) \\ &= \tfrac{1}{2}\varepsilon_0 \otimes \varepsilon_0 + \tfrac{1}{2}\varepsilon_1 \otimes \varepsilon_1 \\ &= \tfrac{1}{2}\varepsilon_{(0,0)} + \tfrac{1}{2}\varepsilon_{(1,1)}, \end{aligned}$$

whence $\lambda \neq \mu \otimes N$ even though $\nu = \mu N$.

The goal of this section is to study under which conditions it is possible to represent a probability measure λ of $\Pi(\mu, \nu)$ in the form $\lambda = \mu \otimes N$, where N is a suitable Markov kernel from $(E, \mathcal{E})$ to $(F, \mathcal{F})$. This justifies the following definition.

(1.4) Definition. Let $(E, \mathcal{E})$ and $(F, \mathcal{F})$ be two measurable space, λ a probability measure on the product space $(E \times F, \mathcal{E} \otimes \mathcal{F})$, and let us denote by μ its marginal relative to the first component. A Markov kernel N from $(E, \mathcal{E})$ to $(F, \mathcal{F})$ shall be called a (**version** of the) **disintegration** of λ **with respect to** μ if $\lambda = \mu \otimes N$.

The definition given above may look rather awkward at first glance, for it is abstract and does not fully show the relations between all the parties involved. The following result repairs the flaws of the definition by showing some different equivalent characterisations of the disintegration.

(1.5) Proposition. *Let $(E,\mathcal{E})$ and $(F,\mathcal{F})$ be two measurable spaces, μ a probability measure on $(E,\mathcal{E})$, λ a probability measure on the product space $(E\times F,\mathcal{E}\otimes\mathcal{F})$, and N a Markov kernel from $(E,\mathcal{E})$ to $(F,\mathcal{F})$. Then, the following conditions are equivalent:*

(a) *μ is the marginal of λ relative to the first component and N is a version of the disintegration of λ with respect to μ.*

(b) *For every bounded measurable function h on $(E\times F,\mathcal{E}\otimes\mathcal{F})$, one has*

$$\int h\,\mathrm{d}\lambda = \int \mu(\mathrm{d}x)\int N(x,\mathrm{d}y)\,h(x,y).$$

(c) *For every bounded measurable function f on $(E,\mathcal{E})$ and every bounded measurable function g on $(F,\mathcal{F})$,*

$$\int f\otimes g\,\mathrm{d}\lambda = \int \mu(\mathrm{d}x)\,f(x)\int N(x,\mathrm{d}y)\,g(y).$$

Proof. The equivalence between conditions (a) and (b) is an immediate consequence of Definition IV(1.15).

Condition (c) is just a special case of condition (b), so it suffices to prove that implication (c) $\Rightarrow$ (b) holds. To this end, assuming that condition (c) holds, it suffices to observe that the class of all bounded measurable real functions h defined on the product space $(E\times F,\mathcal{E}\otimes\mathcal{F})$ such that

$$\int h\,\mathrm{d}\lambda = \int \mu(\mathrm{d}x)\int N(x,\mathrm{d}y)\,h(x,y)$$

by hypothesis contains the class of all functions of the form $f\otimes g$ which is stable under the operation of multiplication, contains the constant 1, and contains all the functions of the form $I_{A\times B}=I_A\otimes I_B$ with $A\in\mathcal{E}$ and $B\in\mathcal{F}$, and the thesis follows from the Monotone Class Theorem (see I(6.4)). □

(1.6) Remark. In the same hypotheses as in the previous proposition, let us now assume that $\mathcal{I}$ be a base for $\mathcal{E}$, and $\mathcal{J}$ be a base for $\mathcal{F}$. Then, for the probability measures λ and $\mu\otimes N$ to coincide it is necessary and sufficient that they coincide on the base of the σ-field $\mathcal{E}\otimes\mathcal{F}$, constituted by all elements of $\mathcal{E}\otimes\mathcal{F}$ of the form $A\times B$, with A element of $\mathcal{I}$ and B element of $\mathcal{J}$. In other words, each of the conditions expressed by the previous criterion are equivalent to the following condition

(d) *For every element A of $\mathcal{I}$ and every element B of $\mathcal{J}$,*

$$\lambda(A\times B) = \int_A N(x,B)\,\mu(\mathrm{d}x).$$

Let us now observe that, in the general setting of Definition (1.4), it is not guaranteed that a version of the disintegration exists, nor that, if it exists, it is unique. We shall discuss the problem of the existence in §3 and §5, where we shall see that existence is guaranteed as soon as some mild additional hypotheses are added. As regard to the uniqueness, the following theorem shows that, as soon as the σ-field $\mathcal{F}$ is 'separable' then a form of *essential uniqueness*, that is a 'uniqueness modulo μ', is ensured.

(1.7) Theorem. *In the same hypotheses as in Definition (1.4), assume that the σ-field $\mathcal{F}$ is separable, that is to say that there exists a countable set of generators for $\mathcal{F}$. Then, two versions N and N' of the disintegration of λ with respect to μ, are equivalent modulo μ, that is to say, the set*

$$\{x \in E : N(x, \cdot\,) \neq N'(x, \cdot\,)\}$$

belongs to $\mathcal{E}$, and it is negligible with respect to μ.

Proof. Let us assume that the σ-field $\mathcal{F}$ be separable, which means that it possesses a countable base, and let us denote it by $\mathcal{J}$. For every element B of $\mathcal{J}$, the positive real functions NI_B and $N'I_B$ on $(E, \mathcal{E})$ both must be equivalent modulo μ, because, by virtue of Remark (1.6), one has

$$\int_A N(x, B)\, \mu(\mathrm{d}x) = \int_A N'(x, B)\, \mu(\mathrm{d}x) \qquad \text{for every } A \text{ in } \mathcal{E}.$$

This is equivalent to say that the set

$$\{x \in E : N(x, B) \neq N'(x, B)\}$$

(which belongs to $\mathcal{E}$) is negligible with respect to μ. On the other hand, the intersection of all these sets, as B ranges over the countable class $\mathcal{J}$, coincides, thanks to the fundamental criterion for the coincidence of two measures, to the set

$$\{x \in E : N(x, \cdot\,) \neq N'(x, \cdot\,)\}$$

and this proves the theorem. □

The following example shows a case of non-uniqueness of the disintegration. It is not surprising that the example is quite complicated, because the request for the σ-field $\mathcal{F}$ to be separable is in fact a decently mild assumption.

(1.8) Example (Non-uniqueness of the disintegration). Let us start by recalling that, in the product of an arbitrary family of measurable spaces, each measurable set belongs to the σ-field generated by a suitable *countable* subfamily of the canonical projections.

For the purpose of this example, we shall call *non-trivial* any measurable space on which there exist two distinct probability measures. Let us also recall that a measure is *diffuse* if the measure of every measurable set constituted by a single element is zero. Let us consider two probability spaces $(E, \mathcal{E}, \mu)$ and $(F, \mathcal{F}, \nu)$, with the following properties:

(a) The measure μ is diffuse.

(b) The measurable space $(F, \mathcal{F})$ is the product of a family, with E as the set of indices, of non-trivial measurable spaces.

(c) The measure ν is the product of a family of probability measures, with E as the set of indices.

On one side, it is evident that a version of the disintegration of the measure $\lambda = \mu \otimes \nu$ with respect to μ is the constant kernel identified by ν. On the other hand, we may also find another kernel from $(E, \mathcal{E})$ to $(F, \mathcal{F})$ which is also a version of the disintegration of λ with respect to μ, but such that, for every element x of E, the measure $N(x, \cdot)$ is different from ν. To this end, let us take, for every x in E, a measure $N(x, \cdot)$ obtained altering, in the expression of ν as a product of measures, *only* the factor with index x. Then, thanks to the result recalled at the beginning of the example, one has that, for every element B of $\mathcal{F}$, the map $x \mapsto N(x, B)$ coincides with the constant $\nu(B)$ outside of a *countable* (whence *μ-negligible*) set. It follows that the family $(N(x, \cdot))_{x \in E}$ of measures on $(F, \mathcal{F})$ is a kernel, and this kernel is a version of the disintegration of λ with respect to μ.

2. Conditional measures with respect to a σ-field

Related to the notion of disintegration is the notion of 'conditional measure' with respect to sub-σ-field which is the object of the present section. Let us start by giving the definition of disintegration.

(2.1) Definition. Let μ be a probability measure on a given measurable space $(E, \mathcal{E})$, let $\mathcal{U}$ be a sub-σ-field of $\mathcal{E}$, and let us denote by μ^0 the restriction of μ to $\mathcal{U}$. A Markov kernel N, from $(E, \mathcal{U})$ to $(E, \mathcal{E})$, shall be called a **conditional measure** (or, equivalently, **conditional probability**) of μ with respect to $\mathcal{U}$ if, for every bounded measurable function f on $(E, \mathcal{U})$ and every bounded measurable function g on $(E, \mathcal{E})$ one has

$$\int fg \, \mathrm{d}\mu = \int fNg \, \mathrm{d}\mu^0,$$

where Ng denotes the transform of g through the kernel N.

(2.2) Remark. Let us observe immediately that for a kernel N from $(E, \mathcal{U})$ to $(E, \mathcal{E})$ to be a conditional measure of μ with respect to $\mathcal{U}$, it is (necessary and) sufficient that the following condition holds:

$$\mu(A \cap B) = \int_A N(x, B) \, \mu^0(\mathrm{d}x) \qquad \text{for every } A \in \mathcal{U} \text{ and } B \text{ in } \mathcal{E}. \tag{2.3}$$

As a matter of fact, the condition is obviously *necessary*, as it is easily shown by choosing $f = I_A$ and $g = I_B$, with A element of $\mathcal{U}$ and B element of $\mathcal{E}$. Thus, it only remains to show that it is also *sufficient*. To this end, observe that, if condition (2.3) holds, then the class $\mathcal{M}_1$ of all bounded measurable functions g on $(E, \mathcal{E})$ such that

$$\int_A g \, \mathrm{d}\mu = \int_A Ng \, \mathrm{d}\mu^0 \qquad \text{for every } A \in \mathcal{U}$$

is monotone, and contains the indicator functions of every B in $\mathcal{E}$, whence coincides with the class of all bounded measurable functions on $(E, \mathcal{E})$ by virtue of the Monotone Class Theorem. Furthermore, thanks to this partial result, the class $\mathcal{M}_2$ of all bounded measurable functions f on $(E, \mathcal{U})$ such that

$$\int fg \, \mathrm{d}\mu = \int fNg \, \mathrm{d}\mu^0 \qquad \text{for every } g \in \mathcal{M}_1$$

is monotone and contains the indicator functions of every A in $\mathcal{U}$, whence coincides with the class of all bounded measurable functions on $(E, \mathcal{U})$ by virtue of the Monotone Class Theorem, and this yields that the kernel N is indeed a conditional measure of μ with respect to $\mathcal{U}$.

The link between the notion of conditional measure and the notion of disintegration is unfolded in the following proposition.

(2.4) Proposition. *Let μ be a probability measure on a given measurable space $(E, \mathcal{E})$, let $\mathcal{U}$ be a sub-σ-field of $\mathcal{E}$, and let us denote by μ^0 the restriction of μ to $\mathcal{U}$. Let N be a Markov kernel from $(E, \mathcal{U})$ to $(E, \mathcal{E})$. Let us also denote by j the measurable map from $(E, \mathcal{E})$ to $(E \times E, \mathcal{U} \otimes \mathcal{E})$ defined, for every element x of E, by $j(x) = (x, x)$.*

Then, the following conditions are equivalent.

(a) *N is a conditional measure of μ with respect to $\mathcal{U}$.*

(b) *N is a version of the disintegration of $j(\mu)$ with respect to μ^0.*

Proof. To prove the equivalence between (a) and (b), let us set $\lambda = j(\mu)$, let f be a bounded measurable function on $(E, \mathcal{U})$ and g be a bounded measurable function on $(E, \mathcal{E})$, and observe that, assuming (a),

$$\int f \otimes g \, \mathrm{d}\lambda = \int fg \, \mathrm{d}\mu = \int fNg \, \mathrm{d}\mu^0.$$

Conversely, assuming (b),

$$\int fg \, \mathrm{d}\mu = \int f \otimes g \, \mathrm{d}\lambda = \int fNg \, \mathrm{d}\mu^0. \qquad \square$$

In the same hypotheses as in Definition (2.1), let us observe that a special important case of conditional measure occurs when the sub-σ-field $\mathcal{U}$ is the σ-field generated by a measurable map φ from $(E, \mathcal{E})$ to a measurable space $(F, \mathcal{F})$.

(2.5) Proposition. *Let $(E, \mathcal{E})$ be a measurable space, μ a probability measure on this space, and φ a measurable map defined on $(E, \mathcal{E})$ with values in a measurable space $(F, \mathcal{F})$. For μ to have a conditional measure with respect to $\sigma(\varphi)$ it is necessary and sufficient that there exists a Markov kernel Q from $(F, \mathcal{F})$ to $(E, \mathcal{E})$ such that, for every bounded measurable function f on $(E, \mathcal{E})$ and every bounded measurable function g on $(F, \mathcal{F})$, one has*

$$\int f \cdot (g \circ \varphi) \, \mathrm{d}\mu = \int Qf \cdot g \, \mathrm{d}\nu$$

where $\nu = \varphi(\mu)$ denotes the image of μ under φ.

Proof. Let us start by observing that the condition is *sufficient.* This is very easy, for the kernel N defined for every x in E by $N(x,\,\cdot\,) = Q(\varphi(x),\,\cdot\,)$ is, by construction, a conditional measure of μ with respect to $\sigma(\varphi)$.

Let us prove that the condition is also *necessary.* To this end, assume that N is a conditional measure of μ with respect to $\sigma(\varphi)$, generated by φ, and observe that, then, for every bounded measurable function f on $(E, \mathcal{E})$, the map Nf, transform of f through N, is $\sigma(\varphi)$-measurable. Hence, by Doob's measurability criterion (see I(2.14)), there exists a measurable real function Qf, defined on $(F, \mathcal{F})$, such that $Nf = Qf \circ \varphi$, and observe that, by virtue of Theorem IV(1.6), the map $f \mapsto Qf$ defines a kernel from $(F, \mathcal{F})$ to $(E, \mathcal{E})$. Furthermore,

$$\begin{aligned} \int f \cdot (g \circ \varphi)\, \mathrm{d}\mu &= \int Nf \cdot (g \circ \varphi)\, \mathrm{d}\mu \\ &= \int (Qf \circ \varphi) \cdot (g \circ \varphi)\, \mathrm{d}\mu \\ &= \int Qf \cdot g\, \mathrm{d}\nu. \end{aligned}$$ □

(2.6) Definition. In the same hypotheses as in the previous proposition, the kernel Q is called a **conditional measure** of the measure μ with respect to the map φ.

Let us complete this section with a simple example.

(2.7) Example. Let $E = \{x_1, x_2, x_3, x_4\}$ and $F = \{y_1, y_2\}$ thought as measurable spaces with their corresponding discrete σ-fields, and let us consider the map φ, from E to F, defined by

$$\varphi(x_1) = \varphi(x_3) = y_1, \qquad \varphi(x_2) = \varphi(x_4) = y_2.$$

Let a_1, a_2, a_3, a_4 be four strictly positive real numbers that add up to 1, set

$$\mu = a_1\varepsilon_{x_1} + a_2\varepsilon_{x_2} + a_3\varepsilon_{x_3} + a_4\varepsilon_{x_4},$$

and let us denote by ν the image of μ under φ, that is to say the measure defined by $\nu = (a_1 + a_3)\varepsilon_{y_1} + (a_2 + a_4)\varepsilon_{y_2}$. It is not difficult to recognise that a version of the disintegration of μ with respect to φ is given by the Markov kernel Q defined by

$$Q(y_1,\,\cdot\,) = \frac{a_1}{a_1 + a_3}\varepsilon_{x_1} + \frac{a_3}{a_1 + a_3}\varepsilon_{x_3}, \quad Q(y_2,\,\cdot\,) = \frac{a_2}{a_2 + a_4}\varepsilon_{x_1} + \frac{a_4}{a_2 + a_4}\varepsilon_{x_4}.$$

3. Regular measurable spaces

Hitherto, we have only studied the properties of disintegrations and disintegrations, without facing the problem of the existence. In general, in fact, there is nothing that ensures the existence of a disintegration with respect to a given measure. In this section, we shall endeavour to repair to this flaw, by showing a class of measurable spaces where the disintegration with respect to any measure is guaranteed. Precisely, let us give the following definition.

(3.1) Definition. A measurable space $(F, \mathcal{F})$ shall be called **regular**, for the concept of disintegration, if, for every probability space $(E, \mathcal{E}, \mu)$, and every probability measure λ, on $(E \times F, \mathcal{E} \otimes \mathcal{F})$, with μ the marginal with respect to the first component, there exists a version of the disintegration of λ with respect to μ.

A first paramount result is the following.

(3.2) Theorem. *The measurable space* $(\mathbb{R}, \mathcal{B}(\mathbb{R}))$ *is regular.*

Proof. Let $(E, \mathcal{E})$ be a measurable space, λ a probability measure on the product space $(E \times \mathbb{R}, \mathcal{E} \otimes \mathcal{B}(\mathbb{R}))$, and let us denote by μ the marginal of λ relative to its first component. To prove the theorem, it suffices to exhibit a version of the disintegration of λ with respect to μ. Since the intervals of the form $(-\infty, r]$, with r rational number, form a base for the Borel σ-field $\mathcal{B}(\mathbb{R})$, it suffices to construct a Markov kernel N, from $(E, \mathcal{E})$ to $(\mathbb{R}, \mathcal{B}(\mathbb{R}))$, such that

$$\int f \otimes I_{(-\infty,r]} \, d\lambda = \int f N I_{(-\infty,r]} \, d\mu \tag{3.3}$$

for every bounded measurable function f on $(E, \mathcal{E})$. To this end, let us start by observing that the map

$$f \mapsto L_r(f) = \int f \otimes I_{(-\infty,r]} \, d\lambda$$

is manifestly a continuous linear functional defined on the space $\mathcal{L}^1(\mu)$ of all the real functions defined on $(E, \mathcal{E})$ which are integrable with respect to the measure μ. Thus, by virtue of Theorem VI(4.4), there exists a bounded measurable function f_r on $(E, \mathcal{E})$ such that

$$L_r(f) = \int f f_r \, d\mu.$$

Thanks to this definition, it is easy to check that each of the sets below has, under the probability μ, measure equal to 1:

$$A = \bigcap\nolimits_{r,s \in \mathbb{Q},\, r<s} \{f_r \le f_s\}, \qquad B = \bigcap\nolimits_{r \in \mathbb{Q}} \{f_r = \inf\nolimits_n f_{r+1/n}\},$$
$$C = \{\sup\nolimits_n f_n = 1,\ \inf\nolimits_n f_{-n} = 0\}.$$

Hence, setting $D = A \cap B \cap C$, it must be $\mu(D) = 1$. For every element x of D, let us denote by $N(x, \cdot\,)$ the probability measure, on $\mathcal{B}(\mathbb{R})$, which is identified by the equality

$$N\big(x, (-\infty, r]\big) = f_r(x) \qquad \text{for every rational number } r.$$

Let us also set $N(x, \cdot\,) = \varepsilon_0$ for every element x of $E \setminus D$. It is thus defined a Markov kernel N that solves the problem. Indeed, for every rational number r, the function $x \mapsto N(x, (-\infty, r])$ is measurable on $(E, \mathcal{E})$, because it is constant on $E \setminus D$, and coincides with the function f_r on D, and the equality (3.3) is satisfied by construction. □

In order to give a more thorough description of regular measurable spaces, let us start by giving the following definition.

(3.4) Definition. Given two measurable spaces $(F, \mathcal{F})$ and $(F_0, \mathcal{F}_0)$, we shall say that $(F_0, \mathcal{F}_0)$ is **subordinate** to $(F, \mathcal{F})$ if there exists a measurable map f from $(F_0, \mathcal{F}_0)$ to $(F, \mathcal{F})$ and a measurable map g from $(F, \mathcal{F})$ to $(F_0, \mathcal{F}_0)$ such that $g \circ f$ coincides with the identity on F_0.

(3.5) Example. Assume that $(F_0, \mathcal{F}_0)$ is a measurable subspace of $(F, \mathcal{F})$; in other words, assume that F_0 is a measurable subset of F and $\mathcal{F}_0$ coincides with the trace of $\mathcal{F}$ onto F_0, that is to say, the σ-field of all elements of $\mathcal{F}$ included in F_0. Then $(F_0, \mathcal{F}_0)$ is subordinate to $(F, \mathcal{F})$. Indeed, to see this it suffices to take as f the canonical injection of F_0 into F, and as g any measurable map that fixes every point in F_0 and that is constant on $F \setminus F_0$.

(3.6) Example. If two measurable spaces are *isomorphic*, that is to say, if there exists a bijective map from one to the other, which is measurable and whose inverse is also measurable, then, obviously, each of these spaces is subordinate to each other.

The interest for subordinate measurable spaces is due to the following result.

(3.7) Proposition. *If $(F, \mathcal{F})$ is a regular measurable space, then such is every subordinate measurable space $(F_0, \mathcal{F}_0)$.*

Proof. Assume that $(F, \mathcal{F})$ is regular, and let f and g be two maps as they are defined in Definition (3.4). Furthermore, let λ be a measure on the product space $(E \times F_0, \mathcal{E} \otimes \mathcal{F}_0)$. Then, the image of the measure λ under the map ϕ defined by $\phi(x, y) = (x, f(x))$ is a measure on $(E \times F, \mathcal{E} \otimes \mathcal{F})$ with marginal relative to the first component equal to μ. Since $(F, \mathcal{F})$ is regular, there exists a version of the disintegration of $\phi(\lambda)$ with respect to μ. Then, the kernel Q defined by $Q(x, \cdot\,) = g(N(x, \cdot\,))$ is a version of the disintegration of λ with respect to μ. □

The next result is a simple consequence of Ionescu-Tulcea's theorem.

(3.8) Proposition. *A measurable space that is the product of a countable family of regular measurable spaces is regular.*

Proof. Thanks to the previous proposition, it suffices to prove the statement when the countable family is an infinite sequence. Let us therefore consider an infinite sequence $((E_n, \mathcal{E}_n))_{n \geq 1}$ of regular measurable spaces, let μ be a probability measure on an arbitrary measurable space $(E_0, \mathcal{E}_0)$, let us set

$$F = \textstyle\prod_{n=0}^{\infty} E_n, \qquad \mathcal{F} = \bigotimes_{n=0}^{\infty} \mathcal{E}_n,$$

and, for every positive integer n, denote by j_n the projection of F onto the product space $E_0 \times \cdots \times E_n$. Let us assume that $j_0(\lambda) = \mu$ and let us prove that there exists a version of the disintegration of λ with respect to μ. To this end, since the spaces $(E_n, \mathcal{E}_n)$ are regular, for every strictly positive index n, let us denote by N_n a version of the disintegration of $j_{n+1}(\lambda)$ with respect to $j_n(\lambda)$. In other words, assume that

$$j_0(\lambda) = \mu, \qquad j_{n+1}(\lambda) = j_n(\lambda) \otimes N_n \qquad \text{for every } n \geq 0.$$

Let Q be Ionescu-Tulcea's kernel associated to the sequence (N_n). From Proposition IV(3.8) it follows that $\lambda = \mu Q$. Furthermore, if p denotes the projection of $(F, \mathcal{F})$ onto the measurable set $(F_0, \mathcal{F}_0)$ where

$$F_0 = \textstyle\prod_{n\geq 1} E_n, \qquad \mathcal{F}_0 = \textstyle\bigotimes_{n\geq 1} \mathcal{E}_n,$$

and if, for every element x of E_0, we denote by $N(x, \cdot)$ the image of $Q(x, \cdot)$ under this map, it is clear that N is a kernel from $(E_0, \mathcal{E}_0)$ to $(F_0, \mathcal{F}_0)$ and that $\lambda = \mu Q = \mu \otimes N$, whence N is a version of the disintegration of λ with respect to μ, and this proves that the product space $(F_0, \mathcal{F}_0)$ is indeed regular, as required. □

(3.9) Corollary. *Every Polish space (equipped with its corresponding Borel σ-field) is a regular measurable space.*

Proof. Recall from VII(1.11) that every Polish space is homeomorphic to a Borel subspace (more precisely a G_δ) of the space $[0,1]^{\mathbb{N}}$. Since $(\mathbb{R}, \mathcal{B}(\mathbb{R}))$ is regular (by Theorem (3.2)), then the measurable space $([0,1], \mathcal{B}([0,1]))$ is also regular, because it is subordinate to $(\mathbb{R}, \mathcal{B}(\mathbb{R}))$, and the conclusion follows applying the previous proposition. □

4. An example of non-existence

This short section is devoted to show that it is not always possible to ensure the existence of a disintegration. To this end, it suffices to exhibit an example of a probability measure μ on a measurable space $(E, \mathcal{E})$ for which there does not exist a conditional measure with respect to a given sub-σ-field $\mathcal{E}_0$.

To achieve this, let us start by giving the following definition.

(4.1) Definition. We shall say that a σ-field $\mathcal{U}$ on a non-empty set E possess the **diagonal property** if the diagonal $\Delta = \{(x, x) : x \in E\}$ belongs to the σ-field $\mathcal{U} \otimes \mathcal{U}$.

The following proposition gives a criterion based on two simple conditions, to establish when a kernel is a conditional measure with respect to a sub-σ-field which has the diagonal property.

(4.2) Proposition. *Let $(E, \mathcal{E})$ be a measurable space, $\mathcal{U}$ a sub-σ-field of $\mathcal{E}$ with the diagonal property and μ a probability measure on $(E, \mathcal{E})$. For a Markov kernel N from $(E, \mathcal{U})$ to $(E, \mathcal{E})$ to be a conditional measure μ with respect to $\mathcal{U}$, it is necessary and sufficient that the two conditions below hold:*

(a) *μ coincides with the measure ρN, where ρ is the restriction of μ to $\mathcal{U}$.*

(b) *The set D of all elements x of E such that $N(x,\,\cdot\,) \neq \varepsilon_x$ belongs to $\mathcal{U}$, and it is negligible with respect to ρ (hence μ).*

Proof. Let us start by proving that these conditions are *necessary.* Let N be a conditional measure of μ with respect to $\mathcal{U}$, and let us denote by ρ the restriction of μ to $\mathcal{U}$. If f is a bounded measurable function on $(E,\mathcal{U})$ and g is a measurable function on $(E,\mathcal{E})$, then, by definition of conditional measure,

$$\int fg\,\mathrm{d}\mu = \int fNg\,\mathrm{d}\rho$$

and, taking $f = 1$, using the duality of kernels, this gives in particular,

$$\langle \mu, g\rangle = \langle \rho, Ng\rangle = \langle \rho N, g\rangle$$

which shows that $\mu = \rho N$, whence condition (a) holds. It remains to prove that condition (b) holds as well. To this end, let us denote by j the measurable map from $(E,\mathcal{U})$ to $(E \times E, \mathcal{U} \otimes \mathcal{E})$ defined by $x \mapsto (x,x)$, and observe that, by hypotheses, the diagonal Δ lies in $\mathcal{U} \otimes \mathcal{U} \subseteq \mathcal{E} \otimes \mathcal{E}$ whence, by virtue of Proposition (2.4) and Proposition (1.5)(b),

$$0 = j(\mu)(\Delta^c) = \int \rho(\mathrm{d}x) \int N(x,\mathrm{d}y) I_{\Delta^c}(x,y),$$

which means that the set

$$\left\{x \in E : \int N(x,\mathrm{d}y)\, I_{\Delta^c}(x,y) \neq 0\right\}$$

is negligible with respect to ρ. On the other hand, since this set coincides with the set of all points x in E such that $N(x,\,\cdot\,) \neq \varepsilon_x$, condition (b) is satisfied.

It remains to prove that the conditions are also *sufficient.* To this end, assume that conditions (a) and (b) hold, and, for every pair bounded measurable functions f, g on $(E,\mathcal{U})$ and $(E,\mathcal{E})$, respectively, observe that, for every x outside of D, one has $N(x,\,\cdot\,) = \varepsilon_x$, whence,

$$g(x) = \int N(x,\mathrm{d}y) g(y) = Ng(x).$$

On the other hand, by hypotheses, Ng is $\mathcal{U}$-measurable, and this yields

$$\int fg\,\mathrm{d}\mu = \int fNg\,\mathrm{d}\mu = \int fNg\,\mathrm{d}\rho,$$

which proves that N is a conditional measure of μ with respect to $\mathcal{U}$. □

(4.3) Remark. As a result of the previous proposition, if a kernel N satisfies condition (b), then every element B of $\mathcal{E}$ can always be sandwiched between two elements of the σ-field $\mathcal{U}$ with the same measure with respect to μ. In fact, for every element x of D^c, one has

$$(NI_B)(x) = N(x,B) = \varepsilon_x(B) = I_B(x).$$

It is thus clear that $I_{D^c}I_B = I_{D^c}NI_B$, and, since the right-hand side of this equality is measurable on $(E,\mathcal{U})$, it follows that the set $D^c \cap B$ must lie in $\mathcal{U}$. Furthermore, the set B is sandwiched between the two sets $D^c \cap B$ and $D \cap B = (D^c \cap B) \cup D$, both belonging to $\mathcal{U}$ and all with the same measure:

$$\rho(D^c \cap B) = \rho\big((D^c \cap B) \cup D\big) = \mu(B).$$

It is thus clear, that in order to have an example of non-existence of a conditional measure with respect to a sub-σ-field, it suffices to construct on the probability measure space $(E, \mathcal{E}, \mu)$, a sub-σ-field $\mathcal{U}$ of $\mathcal{E}$, that has the diagonal property, and where not every element of $\mathcal{E}$ is included between two elements of $\mathcal{U}$ with the same measure. In this case, in fact, we shall construct an example even stronger: That is to say an example where there exist no Markov kernel from $(E,\mathcal{U})$ to $(E,\mathcal{E})$ for which condition (b) of the previous proposition holds.

We are finally ready to discuss the desired example.

(4.4) Example. Take the set E to be the interval $[0,1]$ and $\mathcal{U}$ to be the Borel σ-field of $[0,1]$ (which evidently does have the diagonal property). Let us recall from Example III(8.9) that the Borel σ-field $\mathcal{B}([0,1])$ is included in the trace of the Lebesgue σ-field $\mathfrak{L}$ onto $[0,1]$, and that this bigger σ-field does not include all subsets of $[0,1]$. Take therefore a subset H of $[0,1]$ is not measurable with respect to the Lebesgue measure, but whose outer Lebesgue measure is 1 (that is the maximum possible).

Take $\mathcal{E}$ to be the class constituted by all subsets of $[0,1]$ whose intersection with H is a Borel subset of H, and define the measure μ on $\mathcal{E}$ as follows: Given an element of $\mathcal{E}$, that is to say a part B of $[0,1]$ whose intersection with H is a Borel subset of H, which is equivalent to say that B can be written in the form $B \cap H = A \cap H$, where A is a suitable Borel subset of $[0,1]$, and let us define $\mu(B)$ as equal to the Lebesgue measure of A. Observe that this definition is not ambiguous: In fact, if A' is another Borel subset of $[0,1]$ such that $B \cap H = A' \cap H$, then the symmetric difference $A \triangle A'$, being included in $[0,1] \setminus H$, is negligible with respect to the Lebesgue measure (for H has outer measure equal to 1). It is not difficult to check that μ is indeed a probability measure on $\mathcal{E}$, and that it coincides with the Lebesgue measure on each element of $\mathcal{U}$, that is to say on each Borel subset of $[0,1]$. Thus, H cannot be included between two elements of $\mathcal{U}$ with the same measure under μ.

In conclusion, the space $(E, \mathcal{E}, \mu)$ and the sub-σ-field $\mathcal{U}$ gives the desired example, for there cannot exist any Markov kernel which is the conditional measure of μ with respect to $\mathcal{U}$.

5. Jiřina's theorem

It is due to the Czech mathematician Miloslav Jiřina (1927-2011) a sufficient criterion for the existence of a disintegration that is more general than the one discussed in (3.9). The idea behind Jiřina's result is somewhat related to the one of Theorem (3.2), and it requires a preliminary definition, which is just an abstraction of the corresponding definition in a topological space.

(5.1) Definition. Given a non-empty set E, a class $\mathcal{K}$ of subsets of E is said to be **compact** if, from any countable family of elements of $\mathcal{K}$ whose intersection is empty, it is always possible to extract a finite sub-family whose intersection is also empty.

In order to prove Jiřina's theorem, we are going to need two preliminary results regarding compact class.

(5.2) Lemma. *If $\mathcal{K}$ is a compact class on a non-empty set E, then such is the class $\mathcal{K}_s$ of all finite unions of elements of $\mathcal{K}$.*

Proof. Let (H_n) be a sequence of elements of $\mathcal{K}_s$ whose intersection is empty. For every positive integer n, let $(K_i^n)_{i\in I_n}$ be a finite family of elements of $\mathcal{K}$ such that

$$H_n = \bigcup_{i\in I_n} K_i^n.$$

Setting $I = \prod_n I_n$, one has

$$\emptyset = \bigcap_n H_n = \bigcup_{\sigma\in I} \bigcap_n K_{\sigma(n)}^n.$$

This means that, for every element σ of I, the sequence (K_σ^n) has empty intersection, thus, there exists a positive integer p (which depends on σ) such that

$$\bigcap_{n=1}^p K_{\sigma(n)}^n = \emptyset.$$

In other words, if, for every positive integer p, we set

$$G_p = \big\{\sigma \in I : \bigcap_{n=1}^p K_{\sigma(n)}^n = \emptyset\big\},$$

one has $I = \bigcup_p G_p$. On the other hand, considering on each of the sets I_n the discrete topology, and on the set I the corresponding product topology, this means that the set I is compact, and that the sets G_p are open. As a result, there exists an integer p with $G_p = I$. Hence, for every σ in I,

$$\bigcap_{n=1}^p K_{\sigma(n)}^n = \emptyset.$$

Thus,

$$\bigcap_{n=1}^p H_n = \bigcup_{\sigma\in I} \bigcap_{n=1}^p K_{\sigma(n)}^n = \emptyset,$$

which proves that the class $\mathcal{K}_s$ is compact. □

(5.3) Lemma. *Let F be a non-empty set, $\mathcal{F}_0, \mathcal{F}_1$ a pair of algebras on F, with $\mathcal{F}_0 \subseteq \mathcal{F}_1$, and λ a positive additive function defined on $\mathcal{F}_1$. Assume that there exists a compact class $\mathcal{K}$ of subsets of F with the following property: For every element B of $\mathcal{F}_0$ and every strictly positive real number ε, there exists a pair (A, K) of subsets of F such that:*

$$A \in \mathcal{F}_1, \qquad K \in \mathcal{K}, \qquad A \subseteq K \subseteq B, \qquad \lambda(B \setminus A) < \varepsilon.$$

Then, the restriction of λ to $\mathcal{F}_0$ is countably additive.

Proof. It suffices to prove that, if (B_n) is a decreasing sequence of elements of $\mathcal{F}_0$, with $\bigcap_n B_n = \emptyset$, then $\inf_n \lambda(B_n) = 0$. To this end, let us fix a strictly positive real number ε, and, for every index n, let us consider a pair (A_n, K_n) of subsets of F such that:

$$A_n \in \mathcal{F}_1, \qquad K_n \in \mathcal{K}, \qquad A_n \subseteq K_n \subseteq B_n, \qquad \lambda(B_n \setminus A_n) < 2^{-n}\varepsilon.$$

Then, the sequence (K_n) has empty intersection, and therefore there exists a positive integer p such that $\bigcap_{n \leq p} L_n = \emptyset$, whence $\bigcap_{n=1}^{p} A_n = \emptyset$. thus,

$$B_p \subseteq \textstyle\bigcap_{n \leq p}(B_n \setminus A_n), \quad \text{hence} \quad \lambda(B_p) \leq \sum_{n \leq p} \lambda(B_n \setminus A_n) \leq \varepsilon.$$

Since ε is arbitrary, this is sufficient to conclude that $\inf_n \lambda(B_n) = 0$. □

We are finally ready to prove Jiřina's theorem.

(5.4) Theorem (Jiřina). *Let $(E, \mathcal{E})$ and $(F, \mathcal{F})$ be two measurable spaces, let λ be a probability measure on the product space $(E \times F, \mathcal{E} \otimes \mathcal{F})$, and let us denote by μ, ν the marginals of λ relative to the first and second component, respectively. In order for a disintegration of λ with respect to μ to exist it suffices that the following* Jiřina's condition *holds:*

There exists a countable algebra $\mathcal{F}_0$, which generates the σ-field $\mathcal{F}$, and a compact class $\mathcal{K}$ of subsets of F, such that, for every element B of $\mathcal{F}_0$, and every strictly positive real number ε, it is possible to find a pair (A, K) of elements of F such that

$$A \in \mathcal{F}, \qquad K \in \mathcal{K}, \qquad A \subseteq K \subseteq B, \qquad \nu(B \setminus A) < \varepsilon.$$

Proof. Assume that Jiřina's condition holds. Thanks to Lemma (5.2), we may assume that the class $\mathcal{K}$ is stable under finite union. Then, to every element B of the algebra $\mathcal{F}_0$ it is possible to associate, for every strictly positive integer n, a pair (L_n^B, K_n^B) of subsets of F such that

$$L_n^B \in \mathcal{F}, \qquad K_n^B \in \mathcal{K}, \qquad L_n^B \subseteq K_n^B \subseteq B, \qquad \nu(B \setminus L_n^B) \leq 1/n. \tag{5.5}$$

Using the fact that $\mathcal{K}$ is stable under finite union, we may also assume, with no loss of generality, that the sequence (L_n^B) is increasing, so that

$$\nu\Big\{ I_{L_n^B} \uparrow I_B \Big\} = 1. \tag{5.6}$$

Let us also observe that the algebra generated by $\mathcal{F}_0$ and by the sets of the form L_n^B (with $B \in \mathcal{F}_0$ and $n \geq 1$) continues to be countable: Let us denote it by $\mathcal{F}_1$. Let us now observe that the map

$$f \mapsto J_B(f) = \int f \otimes I_B \, \mathrm{d}\lambda$$

is manifestly a continuous linear functional on the space $\mathcal{L}^1(\mu)$ of all real functions defined on $(E, \mathcal{E})$ which are integrable with respect to the measure μ. Thus, by virtue of Theorem VI(4.4), there exists a bounded measurable function f_B such that $J_B(f) = \int f_B f \, \mathrm{d}\mu$. Furthermore, with no loss of generality, we may also assume that f_B takes values in $[0, 1]$. In fact, we may assume that, when $B = F$, the function f_B coincides with the constant 1 on E. Moreover, for every pair B, B' of disjoint elements of $\mathcal{F}_1$, thanks to the identity

$$I_{B \cup B'} = I_B + I_{B'}$$

and the linearity of J_B, one gets

$$\mu\{f_{B \cup B'} = f_B + f_{B'}\} = 1,$$

that is to say, the set

$$\{f_{B \cup B'} = f_B + f_{B'}\} \tag{5.7}$$

belongs to the σ-field $\mathcal{E}$, and has, under μ, measure 1. In addition, for every element B of $\mathcal{F}_0$, the identity (5.6), implies

$$\mu\Big\{f_{L_n^B} \uparrow f_B\Big\} = 1.$$

In other words, the sets

$$\Big\{f_{L_n^B} \uparrow f_B\Big\} \tag{5.8}$$

belongs to the σ-field $\mathcal{E}$, and has, under μ, measure 1. Let us denote by D the intersection of all subsets of F of the form (5.8), as B ranges over $\mathcal{F}_0$, and all those of the form (5.7), as (B, B') ranges over all pairs of disjoint elements of $\mathcal{F}_1$. Then, D is an element of $\mathcal{E}$ with measure 1 under μ.

For every element x of D, let us denote by λ_x the function, with values in $[0, 1]$, defined on $\mathcal{F}_1$ by:

$$\lambda_x(B) = f_B(x) \qquad \text{for } B \in \mathcal{F}_1.$$

This is clearly an additive function on $\mathcal{F}_1$, and takes the value 1 on F. In addition, thanks to (5.5), Lemma (5.3) yields that the restriction of λ_x to $\mathcal{F}_0$ is countably additive. Thus, Carathéodory-Hahn's extension theorem yields that there exists a unique probability measure ν_x, defined on $\mathcal{F}$, which coincides with λ_x on the algebra $\mathcal{F}_0$. To extend this definition from D to F, choose an element y_0 in F and set

$$N(x, \cdot\,) = \begin{cases} \nu_x & \text{for } x \in D, \\ \varepsilon_{y_0} & \text{otherwise.} \end{cases}$$

Then, for any element B of $\mathcal{F}_0$, and for any element x of E, one has

$$N(x,B) = \begin{cases} \nu_x(B) = \lambda_x(B) = f_B(x) & \text{for } x \in B, \\ I_B(y_0) & \text{otherwise.} \end{cases}$$

It is thus clear that the function NI_B coincides with f_B on D and with a constant otherwise. Hence, for every bounded measurable function f on $(E, \mathcal{E})$, one has $\int f \otimes I_B \, \mathrm{d}\lambda = \int f f_B \, \mathrm{d}\mu = \int f N I_B \, \mathrm{d}\mu$. From the fact that $\mathcal{F}_0$ is a base for the σ-field $\mathcal{F}$, this is sufficient to conclude that N is a version of the disintegration of λ with respect to μ. □

(5.9) Remark. In the same hypotheses as the previous theorem, observe that Jiřina's condition is satisfied, in particular, if we suppose that F is a Polish space and that $\mathcal{F}$ is its corresponding Borel σ-field. In fact, in this case, the σ-field $\mathcal{F}$, being countably generated, admits a base which is a countable algebra; furthermore, the set of all compact subsets of F is also compact in the sense of Definition (5.1), and, thanks to Ulam's theorem (see VII(4.5)) the measure ν is tight, which ultimately guarantees that Jiřina's condition holds. Thus, Jiřina's theorem gives an alternative proof that Polish spaces are regular for the concept of disintegration, but that also gives a more abstract and general sufficient condition for a measurable space to be regular (for the concept of disintegration).

Chapter XI

Lebesgue Measure

This chapter is going to be extremely different from all the others. In fact, in lieu of working in a general abstract setting, or in a still rather abstract topological space, we will now be only working on a specific underlying space, and we shall also look at only one measure! Specifically, this chapter is devoted to studying the Borel-Lebesgue measure on the real line first, and then on the Euclidean space with n dimensions.

The ultimate goal of this chapter is to introduce the necessary results, like integration by parts and integration by substitution, that enable numerical calculations, and we shall see interesting numerical examples of calculations of integrals. After having done this in the unidimensional case, we shall move onto the more general case of functions of several variables, where we shall start by exploiting the properties of the Borel-Lebesgue measure as a product measure to reduce the calculation of integrals to iterate integrals of one-dimensional functions. In order to prove the change of variables in $\mathbb{R}^n$, we shall recall some fundamental theorems of differential calculus, whose proofs are elegant and interesting in their own fashion, and we shall shows that every diffeomorphism in $\mathbb{R}^n$ is *locally primitive.* This is going to be a pivotal concept to prove the change of variable formula. Finally, empowered by the change of variable, we shall complete the chapter discussing the important case of polar coordinates.

In spite of arriving so late in the book, this chapter is much easier to follow, and way less abstract, than any of the previous ones. Some of the reason why this chapter becomes so incredibly simple is largely due to the fact that all the hard work has already been done, and all we are going to do here is to translate in the specific case the results done in all generality before. But the purpose of this chapter is not just to show an example of application of the abstract theory, but to introduce, in a rigorous manner, a measure that is extraordinarily important in Mathematical Analysis, and that every mathematician should learn.

DOI: 10.1201/9781003724766-XI

1. Integration with respect to Borel-Lebesgue measure

This section is devoted to studying integration with respect to Borel-Lebesgue measure on $\mathbb{R}$, and we shall give some rules to enable explicit calculations of integrals. Before doing so, however, it is necessary to dwell a little upon the notation adopted for Borel-Lebesgue measure. In fact, the (Borel-)Lebesgue measure on $\mathbb{R}$ is customary denoted by the symbol $\mathcal{L}^1$, which turns out to be a rather unfortunate choice, because $\mathcal{L}^1$ also denotes the space of all integrable functions and, in particular, the space of $\mathcal{L}^1$-integrable functions would have to be denoted by $\mathcal{L}^1(\mathcal{L}^1)$ which is, at the same time, ugly and unclear. To avoid this repulsive expression, we shall write λ^1 for the Borel-Lebesgue measure on the real line (and λ^n for the n-dimensional Borel-Lebesgue measure), so that the space of all integrable functions with respect to the Borel-Lebesgue measure may be written as $\mathcal{L}^1(\lambda^1)$. This notation tends to be the one adopted when speaking in abstract, and it has the further advantage of preserving the tradition of indicating measures with Greek letters, but it is not the preferable choice in some applications, especially in Mathematical Analysis and Geometric Measure Theory, where the general consensus goes towards $\mathcal{L}^1$.

Let us also mention that, if E is a measurable subset of $\mathbb{R}$, in lieu of using the expression $\mathcal{L}^1(\lambda^1_E)$, we shall always use the symbol $\mathcal{L}^1(E)$ to denote the semi-normed space of all real valued functions defined on E, integrable with respect to the trace of the Borel-Lebesgue measure λ^1 onto E. Likewise, we shall always denote by $L^1(E)$ the quotient space of all equivalence classes, modulo λ^1, of all numeric functions defined on E and integrable with respect to the trace λ^1_E of λ^1 onto E.

As regard to integrals with respect to Borel-Lebesgue measure, instead, if f is an element of $\mathcal{L}^1(E)$, we shall preferably adopt the notation $\int_E f(x)\,\mathrm{d}x$, consisting in always showing the dummy variable x in the integrand, and completely omitting the measure, rather than using more cumbersome notations such as $\int_E f(x)\,\lambda^1(\mathrm{d}x)$, $\int_E f(x)\,\mathrm{d}\lambda^1(x)$, or $\int_E f\,\mathrm{d}\lambda^1$. Notwithstanding this notational design, we shall always retain the right to utilise a notation such as these only if, in some specific context, a cumbersome yet rigorous notation contributes to an increasing in the overall conceptual clarity.

The first important remark about integration with respect to Borel-Lebesgue measure is that it is invariant under translation (see Example I(5.7)). We shall now prove that the invariance under translation of Borel-Lebesgue measure extends to the integral.

(1.1) Proposition. *Let f be a positive measurable function or an integrable function defined on $\mathbb{R}$. For any real number y, let us denote by τ_y the 'translation operator' defined by $\tau_y(x) = x + y$.*

Then, $\int (f \circ \tau_y)\,\mathrm{d}\lambda^1 = \int f\,\mathrm{d}\lambda^1$, that is to say

$$\int f(x+y)\,\mathrm{d}x = \int f(x)\,\mathrm{d}x.$$

Proof. In the special case of the real line, Example I(5.7) may be rephrased by saying that, for every real number y, one has $\tau_y(\lambda^1) = \lambda^1$, where τ_y if the function $x \mapsto x + y$. The conclusion thus follows from Theorem II(4.3). □

Now, if f is a function defined on a part E of $\mathbb{R}$ and if A is a Borel subset of E, the function $I_A f$ can always be thought as a function, defined on $\mathbb{R}$, which coincides with f on A, and with 0 otherwise. We shall say that a function f is **locally integrable** if, for every compact set K included in E, the function $I_K f$ is integrable. The class of all locally integrable numeric functions defined on E shall be denoted by $\mathcal{L}^1_{\text{loc}}(E)$. Similarly, we shall denote by $L^1_{\text{loc}}(E)$ the set of all classes of locally integrable functions defined on E. It is not difficult to show that both these classes are Riesz spaces.

If $f \in \mathcal{L}^1_{\text{loc}}(E)$ is a locally integrable function on E, if K is a bounded and closed interval with extremes a and b, with $a < b$, that is to say, if $K = [a, b]$, in lieu of the notation $\int_K f(x)\,\mathrm{d}x$, we shall use the most common notation

$$\int_a^b f(x)\,\mathrm{d}x.$$

This latter notation may appear ambiguous, for it does not distinguish, for example, between the integral over the closed interval $[a, b]$ and one over the open interval (a, b); on the other hand, keeping into account the fact that the two functions $I_{[a,b]}f$ and $I_{(a,b)}f$ are different only on the set $\{a, b\}$, which is in turn negligible (the Borel-Lebesgue measure is diffuse), it follows that these two functions are equivalent modulo λ^1 and their integrals coincide. Furthermore, for any pair of real numbers a, b with $a \leq b$, such that the interval with extremes a and b is included in E, we shall set by definition

$$\int_b^a f(x)\,\mathrm{d}x = -\int_a^b f(x)\,\mathrm{d}x$$

and also $\int_a^a f(x)\,\mathrm{d}x = 0$, which allows us to use the relation

$$\int_a^b f(x)\,\mathrm{d}x = \int_a^c f(x)\,\mathrm{d}x + \int_c^b f(x)\,\mathrm{d}x$$

for any triple a, b, c of real numbers, such that E contains an interval containing all three of them.

(1.2) Theorem. *Let f be a locally integrable function defined on an interval $[a, b]$, let x_0 be a point of $[a, b]$, and*

$$F(x) = \int_{x_0}^x f(t)\,\mathrm{d}t \qquad \text{for every } x \text{ in } [a, b].$$

Then, the function F is continuous on $[a, b]$.

Proof. Suppose, with no loss of generality, that $x_0 < x$, and let (x_n) be any increasing sequence of elements of $[a, b]$, which converges towards x. Observe that the sequence of integrable functions on $[a, b]$

$$(I_{[x_0,x_n]}f),$$

converges towards the integrable function $I_{[x_0,x)}f$. In addition, the absolute values of the elements of this sequence are bounded from above by $I_{[x_0,x]}|f|$, which is also integrable. Hence, Lebesgue's theorem on dominated convergence yields

$$\lim_n \int_{x_0}^{x_n} f(t)\,\mathrm{d}t = \int_{x_0}^{x} f(t)\,\mathrm{d}t$$

and this is sufficient to prove the theorem. □

We already know that all continuous functions are Borel measurable. Further, they are also locally integrable, because they are bounded on every compact set (thanks to Weierstrass theorem). For these functions, one can also prove the following result, which is extremely important for the explicit calculation of integrals with respect to the Borel-Lebesgue measure.

(1.3) Theorem (Fundamental theorem of calculus). *Let f be a real valued function of one real variable, defined on an interval $[a, b]$, and let x_0 be a point of $[a, b]$. If f is continuous, then the function*

$$F(x) = \int_{x_0}^{x} f(t)\,\mathrm{d}t$$

is differentiable, and its derivative in the point x is $f(x)$.

Proof. Let x be any point of $[a, b]$. With no loss of generality, we may also assume that $x > x_0$. Let us differentiate from first principle the function F. To this end, for any real number h,

$$\frac{F(x+h) - F(x)}{h} = \frac{1}{h}\int_{x}^{x+h} f(t)\,\mathrm{d}t.$$

Since the function f is continuous in every point of E, it is continuous in particular on x, and therefore, for each strictly positive real number ε, there exists a strictly positive real number δ, such that, for every real number t, with $|t - x| < \delta$, we have

$$f(x) - \varepsilon < f(t) < f(x) + \varepsilon. \tag{1.4}$$

In particular, if $|h| < \delta$, then, for every real number t between x and $x + h$, one has $|t - x| < |h| < \delta$ and therefore the inequality (1.4) holds. Now, by means of the properties of monotonicity and linearity of integrals, the chain of inequalities

$$f(x) - \varepsilon < \frac{1}{h}\int_{x}^{x+h} f(t)\,\mathrm{d}t < f(x) + \varepsilon$$

holds, hence

$$\lim_{h\to 0} \frac{F(x+h) - F(x)}{h} = \lim_{h\to 0} \frac{1}{h}\int_{x}^{x+h} f(t)\,\mathrm{d}t = f(x)$$

which proves that the derivative of $F(x)$ at x is $f(x)$. □

In order to understand the importance of the fundamental theorem of calculus, let us recall that, if f is a real function, defined and continuous on an interval $[a, b]$, a real function F, also defined on $[a, b]$, is called a **primitive** of f if it is differentiable and its derivative coincides with f. By means of Lagrange's mean value theorem, if F and G are two such primitives of f, then they only differ by a constant. Also, by means of the fundamental theorem of calculus, one of these primitives can be written in the form $F(x) = \int_{x_0}^{x} f(t)\,\mathrm{d}t$ and any other primitive of f can be written in the form

$$\int_{x_0}^{x} f(t)\,\mathrm{d}t + c.$$

Putting together these two remarks, the following result is straightforward.

(1.5) Corollary. *If f is a real function defined and continuous on the interval $[a, b]$ and if G is any primitive of f, then*

$$\int_{a}^{b} f(t)\,\mathrm{d}t = G(b) - G(a).$$

In the application of this corollary of the fundamental theorem of calculus, it is customary to adopted a specific notation to emphasise the right-hand side of the previous equality. Precisely, the term $G(b) - G(a)$, is often shortened with the notation

$$\Big[G(x)\Big]_{x=a}^{x=b}.$$

This is particularly convenient to highlight the structure of the primitive that is used to calculate the integral.

(1.6) Remark. Among the many possible ways to state that the function $F(x) = \int_{a}^{x} f(t)\,\mathrm{d}t$ is a primitive of $f(x)$, some occasionally state that

$$\frac{\mathrm{d}}{\mathrm{d}x}\int_{x_0}^{x} f(t)\,\mathrm{d}t = f(x).$$

Furthermore, using the notation f' for $(\mathrm{d}f/\,\mathrm{d}x)$, Theorem (1.5) shows that

$$\int_{x_0}^{x} f'(x)\,\mathrm{d}x = f(x) - f(x_0),$$

and this formula holds for any function f differentiable on the interval $[a, b]$, because a primitive of f' is the function f. This justifies why the research of a primitive for a function is often referred to as **integration** (even though it is not carried out calculating an integral). We must also remark that this link between differentiation and integration is the reason why, erroneously and with a clumsy and potentially dangerous language, some people state that 'integration is the inverse of differentiation'. This latter statement is not just unfortunate but a real and proper mistake, because it mixes up the two notions of integral and of primitive, which are two different concepts, and where the latter, rather than the former, may be though as a sort of inverse operation of the derivative, notwithstanding the fact that the derivative is not injective, whence not invertible either. Obviously, being a linear function, the homomorphism theorem ensures that the quotient of the derivative over its kernel is injective. But the fact that this kernel is in some way 'simple' does not in any way allow to identify the derivative with the corresponding injective linear function induced by the homomorphism theorem.

The rest of this section is devoted to some results that may be utilised to simplify numerical calculations. The first of them is referred to as 'integrations by parts', and the second is the theorem of 'change of variables' or 'substitution'.

(1.7) Proposition (Integration by parts). *Let f and g be two continuous functions defined on the same interval I, and let F be a primitive of f and G a primitive of g. If a and b are two elements of I, such that the interval with extremes a and b is all included in I, then*

$$\int_a^b F(x)g(x)\,\mathrm{d}x = \Big[F(x)G(x)\Big]_{x=a}^{x=b} - \int_a^b f(x)G(x)\,\mathrm{d}x.$$

Proof. Observe that the function FG is a primitive of $Fg + fG$, and, using Theorem (1.5), it follows

$$\int_a^b \Big[F(x)g(x) + f(x)G(x)\Big]\,\mathrm{d}x = \Big[F(x)G(x)\Big]_{x=a}^{x=b}$$

and the thesis follows using the linearity of integrals. □

Integration by parts is indeed quite interesting, for it allows to transform an integral of a function Fg into an integral of the function fG, that is to say, it allows to differentiate the function F, providing to replace g with a primitive G. This is useful when the primitive of G is easy to find and one would benefit from differentiating the function f, e.g. when $f(x) = x^n$ and its derivative, $f'(x) = nx^{n-1}$, would have a smaller degree.

(1.8) Example. We wish to calculate the integral $\displaystyle\int_0^1 x^2 e^x\,\mathrm{d}x$.

This can be done using integration by parts. To this end, let us choose two suitable function $F(x)$ and $g(x)$ to utilise. Let us observe that the best choice would be to set $F(x) = x^2$ and $g(x) = e^x$, because, in this way, one gets $f(x) = 2x$ and $G(x) = e^x$ and the degree of $F(x)$ is reduced. This gives

$$\int_0^1 x^2 e^x\,\mathrm{d}x = \Big[x^2 e^x\Big]_{x=0}^{x=1} - \int_0^1 2xe^x\,\mathrm{d}x = e - 2\int_0^1 xe^x\,\mathrm{d}x.$$

Now, to calculate the integral on the right-hand side of the previous equality, it is convenient to use integration by parts again, this time choosing $F(x) = x$

and $g(x) = e^x$. This gives therefore $f(x) = 1$ and $G(x) = e^x$, whence

$$\int_0^1 xe^x \, \mathrm{d}x = \Big[xe^x\Big]_{x=0}^{x=1} - \int_0^1 e^x \, \mathrm{d}x = e - \Big[e^x\Big]_{x=0}^{x=1} = e - (e-1) = 1.$$

Putting everything together, this gives finally

$$\int_0^1 x^2 e^x \, \mathrm{d}x = e - 2\int_0^1 xe^x \, \mathrm{d}x = e - 2 \cdot 1 = e - 2.$$

On a completely different level, substitution is perhaps the main tool not only on a theoretical level, but also when it comes to numerical calculations.

(1.9) Proposition (Integration by substitution). *Let I be an interval of the real line and let φ be a differentiable monotone function defined on I. For every real function f, integrable on the image of I under φ, and for each pair of real numbers α and β in I, one has therefore*

$$\int_a^b f(x) \, \mathrm{d}x = \int_\alpha^\beta f\big(\varphi(u)\big)\varphi'(u) \, \mathrm{d}u \tag{1.10}$$

where $a = \varphi(\alpha)$ and $b = \varphi(\beta)$.

Proof. Let us start by observing that, if (1.10) holds for the indicator function of every Borel subset of $[a,b]$, then it also holds when f is a simple function on $[a,b]$ by linearity of integrals, hence for every positive Borel function on $[a,b]$ by Beppo Levi's theorem, hence for every integrable function f on $[a,b]$ by linearity of integrals again. Thus it suffices to prove the formula only when f is the indicator function of a Borel subset of $[a,b]$.

To prove (1.10) in this special case, observe that, since φ is monotone, it is either increasing or decreasing: Assume that φ is increasing, for the proof is analogous when φ is decreasing. Under this additional hypothesis, the function φ' is positive. Furthermore, since φ' is continuous, it is also Borel measurable. Thus, let us set

$$\lambda(A) = \int_a^b I_A(x) \, \mathrm{d}x, \qquad \mu(A) = \int_\alpha^\beta I_A\big(\varphi(u)\big)\varphi'(u) \, \mathrm{d}u$$

for every Borel set A included in the interval $[a,b]$, image of $[\alpha,\beta]$ under φ. By definition, λ and μ are two Borel measures on $[a,b]$. In particular, on every sub-interval $[c,d]$ of $[a,b]$, one has $\lambda([c,d]) = \int_a^b I_{[c,d]}(x) \, \mathrm{d}x = d - c$, and since the image of any interval through a continuous function is an interval,

$$\begin{aligned}
\mu\big([c,d]\big) &= \int_\alpha^\beta I_{[c,d]}\big(\varphi(u)\big)\varphi'(u) \, \mathrm{d}u \\
&= \int_\alpha^\beta I_{[\varphi^{-1}(c),\varphi^{-1}(d)]}(u)\varphi'(u) \, \mathrm{d}u \\
&= \int_{\varphi^{-1}(c)}^{\varphi^{-1}(d)} \varphi'(y) \, \mathrm{d}u.
\end{aligned}$$

Furthermore, since φ is a primitive of φ, thanks to Theorem (1.5), one has

$$\mu([c,d]) = \varphi(\varphi^{-1}(d)) - \varphi(\varphi^{-1}(c)) = d - c.$$

Since the collection of all sub-intervals of $[a,b]$ form a base for its Borel σ-field, the fundamental criterion for the coincidence of two measures enables to infer that the measures λ and μ coincide. □

From a strictly formal point of view, when the integration by substitution is employed to calculate an integral, one may think of considering the substitution consisting in setting $x = \varphi(u)$, and correspondingly $\mathrm{d}x = \varphi'(u)\,\mathrm{d}u$ and also $a = \varphi(\alpha)$ and $b = \varphi(\beta)$. The formulae involving the symbols $\mathrm{d}x$ and $\mathrm{d}u$, however, are only formal shortcuts and have no real meaning outside the sign of integral. The following examples show how these symbols can be utilised in calculations.

(1.11) Example. We wish to calculate the integral $\displaystyle\int_1^{\sqrt{2}} 2xe^{x^2}\,\mathrm{d}x$.

Let us set $y = x^2$ and therefore $\mathrm{d}y = 2x\,\mathrm{d}x$. Thus one has

$$\int_1^{\sqrt{2}} e^{x^2}\,2x\,\mathrm{d}x = \int_{1^2}^{(\sqrt{2})^2} e^y\,\mathrm{d}y = \int_1^2 e^y\,\mathrm{d}y = \Big[\,e^y\,\Big]_{y=0}^{y=1} = e(e-1).$$

(1.12) Example. We wish to calculate the integral $\displaystyle\int_0^1 \sqrt{1-x^2}\,\mathrm{d}x$.

To calculate this integral let us set $x = \sin t$, which yields $\mathrm{d}x = \cos t\,\mathrm{d}t$. For the extremes of integration, observe that $\sin 0 = 0$ and $\sin \frac{1}{2}\pi = 1$, whence the integral becomes

$$\begin{aligned}\int_0^1 \sqrt{1-x^2}\,\mathrm{d}x &= \int_0^{\frac{1}{2}\pi} \sqrt{1-\sin^2 t}\,\cos t\,\mathrm{d}t = \int_0^{\frac{1}{2}\pi} \cos^2 t\,\mathrm{d}t \\ &= \tfrac{1}{2}\int_0^{\frac{1}{2}\pi} (1-\cos 2t)\,\mathrm{d}t = \tfrac{1}{4}\pi - \tfrac{1}{2}\int_0^{\frac{1}{2}\pi} \cos 2t\,\mathrm{d}t.\end{aligned}$$

To complete the calculation, it suffices to calculate the integral on the right-hand side. This is also easy to calculate using substitution. Indeed, setting $s = 2t$, whence $\mathrm{d}s = 2\,\mathrm{d}t$, and therefore

$$\int_0^{\frac{1}{2}\pi} \cos 2t\,\mathrm{d}t = \tfrac{1}{2}\int_0^{\pi} \cos s\,\mathrm{d}s = 0.$$

Thus, in the end,

$$\int_0^1 \sqrt{1-x^2}\,\mathrm{d}x = \tfrac{1}{4}\pi.$$

Hitherto, we have only studied integrable functions on a bounded closed interval of the real line, showing in particular two methods to calculate integrals. The rest of this section shows how to work with an unbounded interval. Let us start by the following.

(1.13) Proposition. *Let f be a positive function or an integrable function, defined on a subset E of $\mathbb{R}$ which contains an interval of the form $[c, +\infty)$. Then,*

$$\int_c^{+\infty} f(x)\,\mathrm{d}x = \lim_{b\to+\infty} \int_c^b f(x)\,\mathrm{d}x.$$

Proof. Let (b_n) be any increasing, unbounded sequence of positive real numbers all greater than c. Let us start by assuming that f is positive. In this particular case, let us notice that the function $I_{[c,+\infty)}f$ is the limit of the sequence $\left(I_{[c,b_n)}f\right)$ and the thesis of the theorem follows immediately from Beppo Levi's theorem. Now suppose that the function f is integrable. In this case, observe that the sequence $\left(I_{[c,b_n)}f\right)$ is made of integrable functions and each element is bounded from above, in absolute value, by the integrable function $I_{[c,+\infty)}|f|$ and the thesis follows, this time, from Lebesgue's theorem on dominated convergence. □

Alongside the latter, there is an equivalent result, which holds for functions defined on an interval of the form $(-\infty, c]$, the proof of which follows from the previous proposition applying the substitution given by setting $x = -u$. Precisely, the following holds.

(1.14) Proposition. *Let f be a positive functions or an integrable function, defined on a subset E of $\mathbb{R}$ which contains an interval of the form $(-\infty, c]$. Then,*

$$\int_{-\infty}^{c} f(x)\,\mathrm{d}x = \lim_{a\to-\infty} \int_a^c f(x)\,\mathrm{d}x.$$

Putting together the previous two propositions, the following becomes obvious.

(1.15) Proposition. *Let f be positive or integrable function defined on the whole real line. Then, for each real number c, one has*

$$\int_{-\infty}^{+\infty} f(x)\,\mathrm{d}x = \lim_{a\to-\infty} \int_a^c f(x)\,\mathrm{d}x + \lim_{b\to+\infty} \int_c^b f(x)\,\mathrm{d}x.$$

Let us observe that the fact that the limits that appear on the right-hand side of the formulae in the Propositions (1.13), (1.14), and (1.15), exist and are finite, is not sufficient to guarantee that the function f is integrable. This justifies the following definition.

(1.16) Definition. In the same hypotheses as in Propositions (1.13), (1.14), and (1.15), respectively, the function f shall be called integrable **improperly** if the limit that appears on the right-hand side of the formula exists and is finite, and the integral thereby defined by the formula shall be called an **improper** integral.

The following example shows that, as we have mentioned before, a function may be improperly integrable without being integrable.

(1.17) Example. The function

$$f(x) = \begin{cases} \dfrac{\sin x}{x} & \text{if } x \geq 0, \\ 1 & \text{if } x = 0, \end{cases}$$

is continuous in $[0, +\infty)$, and it is integrable improperly. In fact, integrating by parts, we find

$$\int_0^{+\infty} \frac{\sin x}{x}\, \mathrm{d}x = \left[-\frac{\cos x}{x}\right]_{x=0}^{x\uparrow+\infty} - \int_0^{+\infty} \frac{\cos x}{x^2}\, \mathrm{d}x = 1 - \int_0^{+\infty} \frac{\cos x}{x^2}\, \mathrm{d}x.$$

Now, the integral on the right-hand side of the previous chain of equalities is convergent, because the integrand, i.e. the function $\cos x/x^2$, is dominated, in absolute value, by the function $1/x^2$, which is integrable in $(0, +\infty)$. Indeed,

$$\left|\int_0^{+\infty} \frac{\cos x}{x^2}\, \mathrm{d}x\right| \leq \int_0^{+\infty} \left|\frac{\cos x}{x^2}\right| \mathrm{d}x \leq \int_0^{+\infty} \frac{1}{x^2}\, \mathrm{d}x.$$

This proves that the integral of f exists improperly in $[0, +\infty)$.

On the other hand, in n is a positive integer, we also have

$$\begin{aligned}
\int_0^{2n\pi} \left|\frac{\sin x}{x}\right| \mathrm{d}x &= \sum_{k=1}^{n} \int_{2(k-1)\pi}^{2k\pi} \left|\frac{\sin x}{x}\right| \mathrm{d}x \\
&\geq \sum_{k=1}^{n} \frac{1}{2k\pi} \int_{2(k-1)\pi}^{2k\pi} |\sin x|\, \mathrm{d}x \\
&= \sum_{k=1}^{n} \frac{1}{2k\pi} \int_0^{2\pi} |\sin x|\, \mathrm{d}x \\
&= \frac{2}{\pi} \sum_{k=1}^{n} \frac{1}{k}.
\end{aligned}$$

Taking the limit as n approaches ∞, observe that the harmonic series on the right-hand side is divergent, hence $\int_0^{+\infty} f(x)\, \mathrm{d}x = +\infty$. As a result, the function f is not integrable, or, what is the same, $f \notin \mathcal{L}^1(0, +\infty)$.

2. The Borel-Lebesgue measure in $\mathbb{R}^n$

The one-dimensional Borel-Lebesgue measure λ^1 in $\mathbb{R}$ has been defined as the only measure on the Borel σ-field $\mathcal{B}(\mathbb{R})$ such that

$$\lambda^1\big([a,b]\big) = b - a,$$

for every pair a, b of real numbers, with $a < b$. Using the product measure, we can now define the n-dimensional measure to the Borel σ-field of $\mathbb{R}^n$, for any $n \geq 1$. To this end, observe that, by means of Theorem I(7.3), the Borel σ-field $\mathcal{B}(\mathbb{R}^n)$ coincides with $\mathcal{B}(\mathbb{R})^{\otimes n}$, the product of n copies of the σ-field $\mathcal{B}(\mathbb{R})$. This observation justifies the following definition.

(2.1) Definition. On the measurable space $(\mathbb{R}^n, \mathcal{B}(\mathbb{R}^n))$, product of n copies of $(\mathbb{R}, \mathcal{B}(\mathbb{R}))$, the **$n$-dimensional Borel-Lebesgue measure**, is denoted by λ^n, and it is defined by

$$\lambda^n = \lambda^{n-1} \otimes \lambda^1 = \lambda^1 \otimes \cdots \otimes \lambda^1 \qquad (n \text{ times}).$$

Recall that the measure λ^n constructed in Definition (2.1) is the only measure on the measurable space $(\mathbb{R}^n, \mathcal{B}(\mathbb{R}^n))$ such that

$$\lambda^n\big(\textstyle\prod_{i=1}^n [a_i, b_i]\big) = \prod_{i=1}^n \lambda^1\big([a_i, b_i]\big) = \prod_{i=1}^n (b_i - a_i).$$

If k is any whole number between 1 and $n-1$, it is easily shown that

$$\lambda^n = \lambda^{n-k} \otimes \lambda^k.$$

It is also easy to check that the set $\mathcal{Q}$ of all products of closed intervals of the form $[a, b]$, with $a < b$, is stable under finite intersection, and generates the Borel σ-field of $\mathbb{R}^n$, hence it is a base for $\mathcal{B}(\mathbb{R}^n)$. Every element of such a base shall be called a **closed cuboid**. Similarly, any product of intervals, not necessarily closed, shall be called a **cuboid** in $\mathbb{R}^n$. The set of all (non necessarily closed) cuboids in $\mathbb{R}^n$ is clearly another base for the Borel σ-field.

Sometimes, the set of all cuboids might be too general and one may need a slightly more special type of base for the Borel σ-field of $\mathbb{R}^n$. To this end, if $a = (a_1, \ldots, a_n)$ is an element of $\mathbb{R}^n$ and δ is a positive real number, set

$$Q(a, \delta) = \Big\{x \in \mathbb{R}^n : a_i \leq x_i < a_i + \delta \text{ for all } i = 1, \ldots, n\Big\} = \prod_{i=1}^n [a_i, a_i + \delta).$$

This set shall be called the **δ-cube** with corner at a.

Other times, cuboids or δ-cubes are just too many, and one would rather take a countable base. A very useful countable base can be constructed by taking advantage of the special structure of dyadic numbers, which form a dense subset of the real numbers. To this end, let us set

$$\mathcal{D}_k = \left\{Q(2^{-k}q, 2^{-k}) : q \in \mathbb{Z}^n\right\}.$$

Every element of $\mathcal{D}_k$ is Borel measurable, and its Borel-Lebesgue measure is 2^{-nk}. Furthermore, all elements of $\mathcal{D}_k$ are mutually disjoint, and their union coincides with $\mathbb{R}^n$. Moreover, if $h < k$, $Q \in \mathcal{D}_h$ and $Q' \in \mathcal{D}_k$, then either $Q' \subseteq Q$ or $Q \cap Q' = \emptyset$.

Thus, if we set $\mathcal{D} = \bigcup_k \mathcal{D}_k$, this is a countable union of countable sets, hence it is countable, and, for every pair Q, Q' of elements of $\mathcal{D}$, either they are disjoint, or they intersect and one of the two sets is included in the other. Either way, by adding $\emptyset$ to $\mathcal{D}$, this proves that $\mathcal{D}$ is stable under finite intersection, hence it is a countable base of $\mathcal{B}(\mathbb{R}^n)$.

(2.2) Proposition. *The following properties hold:*

(a) (Translation invariance) *For any $A \in \mathcal{B}(\mathbb{R}^n)$ and every $x \in \mathbb{R}^n$, one has $\lambda^n(A + x) = \lambda^n(A)$, where*

$$A + x = \left\{y + x : y \in A\right\}.$$

(b) *If μ is any translation invariant measure, defined on $(\mathbb{R}^n, \mathcal{B}(\mathbb{R}^n))$, such that $\mu(K) < \infty$, for every compact set K in $\mathbb{R}^n$, then there exists a positive real number c_μ such that*

$$\mu(A) = c_\mu \lambda^n(A) \qquad \text{for every element } A \text{ in } \mathcal{B}(\mathbb{R}^n).$$

(c) (Rotation invariance) *For any orthogonal map R on $\mathbb{R}^n$, one has*

$$\lambda^n\big(R(A)\big) = \lambda^n(A) \qquad \text{for every element } A \text{ in } \mathcal{B}(\mathbb{R}^n).$$

(d) *For any linear transformation $T : \mathbb{R}^n \to \mathbb{R}^n$, one has*

$$\lambda^n\big(T(A)\big) = |\det T|\lambda^n(A) \qquad \text{for every element } A \text{ in } \mathcal{B}(\mathbb{R}^n).$$

Proof. To prove (a), it suffices to observe that the two measures

$$A \mapsto \lambda^n(A), \qquad A \mapsto \lambda^n(A + x)$$

defined on the measurable space $(\mathbb{R}^n, \mathcal{B}(\mathbb{R}^n))$ agree on the base $\mathcal{Q}$ of all cuboids, and then apply the fundamental criterion for coincidence of two measures (see I(5.6)).

To prove (b), let us start by observing that, since μ is translation invariant, for any positive integer k, all elements of $\mathcal{D}_k$ have the same measure under μ. Thus, let us take any set Q_0 in $\mathcal{D}_0$, and set $\mu(Q_0) = c_\mu$. Since Q_0 is bounded, it is included in a compact set, whence its measure is finite. This means that the constant c_μ is finite. Now observe that, for every index k, the set Q_0 is the disjoint union of 2^{nk} elements of $\mathcal{D}_k$, which, as we have mentioned before, all share the same measure under μ. Thus,

$$\mu(Q) = 2^{-nk}c_\mu = c_\mu\lambda^n(Q) \qquad \text{for every index } k \text{ and every } Q \text{ in } \mathcal{D}_k.$$

Since the measures $A \mapsto \mu(A)$ and $A \mapsto c_\mu\lambda^n(A)$ agree on the base $\mathcal{D}$, the fundamental criterion for coincidence of two measures shows that these two measures are indeed identical.

Let us now prove (c). To this end, observe that, for any element x of $\mathbb{R}^n$, one has $R(A+x) = R(A)+R(x)$, hence the measure $A \mapsto \lambda^n(R(A))$ is translation invariant. Thus, by means of (b), there exists a positive constant c such that $\lambda^n(R(A)) = c\lambda^n(A)$ for every element A of $\mathcal{B}(\mathbb{R}^n)$. Finally, by choosing A equal to the unit ball with centre in the origin, we get $c = 1$.

Finally, let us prove (d). To this end, let us start by recalling that any linear transformation T can always be written in the form $T = RS$, where R is orthogonal, and S is symmetric and positive definite. Then, notice that, on one hand, $|\det T| = |\det(R)\det(S)| = \det S$, because the determinant of an orthogonal transformation is always ± 1; and, on the other hand, thanks to (c), we know that $\lambda^n(T(A)) = \lambda^n(R(S(A))) = \lambda^n(S(A))$.

Therefore, all is reduced to proving that $\lambda^n(S(A)) = \det(S)\lambda^n(A)$ for any symmetric and positive definite transformation S. To prove this special case of (d), it suffices to observe that, for every point x of $\mathbb{R}^n$, one has $S(A+x) = S(A) + S(x)$, which, thanks to (a), yields that the measure $A \mapsto \lambda^n(S(A))$ is translation invariant, whence, by means of (b), there exist a positive real number c such that $\lambda^n(S(A)) = c\lambda^n(A)$, for every Borel set A in $\mathbb{R}^n$. To find the value of the constant c, it suffices to choose a suitable Borel measurable set A: Consider an orthonormal base $\{v_1, \ldots, v_n\}$ of eigenvectors for S, and denote by $\alpha_1, \ldots, \alpha_n$ the corresponding eigenvalues. Since S is positive definite, every eigenvalue is positive. Take the set

$$A = \Big\{\textstyle\sum_{i=1}^n x_i v_i : 0 \le x_i \le 1 \text{ for } i = 1, \ldots, n\Big\},$$

and observe that

$$S(A) = \Big\{\textstyle\sum_{i=1}^n \alpha_i x_i v_i : 0 \le x_i \le 1 \text{ for } i = 1, \ldots, n\Big\}.$$

Furthermore, if U denotes the orthonormal transformation that maps the orthonormal base $\{v_1, \ldots, v_n\}$ into the canonical base of $\mathbb{R}^n$, then, clearly, one has $U(A) = [0,1]^n$ and $U(S(A)) = \prod_{i=1}^n [0, \alpha_i]$. Thanks to (c), we have $\lambda^n(A) = \lambda^n(U(A)) = 1$, and $\lambda^n(S(A)) = \lambda^n(U(S(A))) = \alpha_1 \cdots \alpha_n$. Since $\det S = \alpha_1 \cdots \alpha_n$, this fully proves the proposition. □

Prior to discussing integrals with respect to the n-dimensional Borel-Lebesgue measure, it is necessary to dwell on the notation and the language used for integrals with respect to this measure. If Ω is a connected Borel set of $\mathbb{R}^n$, which we shall refer to as a **domain**, we shall say that a real function f, defined on Ω, is integrable **on** Ω if the function $I_\Omega f$, which can be thought as the function defined on the whole space $\mathbb{R}^n$ that coincides with f on Ω and is zero on the complement of Ω, is integrable with respect to λ^n. In this case, the integral of f on Ω shall be denoted by $\int_\Omega f(x)\,\mathrm{d}x$. Generally, the notation $\int_\Omega f\,\mathrm{d}\lambda^n$ shall not be employed, but sometimes the notation $\int_\Omega f(x)\,\lambda^n(\mathrm{d}x)$ may be adopted in order to remind the fact that the measure is the Borel-Lebesgue measure. If $x = (x_1, \ldots, x_n)$, we may also use the notation $\int \cdots \int_\Omega f(x_1, \ldots, x_n)\,\mathrm{d}x_1 \cdots \mathrm{d}x_n$, where the integral sign is repeated n times. (Typically, this happens only when n is small enough, that is to say, for $n = 2$, 3 or 4.) An integral such as those shall be referred to as a **multiple integral**.

In particular, if $n = 2$ it shall be called a **double integral** and if $n = 3$ it shall be called a **triple integral**, and they shall be written as

$$\iint_\Omega f(x,y)\,\mathrm{d}x\,\mathrm{d}y \quad \text{and} \quad \iiint_\Omega f(x,y,z)\,\mathrm{d}x\,\mathrm{d}y\,\mathrm{d}z,$$

respectively.

Let us stress the fact that a multiple integral is an integral with respect to the n-dimensional Borel-Lebesgue measure, that is to say, to the product of n copies of the 1-dimensional Borel-Lebesgue measure, and we already know several techniques (namely integration by substitution, integration by parts) to calculate integrals with respect to the latter. Using Fubini's theorem, we can now link these techniques that we already know for 1-dimensional integrals to multiple integrals, and use them to calculate multiple integrals.

To this end, let us start by assuming that Ω is of the form $E \times F$, where E is a Borel subset of $\mathbb{R}^k$ and F is a Borel subset of $\mathbb{R}^{n-k}$, and observe that any element z of $\mathbb{R}^n$ may always be identified with a pair (x, y), where x is an element of $\mathbb{R}^k$ and y an element of $\mathbb{R}^{n-k}$. Thus, the integral with respect to λ^n can be identified with the 'double integral' with respect to the product measure $\lambda^k \otimes \lambda^{n-k}$, which can also be written as $\iint f(x,y)\,\mathrm{d}x\,\mathrm{d}y$. Thus, applying Fubini's theorem, we get

$$\iint_{E\times F} f(x,y)\,\mathrm{d}x\,\mathrm{d}y = \int_F \left[\int_E f(x,y)\,\mathrm{d}x\right] \mathrm{d}y = \int_E \left[\int_F f(x,y)\,\mathrm{d}y\right] \mathrm{d}x.$$

This formula shows a possible way to 'reduce' a multiple integral into a sequence of two 'iterated' integrals in lower dimensions. Clearly, the actual problem consists in being able to write the domain Ω in the form $E \times F$, and also to be able to integrate the function f on at least one of the two sets E or F. Fortunately, however, it often happens to be able to integrate over a special type of domains where this is quite simple to be done. Even though this is a special case, this type of domains are extremely common.

(2.3) Definition. A domain Ω in $\mathbb{R}^n$ shall be called **normal** if there exists a domain E in $\mathbb{R}^{n-1}$ and two real function g_1 and g_2, defined on E, such that

$$\Omega = \Big\{(x, y) \in \mathbb{R}^{n-1} \times \mathbb{R} : x \in E,\, g_1(x) \le y \le g_2(x)\Big\}.$$

A direct consequence of Fubini's theorem is that, if Ω is that, if Ω is normal, and if f is a real function on Ω, integrable on Ω, then

$$\iint_\Omega f(x, y)\,\mathrm{d}x\,\mathrm{d}y = \int_E \left[\int_{g_1(x)}^{g_2(x)} f(x, y)\,\mathrm{d}y\right] \mathrm{d}x.$$

Using this formula, is now possible to reduce the dimension of integration by 1, and this makes the integral simpler to handle. In many practical cases, this process can be repeated inductively, transforming an integral in $\mathbb{R}^n$ into n iterated integrals in one dimension.

(2.4) Example. Let $f(x, y) = xy^2$, and consider the set Ω of all points in $\mathbb{R}^2$ that satisfy the relations $0 \le x \le 1$ and $x^2 \le y \le x$, which is, clearly, a normal domain. Applying the method discussed above, we get:

$$\begin{aligned}\iint_\Omega f(x, y)\,\mathrm{d}x\,\mathrm{d}y &= \int_0^1 \left[\int_{x^2}^{x} xy^2\,\mathrm{d}y\right] \mathrm{d}x = \int_0^1 x\left[\tfrac{1}{3}y^3\right]_{y=x^2}^{y=x} \mathrm{d}x\\ &= \tfrac{1}{3}\int_0^1 \left(x^4 - x^7\right) \mathrm{d}x = \tfrac{1}{3}\left[\tfrac{1}{5}x^5 - \tfrac{1}{8}x^8\right]_{x=0}^{x=1} = \tfrac{1}{40}.\end{aligned}$$

(2.5) Example. Let

$$f(x, y) = \begin{cases} 2e^{-y}e^{-2x} & \text{if } x > 0 \text{ and } y > 0,\\ 0 & \text{otherwise,}\end{cases}$$

and let us calculate the integral over the domain $\Omega = \{(x, y) : y \le x\}$, which is clearly normal. Applying the method discussed above, we get

$$\begin{aligned}\iint_\Omega f(x, y)\,\mathrm{d}x\,\mathrm{d}y &= \int_0^{+\infty} \left[\int_0^x 2e^{-y}e^{-2x}\,\mathrm{d}y\right] \mathrm{d}x\\ &= \int_0^{+\infty} 2e^{-2x} \cdot \left[-e^{-y}\right]_{y=0}^{y=x} \mathrm{d}x\\ &= \int_0^{+\infty} 2e^{-2x}\left(1 - e^{-x}\right) \mathrm{d}x\\ &= \int_0^{+\infty} 2e^{-2x}\,\mathrm{d}x - \int_0^{+\infty} 2e^{-3x}\,\mathrm{d}x\\ &= \left[-e^{-2x}\right]_{x=0}^{x\uparrow+\infty} - \left[-\tfrac{2}{3}e^{-3x}\right]_{x=0}^{x\uparrow+\infty}\\ &= 1 - \tfrac{2}{3} = \tfrac{1}{3}.\end{aligned}$$

3. The implicit and the inverse function theorems

The purpose of this section is to recall the main tools from differential calculus that are going to be necessary to state and prove the formula for the change of variables in $\mathbb{R}^n$. The definitions and the theorems that we are about to introduce are slightly more general than what we are actually going to need; the slight increase in generality, however, aims to give a clearer understanding of these classical notions, and all the proofs have been carried out in full not just with the aim to be complete and self-contained, but rather to fully show the beauty and the elegance of the proofs themselves.

To this end, let us start by briefly recalling that, if E is a subset of $\mathbb{R}^n$, a function f, defined on E, with values in $\mathbb{R}^m$, can be seen as an m-tuple of real function $f_1, \ldots, f_m$, defined on E, such that,

$$f(x) = \big(f_1(x), \ldots, f_m(x)\big) \qquad \text{for every element } x \text{ of } E.$$

The function $f_1, \ldots, f_m$ are the components of f. In other words, a function with values in $\mathbb{R}^m$ may always be seen as an m-tuple of real functions all defined over the same domain.

With the aim to fix the notation and the terminology, let us now briefly recall the definition of differentiable function. It is expected however that the reader be already familiar with the basic notions of differential calculus.

(3.1) Definition. Let E be an open set in $\mathbb{R}^n$, let f be a function defined on the set E with values in $\mathbb{R}^m$, and let x^* be a point in E. We shall say that f is **differentiable at** x^* if there exists a (necessarily unique) linear transformation from $\mathbb{R}^n$ to $\mathbb{R}^m$, which is denoted by $Df(x^*)$ and is called the **differential** of f at x^*, such that

$$f(x) = f(x^*) + Df(x^*)(x - x^*) + o(x - x^*) \qquad \text{for all } x \in E,$$

where $o(x - x^*)$ denotes an infinitesimal of order greater than 1 as $x \to x^*$. If, in addition, the function f is differentiable at every point $x \in E$, we shall simply say that f is a **differentiable function** on E.

(3.2) Remark. If E is an open set in $\mathbb{R}^n$, x^* an element of E, f a map from E to $\mathbb{R}^m$, differentiable at x^*, and $f_1, \ldots, f_m$ the components of f, then, the differential, $Df(x^*)$, can be represented by an $m \times n$ matrix whose rows coincide with the gradients $\nabla f_1(x^*), \ldots, \nabla f_m(x^*)$ of the components of f. Such a matrix is called the **Jacobian matrix** of f at x^*. In other words, the Jacobian matrix of f at x^* is matrix whose components are the partial derivatives

$$\frac{\partial f_i}{\partial x_j}(x^*),$$

with $i = 1, \ldots, m$ and $j = 1, \ldots, n$. In the sequel, the differential of a function at a point shall always be identified with its Jacobian matrix through this representation, and the value of the differential at a point shall always be identified with the matrix multiplication of the Jacobian matrix with the (column) vector constituted by the coordinates of the point.

(3.3) Definition. In the same hypotheses as in (3.1), assume that the function f is differentiable on E. If the real functions $\nabla f_1, \dots, \nabla f_m$, all defined on E, with values in $\mathbb{R}^n$, are continuous on E, the function f shall be called **continuously differentiable**.

The following proposition will be useful to compute the Jacobian matrix of a composite function. Its proof is just a mere application of the definition of differentiable function.

(3.4) Proposition (Chain rule). *If E is an open set in $\mathbb{R}^n$, F is an open set in $\mathbb{R}^m$, f a differentiable function from E to $\mathbb{R}^m$, and g a differentiable function from F to $\mathbb{R}^p$, with $f(E) \subseteq F$, then the composite function $h = g \circ f$, from E to $\mathbb{R}^p$, is also differentiable, and the Jacobian matrix of h at a point x in E can be expressed in terms of the Jacobian matrices of f and g as follows:*

$$Dh(x) = Dg\big(f(x)\big)Df(x).$$

Proof. Set $y = f(x)$, $A = Df(x)$ and $B = Dg(y)$. We have

$$f(x+\varepsilon) = f(x) + Df(x)\varepsilon + o(\varepsilon) = y + A\varepsilon + o(\varepsilon)$$

and

$$\begin{aligned} h(x+\varepsilon) &= g\big(f(x+\varepsilon)\big) \\ &= g\big(y + A\varepsilon + o(\varepsilon)\big) \\ &= g(y) + Dg(y)\big(A\varepsilon + o(\varepsilon)\big) + o\big(A\varepsilon + o(\varepsilon)\big) \\ &= g\big(f(x)\big) + BA\varepsilon + Bo(\varepsilon) + o\big(A\varepsilon + o(\varepsilon)\big) \\ &= h(x) + BA\varepsilon + o(\varepsilon). \end{aligned}$$

Comparing the above equality with

$$h(x+\varepsilon) = h(x) + Dh(x)\varepsilon + o(\varepsilon)$$

we conclude immediately that

$$Dh(x) = BA = Dg\big(f(x)\big)Jf(x). \qquad \square$$

Let us now further restrict the hypotheses of Definition (3.1) by assuming that f be a continuously differentiable function defined on an open set E of $\mathbb{R}^n$ with values in $\mathbb{R}^n$. Then, the Jacobian matrix, $Df(x)$, that by hypothesis exists for every x in E, is a square matrix, and it is possible to consider its determinant. In particular, the positive real function $x \mapsto |\det(Df(x))|$, which shall be denoted by Jf, is well defined for every x in E. This function shall be called the **Jacobian** of f.

Observe that, since the partial derivatives of the components of f are all continuous, the Jacobian of f is also continuous. Furthermore, from the chain rule, it turns out that the Jacobian of the composition $h = g \circ f$ of two continuously differentiable function f and g can be written in the form

$$Jh(x) = Jg\big(f(x)\big)Jf(x) \qquad \text{for every } x \in E.$$

Before we may state the Implicit Function Theorem, we are going to need to introduce some extra notation.

(3.5) Notation. If E is the product $A \times B$, with $A \subseteq \mathbb{R}^p$ and $B \subseteq \mathbb{R}^{n-p}$, a function f from E to $\mathbb{R}^m$ may be seen as a function of two vector variables, the first, x, defined over A and the second, y, defined over B. In particular, if (x^*, y^*) is a point in E, we can then consider the two functions

$$x \mapsto f(x, y^*) \qquad \text{and} \qquad y \mapsto f(x^*, y),$$

the first defined over A, and the second defined over B. If the first function is differentiable at x^*, its Jacobian shall be denoted by $D_x f(x^*, y^*)$. Similarly, if the second is differentiable at y^*, its Jacobian shall be denoted by $D_y f(x^*, y^*)$. Let us also observe that, in the special case when $m = 1$, this notations become $\nabla_x f(x^*, y^*)$ and $\nabla_y f(x^*, y^*)$, respectively.

With the definitions and the notations introduced so far, we can now state the following result, due to the Italian mathematician Ulisse Dini (1845–1918), and is usually referred to as the 'Implicit Function Theorem'. This theorem is in fact so astounding that it may be considered as a proper milestone of differential calculus. The proof that we are going to give is due to C. Carathéodory, and it is particularly remarkable, because it is at the same time clever and elementary, going straight to the core of the foundations of the theorem itself.

(3.6) Theorem (Dini's Implicit Function Theorem). *Let E be an open set in $\mathbb{R}^p \times \mathbb{R}^m$, and f a continuously differentiable function from E to $\mathbb{R}^m$. Let (x^*, y^*) be a point in E and suppose that*

$$f(x^*, y^*) = 0 \qquad \text{and} \qquad D_y f(x^*, y^*) : \mathbb{R}^m \to \mathbb{R}^m \text{ is non-singular.}$$

Then, there exist a neighbourhood U of x^* in $\mathbb{R}^p$, a neighbourhood V of y^* in $\mathbb{R}^m$, and a mapping φ from U to V, such that $U \times V \subseteq E$ and

$$f\big(x, \varphi(x)\big) = 0 \qquad \text{for all } x \in U.$$

Furthermore, the function φ is differentiable and

$$D\varphi(x) = -D_y f\big(x, \varphi(x)\big)^{-1} D_x f\big(x, \varphi(x)\big) \qquad \text{for all } x \in U.$$

Proof. In the hypotheses of the theorem, by replacing, if necessary, the function f with $(x, y) \mapsto f(x + x^*, y + y^*) - f(x^*, y^*)$, we may assume, with no loss of generality, that $x^* = 0$ and $y^* = 0$. Let us write

$$f(x, y) = \big(f_1(x, y), \ldots, f_m(x, y)\big),$$

where $f_1, \dots, f_m$ denote the components of the function f. Since the map $(x, y) \mapsto Df(x, y)$ is continuous by assumption, there exist two neighbourhoods U_0 and V_0 of the origins in $\mathbb{R}^p$ and $\mathbb{R}^m$, respectively, such that the matrix, $M(x, y_1, \dots, y_m)$, with rows $\nabla_y f_1(x, y_1), \dots, \nabla_y f_m(x, y_m)$, turns out to be invertible for every $x \in U_0$, and $y_1, \dots, y_m \in V_0$.

Let us now prove that, for every point x of U_0, there exists at most one point y in V_0 such that $f(x, y) = 0$. By contradiction, assume that there existed two points y, z of V_0, with $y \neq z$, such that $f(x, y) = f(x, z) = 0$. By virtue of the mean value theorem, this would imply the existence of m points $y_1, \dots, y_m$, all lying in the line segment between y and z, such that

$$f_i(x, z) - f_i(x, y) = \nabla_y f(x, y_i) \cdot (z - y) = 0 \qquad \text{for all } i = 1, \dots, m.$$

On the other hand, since the matrix $M(x, y_1, \dots, y_m)$ is non-singular, this gives $y = z$, contradiction.

Let $B_r = \overline{B(0, r)}$ be a closed ball with centre 0 and a suitable radius r to ensure that $B_r \subseteq V_0$. Since $f(0, 0) = 0$, we have

$$f(0, y) \neq 0 \text{ for } y \in \partial B_r = \{y \in \mathbb{R}^m : \|y\| = r\},$$

and since the function f is by hypothesis continuous on $U_0 \times V_0$, there exists a strictly positive real number ε such that $\|f(0, y)\| \geq \varepsilon$, for every $y \in \partial B_r$. Hence, the function $g(x, y) = \|f(x, y)\|^2$ satisfies the properties

$$g(0, y) \geq \varepsilon > 0 \text{ for all } y \in \partial B_r \text{ and } g(0, 0) = 0.$$

In addition, since the function g is continuous, there exists a neighbourhood $U \subseteq U_0$ of 0 in $\mathbb{R}^p$ such that

$$g(x, y) \geq \tfrac{1}{2}\varepsilon \text{ and } g(x, 0) \leq \tfrac{1}{2}\varepsilon \text{ for all } x \in U \text{ and } y \in \partial B(0, r).$$

Since $g(0, 0) = 0$, and g cannot vanish in other points, it follows that, for a fixed point x in U, the function $y \mapsto g(x, y)$ achieves a strictly positive minimum on B_r at the point $\varphi(x)$ in the interior of B_r, and we have

$$D_y g\big(x, \varphi(x)\big) = 2 D_y f\big(x, \varphi(x)\big) f\big(x, \varphi(x)\big) = 0,$$

since the matrix $D_y f\big(x, \varphi(x)\big)$ is non-singular, we conclude that

$$f\big(x, \varphi(x)\big) = 0.$$

It remains to prove that the function φ is differentiable. To this end, let us start by writing $\Delta y = \varphi(x + \Delta x) - \varphi(x)$. By virtue of the mean value theorem, this gives

$$0 = D_x f(\tilde{x}, \tilde{y}) \Delta x + D_y f(\tilde{x}, \tilde{y}) \Delta y$$

for some point $(\tilde{x}, \tilde{y})$ lying on the line segment between the point with coordinates $\big(x, \varphi(x)\big)$ and the point with coordinates $\big(x + \Delta x, \varphi(x + \Delta x)\big)$. This implies that, as $\|\Delta x\|$ goes to zero, so does $\|\Delta y\|$, proving that the function $\varphi(x)$ is continuous. In fact, the function $\varphi(x)$ is actually differentiable. To show this, from the fact that f is differentiable, one has

$$\begin{aligned} 0 &= f\big(x + \Delta x, \varphi(x + \Delta x)\big) - f\big(x, \varphi(x)\big) \\ &= D_x f\big(x, \varphi(x)\big)\Delta x + D_y f\big(x, \varphi(x)\big)\Delta y + o\big((\Delta x, \Delta y)\big), \end{aligned}$$

and since $o\big((\Delta x, \Delta y)\big) = o(\Delta x)$, by continuity of $\varphi(x)$, we have

$$\Delta y = -D_y f\big(x, \varphi(x)\big)^{-1} D_x f\big(x, \varphi(x)\big) + o(\Delta x).$$

This proves that $\varphi(x)$ is differentiable at x with

$$D\varphi(x) = -D_y f\big(x, \varphi(x)\big)^{-1} D_x\big(x, \varphi(x)\big)$$

and the theorem is completely proved. □

A simple consequence of the previous theorem is another milestone of differential calculus. To this end, let us give the following definition.

(3.7) Definition. Let U and V be two open sets in $\mathbb{R}^n$. A mapping f, defined on U, with values in V, shall be called a **diffeomorphism**, if it is bijective, and both the maps f and f^{-1} are continuously differentiable. Observe that the inverse of a diffeomorphism and the composition of diffeomorphisms are again diffeomorphisms.

If f is a diffeomorphism from U to V, then, for every open set W included in U, the mapping $x \mapsto f(x)$ from W to $f(W)$ is again a diffeomorphism which is called the diffeomorphism **induced** on f by restriction to W.

Applying the chain rule in the special case of a diffeomorphism, it turns out that the differential $Df(x)$ is non-singular, for every point x of U, and that $D(f^{-1})(y) = Df(x)^{-1}$ whenever $y = f(x)$ or, equivalently, $x = f^{-1}(y)$. In particular, the Jacobian Jf of a diffeomorphism f from U to V never vanishes in U.

(3.8) Theorem (Inverse Function Theorem). *Let E be an open set in $\mathbb{R}^n$, f a continuously differentiable mapping defined on the set E with values in $\mathbb{R}^m$, and x^* an element of E. If $Df(x^*)$ is non-singular, then there exist two open neighbourhoods U of x^*, and V of $f(x^*)$, such that f is a diffeomorphism from U to V, and*

$$Df^{-1}(y) = Df(x)^{-1} \qquad \textit{for every } (x, y) \in U \times V \textit{ such that } y = f(x).$$

Proof. In the hypotheses of the theorem, we wish to locally change the subject of the formula $y = f(x)$ to x, that is to say, to find two neighbourhoods U of x^* and V of $f(x^*)$, and a function φ defined on U, with values in V, such that $f(\varphi(y)) = y$ for every element y of V.

To this end, let us define the function $g(x, y) = f(x) - y$. Then, the function g implicitly defines x as a function of y. Observe that

$$g\big(x^*, f(x^*)\big) = f(x^*) - f(x^*) = 0,$$

the function g is continuously differentiable in its domain, $E \times \mathbb{R}$, and

$$D_x g\big(x^*, f(x^*)\big) = Df(x^*)$$

is non-singular by assumption.

Thus, by virtue of Theorem (3.6), there exist a neighbourhood U of x^* in E, a neighbourhood V of $f(x^*)$, and a function φ from V to U such that $g(\varphi(y), y) = 0$ for every $y \in V$, which means such that $f(\varphi(y)) = y$. To prove that φ is the required inverse, it suffices to note that, if x is an element of U, there exists an element y of V such that $x = \varphi(y)$. Thus,

$$\varphi\big(f(x)\big) = \varphi\big(f\big(\varphi(y)\big)\big) = \varphi(y) = x,$$

and the proof is completed, observing that the desired formula for the derivative of the inverse function follows directly from the chain rule. □

Having now proved the two capstones of differential calculus, let us now show some simple examples of diffeomorphism, and let us use them to introduce the tool that is going to be pivotal to prove the change of variable.

(3.9) Example. Let (a_{ij}) be an $n \times n$ matrix with coefficients in $\mathbb{R}$. Then, the linear map L, defined on $\mathbb{R}^n$ by

$$L(x_1, \dots, x_n) = \Big(\textstyle\sum_{j=1}^n a_{1j}x_j, \dots, \sum_{j=1}^n a_{nj}x_j\Big),$$

is a diffeomorphism of $\mathbb{R}^n$ if and only if $\det(a_{ij}) \neq 0$. If A denotes the matrix (a_{ij}), this map may also be written, more concisely, as $L(x) = Ax$, where $x = (x_1, \dots, x_n) \in \mathbb{R}^n$ and by Ax we denote the usual matrix row-by-column product. Observe that the Jacobian matrix of this function coincides with the matrix A.

(3.10) Definition. Let σ be a permutation of the indices $\{1, \dots, n\}$. The function L_σ defined on $\mathbb{R}^n$ by

$$L_\sigma(x_1, \dots, x_n) = \big(x_{\sigma(1)}, \dots, x_{\sigma(n)}\big)$$

is called a **permutation diffeomorphism**.

It is not difficult to check that a permutation diffeomorphism is indeed a diffeomorphism of $\mathbb{R}^n$. In fact, this is a special type of linear map, whose associated matrix is $(\delta_{i\sigma(j)})$, and whose determinant is $(-1)^{\varepsilon(\sigma)}$ where $\varepsilon(\sigma)$ is 0 if the permutation is even, and 1 if the permutation is odd.

(3.11) Example. A special type of permutation diffeomorphism may be obtained requesting that σ be a cycle, that is to say a permutation that only swaps two indices, say i and j. This map shall be denoted by L_{ij}, and it is defined, when $\{e_1, \ldots, e_n\}$ denotes the canonical base in $\mathbb{R}^n$, by

$$L_{ij}(e_i) = e_j, \quad L_{ij}(e_j) = e_i, \quad \text{and} \quad L_{ij}(e_k) = e_k \quad \text{if } k \neq i, j.$$

Another important type of diffeomorphism is shown in the following definition. To write it more concisely, if we assume that $n > 1$, the space $\mathbb{R}^n$ can be decomposed into the product $\mathbb{R}^{n-1} \times \mathbb{R}$ and, correspondingly every point x in $\mathbb{R}^n$ may be written in the form

$$x = (\hat{x}, x_n), \qquad \text{where } \hat{x} \in \mathbb{R}^{n-1} \text{ and } x_n \in \mathbb{R}.$$

Adopting this notation, we may now give the following definition.

(3.12) Definition. Let U and V be two open sets in $\mathbb{R}^n$, and assume that $n > 1$. A map f from U to V is called a **shear** if it is of the form

$$f(\hat{x}, x_n) = \big(\hat{x}, f_n(\hat{x}, x_n)\big) \qquad \text{for every } (\hat{x}, x_n) \text{ in } U.$$

If $n = 1$, any map may be considered as a (degenerate) shear.

Observe that, for a shear f to be continuously differentiable, it is necessary and sufficient that the last component be continuously differentiable. In this case, it is easy to check that the Jacobian of f coincides with the absolute value of the partial derivative of the last component, f_n, with respect to the nth coordinate, x_n, that is to say, $Jf = |\partial f_n / \partial x_n|$. In particular, for a shear f to be a diffeomorphism it is necessary and sufficient that its last component is differentiable and the corresponding partial derivative is non-zero.

(3.13) Definition. Let U and V be two open sets in $\mathbb{R}^n$. A diffeomorphism f, from U to V, is said to be **primitive** if it is the composition of a finite sequence of diffeomorphisms that are either a permutation diffeomorphism, or a shear diffeomorphism.

We are now in possession of all the ingredients necessary to state and prove the result that we are going to utilise to prove the change of variable in $\mathbb{R}^n$. In order to do so in a clear and concise manner, let us convene that, every time that, given a finite sequence $f_1, \ldots, f_k$ of diffeomorphisms, we consider the composition $f_1 \circ \cdots \circ f_k$, the latter expression be well-defined, that is to say, for every index i, with $1 \leq i \leq k$, the set of values of f_i coincides with the definition set of f_{i+1}. In this case, the map $f_1 \circ \cdots \circ f_k$ is itself a diffeomorphism.

(3.14) Theorem. *Let U and V be two open sets in $\mathbb{R}^n$. Then, every diffeomorphism f from U to V is* locally primitive, *that is to say, for every point x_0 of U, there exists an open neighbourhood W of x_0 in U such that the restriction of f to W is primitive.*

Proof. When $n = 1$ there is nothing to prove because, in this case, every diffeomorphism is, by definition, a shear. Thus, we may assume $n > 1$, and prove the statement by induction. To this end, assume that the result is true for every integer up to $n-1$, and let us prove it when the dimension of the space is n. Taking advantage of the decomposition of $\mathbb{R}^n$ into the product $\mathbb{R}^{n-1} \times \mathbb{R}$, let $w = (\hat{w}, w_n)$ be a given point in U, and let us assume, with no loss of generality, that $(\partial f_n/\partial x_n)(w) \neq 0$. In fact, since f is a diffeomorphism, then the Jacobian $Jf(w)$ of f at w cannot be zero, and therefore there must exist a pair of indices (i, j) such that $(\partial f_i/\partial x_j)(w) \neq 0$. As a consequence, the composition $L_{in} \circ f \circ L_{jn}$ enjoys the property required.

With this additional hypothesis, we may write the map f in the form

$$f(\hat{x}, x_n) = \big(f_0(\hat{x}, x_n), f_n(\hat{x}, x_n)\big),$$

where $f_0(\hat{x}, x_n) = (f_1(\hat{x}, x_n), \ldots, f_{n-1}(\hat{x}, x_n))$, for every $(\hat{x}, x_n)$ in U. Let us consider the shear g, from U to $\mathbb{R}^n$, defined by

$$g(\hat{x}, x_n) = \big(\hat{x}, f_n(\hat{x}, x_n)\big).$$

By construction, the Jacobian of g is $(\partial f_n/\partial x_n)$, and, by assumption, this Jacobian does not vanish at w. By virtue of the Inverse Function Theorem, there exists a neighbourhood W of w in U and a neighbourhood T of $g(w)$, which can always be taken of the form $T \times I$, where T is an open subset of $\mathbb{R}^{n-1}$ and I an open interval of $\mathbb{R}$, such that the function g is invertible, and its inverse is of the form

$$(\hat{u}, u_n) \mapsto \big(\hat{u}, \varphi(\hat{u}, u_n)\big) \qquad \text{for } \hat{u} \text{ in } T \text{ and } u_n \text{ in } I.$$

For every point $(\hat{u}, u_n)$ in $T \times I$, let us now define

$$h(\hat{u}, u_n) = \big(f_0\big(\hat{u}, \varphi(\hat{u}, u_n)\big), u_n\big)$$

and observe that the restriction of f to W coincides with $h \circ g$. In fact, if $x = (\hat{x}, x_n)$ is a point in W, then,

$$\begin{aligned}(h \circ g)(\hat{x}, x_n) &= h\big(\hat{x}, f_n(\hat{x}, x_n)\big)\\ &= \big(f_0\big(\hat{x}, \varphi(\hat{x}, f_n(\hat{x}, x_n))\big), f_n(\hat{x}, x_n)\big)\\ &= \big(f_0(\hat{x}, x_n), f_n(\hat{x}, x_n)\big)\\ &= f(\hat{x}, x_n),\end{aligned}$$

where the penultimate equality follows from the fact that $\varphi(\hat{x}, \cdot\,)$ is the local inverse of $f_n(\hat{x}, \cdot\,)$.

Let us now observe that, for every t in I, the function

$$\hat{u} \mapsto f_0\big(\hat{u}, \varphi(\hat{u}, t)\big),$$

defined on T, is a diffeomorphism, hence, by inductive hypothesis, reducing, if necessary, the neighbourhood T, it can be written in the form

$$g_1^t \circ \cdots \circ g_n^t,$$

where $g_1^t, \ldots, g_k^t$ are either permutations or shear diffeomorphisms. In order to complete the proof, it suffices to transform these functions, which are defined on the $(n-1)$-dimensional space, into functions defined on the n-dimensional space, ensuring that the type of these diffeomorphisms is not altered. To this end, set

$$g_j(\hat{u}, u_n) = L\big(g^{u_n}(\hat{u}), u_n\big),$$

where L denotes the permutation diffeomorphism that swaps around the indices $n-1$ and n; observe that, for every index j, the map g_j is a permutation diffeomorphism if such is g_j^t and a shear diffeomorphism if such is g_j^t, and that the restriction of $f = h \circ g$ to W can now be written in the form

$$(L \circ g_1) \circ \cdots (L \circ g_k) \circ g,$$

whence f is locally primitive. □

4. Change of variables in $\mathbb{R}^n$

The formula for the change of variable is a consequence of Theorem (3.14) and is now easy to prove.

(4.1) Theorem (Change of variable in $\mathbb{R}^n$). *Let U and V be two open subsets in $\mathbb{R}^n$, and let f be a diffeomorphism of U onto V. Then, each Borel subset A of U satisfies*

$$\lambda^n\big(f(A)\big) = \int_A Jf(x)\,\mathrm{d}x, \tag{4.2}$$

and every Borel measurable function g, defined on V, satisfies

$$\int_V g(y)\,\mathrm{d}y = \int_U g\big(f(x)\big) \cdot Jf(x)\,\mathrm{d}x, \tag{4.3}$$

in the sense that, if either of the integrals in the previous equality exists, then both exist and the equality holds.

Proof. Let us start by observing that the relation (4.2) coincides with (4.3) when g is the indicator function of a Borel subset A of U. As a result, if the relation (4.2) is proved, then (4.3) holds for the indicator functions of every Borel subset of U, hence for every Borel simple function defined on U, hence for every positive Borel function defined on U, hence for every integrable function on U. Thus, to prove the theorem, it suffices to prove (4.2).

Let us set $\mu = Jf \cdot \lambda_U^n$, where λ_U^n is the trace of the Borel-Lebesgue λ^n onto the Borel measurable set U. Then, proving (4.2) is equivalent to proving that the measure $f(\mu)$, image of μ under the diffeomorphism f, coincides with the trace of the Borel-Lebesgue measure λ^n onto V.

To this end, observe that, if (4.3) is true for the diffeomorphisms f_1, f_2, then it is also true for the composite diffeomorphism $f = f_2 \circ f_1$. In fact, if A is a Borel subset of U, one has

$$\begin{aligned}\lambda^n\big(f(A)\big) = \lambda^n\big(f_2(f_1(A))\big) &= \textstyle\int_{f_1(A)} Jf_2(w)\,\mathrm{d}w \\ &= \textstyle\int_A Jf_2\big(f_1(x)\big) \cdot Jf_1(x)\,\mathrm{d}x \\ &= \textstyle\int_A Jf(x)\,\mathrm{d}x.\end{aligned}$$

Finally, given a diffeomorphism f from U to V, and assuming that in each point x of U there exists an open neighbourhood U_x, included in U, such that the theorem is true for the restriction of the diffeomorphism f to U_x, then the theorem is also true for the diffeomorphism f. Indeed, the hypothesis is equivalent to say that, in every point x of U, the measure $f(\mu)$ coincides locally with the trace of the Borel-Lebesgue measure λ^n on V. The conclusion, thus, follows from the localisation principle (see Theorem VII(6.4)).

From the previous remarks, and from the fact that, by virtue of (3.14), every diffeomorphism is locally primitive, it follows that it is sufficient to prove the theorem in the two special cases when f is a permutation diffeomorphism, and when f is a shear diffeomorphism.

In the first case, f is the diffeomorphism induced, by restriction, by a linear transformation, whence the theorem follows from (2.2)(d). Thus, the whole theorem is now reduced to proving that (4.2) holds when f is a shear diffeomorphism. Furthermore, in this special case, from the fact that every point x in U has a fundamental system of neighbourhoods composed only of open rectangles, it suffices to prove the theorem when U is an open rectangle, that is to say, a Borel set of $\mathbb{R}^n$ of the form $\prod_{i=1}^n U_i$, where $U_1, \dots, U_n$ are open intervals. By induction, observe that, when $n = 1$, the theorem reduces to (1.9). Thus, we may assume that $n > 1$. In this case, let us set

$$\hat{U} = U_1 \times \cdots \times U_{n-1}, \qquad \hat{x} = (x_1, \dots, x_{n-1}).$$

Then, the shear diffeomorphism f may be written in the form

$$f(\hat{x}, x_n) = \big(\hat{x}, f_n(\hat{x}, x_n)\big).$$

Let us also observe that, then, the map $x_n \mapsto f_n(\hat{x}, x_n)$ is a diffeomorphism from U_n to an open interval of $\mathbb{R}$, which depends on the coordinate $\hat{x}$. Let us denote this interval by $V_n(\hat{x})$. Then, the open set $V = f(U)$ can now be described as the set of all points of the form $(\hat{x}, w)$ with

$$\hat{x} \in \hat{U}, \qquad w \in V_n(\hat{x}).$$

Thus, for every Borel subset A of U, from Fubini's theorem, one gets

$$\begin{aligned}\lambda^n(A) &= \int_{\hat{U}} \mathrm{d}\hat{x} \int_{V_n(\hat{x})} \mathrm{d}w \\ &= \int_{\hat{U}} \mathrm{d}\hat{x} \int_{U_n} |\partial f_n/\partial x_n| \, \mathrm{d}x_n \\ &= \int_{\hat{U}} \mathrm{d}\hat{x} \int_{U_n} Jf(\hat{x}, x_n) \, \mathrm{d}x_n,\end{aligned}$$

where the penultimate equality follows by applying Proposition (1.9) to perform the change of variables $w = f_n(\hat{x}, x_n)$. This proves that (4.2) holds for shear diffeomorphisms, and the proof of the theorem is now complete. □

5. Integration in polar coordinates

After working this hard to prove the change of variable, we may now finally harvest the fruits of our labour. In this section, we shall give a limited account of the power of change of variable, focusing on the special case of the Euclidean plane, $\mathbb{R}^2$. In this special case, it is well-known that every point of the unit circle, that is to say, every point (x, y) such that $x^2 + y^2 = 1$, can be written in the form $(\cos\theta, \sin\theta)$, where θ is a suitable real number, which is uniquely determined up to an integer multiple of 2π. In other words, this result may be expressed by saying that the map $\theta \mapsto (\cos\theta, \sin\theta)$ is periodical, with minimal period equal to 2π, and maps the real line, $\mathbb{R}$, onto the unit circle of $\mathbb{R}^2$. More generally, if r is a strictly positive real number, the map $\theta \mapsto (r\cos\theta, r\sin\theta)$ is periodical, with minimal period equal to 2π, and maps the real line, $\mathbb{R}$, onto the circle with centre in the origin and radius r. As a result, if (x, y) is any point of $\mathbb{R}^2$, different from the origin, by setting $r = \sqrt{x^2 + y^2}$, one may always write

$$x = r\cos\theta, \qquad y = r\sin\theta,$$

where the real number θ is uniquely determined up to an integer multiple of 2π. Thus, the restriction of the map

$$(r, \theta) \mapsto (r\cos\theta, r\sin\theta)$$

to the open set U of $\mathbb{R}^2$ of all pairs (r, θ) of real numbers such that

$$r > 0, \qquad 0 < \theta < 2\pi$$

is a diffeomorphism, F, of this set onto the plane $\mathbb{R}^2$ deprived of the half-line $\{(r, 0) : r > 0\}$, which is called the **polar coordinates** in $\mathbb{R}^2$. Indeed, this map is differentiable, and its Jacobian matrix is

$$DF(r, \theta) = \begin{pmatrix} \cos\theta & -r\sin\theta \\ \sin\theta & r\cos\theta \end{pmatrix}$$

and its Jacobian is

$$JF(r,\theta) = \left|\det\begin{pmatrix}\cos\theta & -r\sin\theta\\ \sin\theta & r\cos\theta\end{pmatrix}\right| = r\cos^2\theta + r\sin^2\theta = r.$$

From the change of variable in $\mathbb{R}^2$, then, the following result holds, for its proof is obvious upon observing that the set $F(U)$ is equivalent, modulo the Borel-Lebesgue measure in $\mathbb{R}^2$, to the whole plane $\mathbb{R}^2$.

(5.1) Proposition. *Let f be a Borel measurable function on $\mathbb{R}^2$. Then,*

$$\iint_{\mathbb{R}^2} f(x,y)\,\mathrm{d}x\,\mathrm{d}y = \iint_U f(r\cos\theta, r\sin\theta)\cdot r\,\mathrm{d}r\,\mathrm{d}\theta, \tag{5.2}$$

in the sense that, if either of the integral in the previous equality exists, then both exist and the equality holds.

Let us observe immediately that the request, in the previous proposition, that the function f is defined on the whole plane $\mathbb{R}^2$ is not restrictive. Indeed, if D is a Borel measurable subset of the plane $\mathbb{R}^2$, and if f is a numeric function defined on D, then the function f may be extended to the whole line by setting $f(x) = 0$ for every $x \in D^c$. Still indicating with f this extension, it is now clear that f coincides with the original function on D and with 0 otherwise, thus its integral over $\mathbb{R}^2$ coincides with its integral over D.

(5.3) Remark. A mnemonic formula often adopted to remember (5.2) is

$$\mathrm{d}x\,\mathrm{d}y = r\,\mathrm{d}r\,\mathrm{d}\theta.$$

Albeit incorrect, this formal equality is able to capture the essence of how the two-dimensional Borel-Lebesgue measure (viz. $\mathrm{d}x\,\mathrm{d}y$) on the space $\mathbb{R}^2$ is transformed when the polar coordinates are deployed (viz. $r\,\mathrm{d}r\,\mathrm{d}\theta$), but in the end it is just the extraction of a piece of (5.2), which is instead a fully rigorous formula, and there is no harm in utilising it.

(5.4) Example. Let $f(x,y) = xy^2$, and consider the set Ω of all points in $\mathbb{R}^2$ that satisfy the relations $x \geq 0$, $y \geq 0$, and $x^2 + y^2 \leq 9$. To calculate the integral

$$\iint_\Omega f(x,y)\,\mathrm{d}x\,\mathrm{d}y$$

let us pass to polar coordinates. This gives

$$f(r\cos\theta, r\sin\theta) = r\cos\theta\cdot r^2\sin^2\theta = r^3\sin^2\theta\cos\theta.$$

Furthermore, the domain Ω, which corresponds to the part of the circle with centre in the origin and radius 3 with the first quadrant, that is to say the set $\mathbb{R}^2_+$, may be written in polar coordinates as $[0,3] \times [0, \frac{1}{2}\pi]$, whence

$$\begin{aligned}
\iint_\Omega xy^2 \,\mathrm{d}x\,\mathrm{d}y &= \int_0^3 \int_0^{\frac{1}{2}\pi} r^3 \sin^2\theta \cos\theta \cdot r\,\mathrm{d}r\,\mathrm{d}\theta \\
&= \left(\int_0^3 r^4 \,\mathrm{d}r \right) \cdot \left(\int_0^{\frac{1}{2}\pi} \sin^2\theta \cos\theta \,\mathrm{d}\theta \right) \\
&= \left[\tfrac{1}{5} r^5 \right]_{r=0}^{r=3} \cdot \left[\tfrac{1}{3} \sin^3\theta \right]_{\theta=0}^{\theta=\frac{1}{2}\pi} = \tfrac{81}{5}.
\end{aligned}$$

Sometimes a curve with equation $F(x,y) = 0$ may not have an explicit equation of the form $y = f(x)$, but it may happen that, in polar coordinates, the equation of the curve is explicit, and this would simplify enormously the calculation of the area enclosed by the curve. Precisely, let us assume that a curve in $\mathbb{R}^2$ is given, and assume that, passing to polar coordinates, this curve may be written in the form

$$r = f(\theta) \qquad \text{for every } \theta \text{ in } [0, 2\pi].$$

Let us further assume that $f(0) = f(2\pi)$, which corresponds to assuming that the curve is closed. Indicating by A the region enclosed by the curve, that is to say, the set of all points with polar coordinates (r,θ) such that $r \le f(\theta)$, it is now simple to calculate the area of A. Indeed, applying the formula for the integration in polar coordinates yields

$$\begin{aligned}
\lambda^2(A) &= \iint_A \mathrm{d}x\,\mathrm{d}y = \int_0^{2\pi} \int_0^{f(\theta)} r\,\mathrm{d}r\,\mathrm{d}\theta \\
&= \int_0^{2\pi} \left[\tfrac{1}{2} r^2 \right]_{r=0}^{r=f(\theta)} \mathrm{d}\theta = \tfrac{1}{2} \int_0^{2\pi} f(\theta)^2 \,\mathrm{d}\theta.
\end{aligned}$$

Thus, the area of the set A enclosed by the curve with polar equation $r = f(\theta)$, is given by the formula

$$\lambda^2(A) = \int_0^{2\pi} f(\theta)^2 \,\mathrm{d}\theta.$$

(5.5) Example. A **cardioid** is a closed curve with equation

$$a^2(x^2 + y^2) = (x^2 + y^2 - ax)^2,$$

where a is a strictly positive real constant. Its shape is shown in Figure XI.1.

To calculate the area enclosed by this curve, that is to say the measure of the set A of all points (x,y) with

$$a^2(x^2 + y^2) \le (x^2 + y^2 - ax)^2,$$

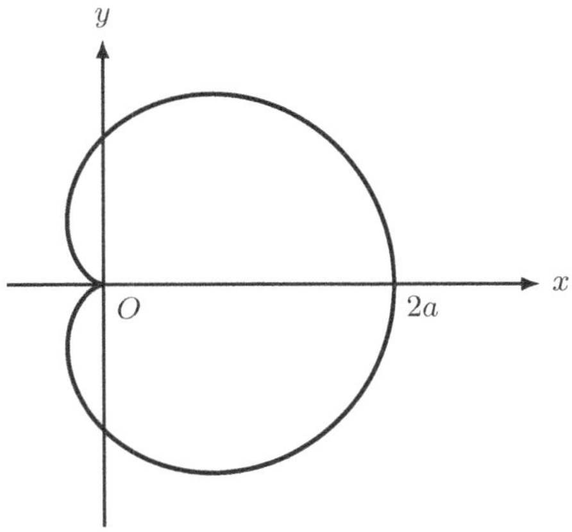

Figure XI.1. The graph of the cardioid with equation $a^2(x^2+y^2) = (x^2+y^2-ax)^2$.

it is convenient to pass to polar coordinates. In fact, applying the transformation $x = r\cos\theta$ and $y = r\sin\theta$, the previous inequality becomes

$$r \le a(1+\cos\theta) \qquad \text{for every } \theta \text{ in } [0, 2\pi].$$

It is now easy to calculate the area enclosed by the cardioid. Indeed,

$$\begin{aligned}\lambda^2(A) &= \tfrac{1}{2}\int_0^{2\pi} a^2(1+\cos\theta)^2 \, d\theta \\ &= \tfrac{1}{2}a^2 \int_0^{2\pi} \left(\tfrac{3}{2} + 2\cos\theta + \tfrac{1}{2}\cos 2\theta\right) d\theta = \tfrac{3}{4}\pi a^2\end{aligned}$$

Using polar coordinates, it is also possible to calculate some special types of uni-dimensional integrals that cannot be calculated directly. The stereotype of this type of calculation is the so-called ‘Gaussian’ integral, which is the object of the next example.

(5.6) Example. The function $f(x) = e^{-ax^2}$ is called the **Gaussian** function or **bell function**, because its graph resembles a bell (see Figure XI.2) and for this reason is often referred to as the **bell curve**. We wish to calculate the area under the bell curve, that is to say the integral

$$\int_{-\infty}^{+\infty} e^{-ax^2} \, dx.$$

This is called a **Gaussian integral**. To this end, let us start by observing that

$$\begin{aligned}\left(\int_{-\infty}^{+\infty} e^{-ax^2} \, dx\right)^2 &= \left(\int_{-\infty}^{+\infty} e^{-ax^2} \, dx\right) \cdot \left(\int_{-\infty}^{+\infty} e^{-ay^2} \, dy\right) \\ &= \iint_{\mathbb{R}^2} e^{-a(x^2+y^2)} \, dx \, dy.\end{aligned}$$

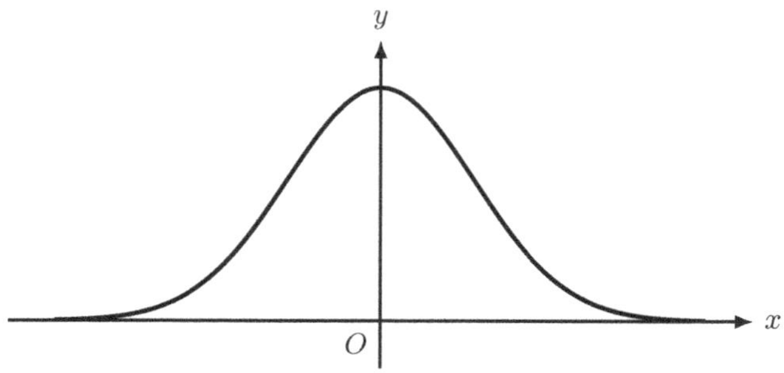

Figure XI.2. The graph of bell curve with equation $f(x) = e^{-ax^2}$.

Thus, to calculate the Gaussian integral, it suffices to calculate the double integral that is shown on the right-hand side of the previous chain of equalities. This is easily done by passing to polar coordinates. Indeed, by making this transformation gives

$$\begin{aligned}\iint_{\mathbb{R}^2} e^{-a(x^2+y^2)}\,\mathrm{d}x\,\mathrm{d}y &= \int_0^{2\pi}\int_0^{+\infty} e^{-ar^2}\cdot r\,\mathrm{d}r\,\mathrm{d}\theta\\ &= \int_0^{2\pi}\mathrm{d}\theta\int_0^{+\infty} re^{-ar^2}\,\mathrm{d}r\\ &= 2\pi\cdot\left[\tfrac{1}{2a}e^{-ar^2}\right]_{r=0}^{r\uparrow+\infty} = \tfrac{\pi}{a},\end{aligned}$$

and therefore

$$\int_{-\infty}^{+\infty} e^{-ax^2}\,\mathrm{d}x = \sqrt{\tfrac{\pi}{a}}.$$

Chapter XII

Hausdorff Measures

The Borel-Lebesgue measure in $\mathbb{R}^n$ is unable to distinguish between different Borel sets of $\mathbb{R}^n$ that are negligible. For instance, a curve, a hyper-plane, or a k-dimensional surface in $\mathbb{R}^n$ are all negligible sets, under the Borel-Lebesgue measure. The family $\mathcal{H}^s$ (with $s \geq 0$) of the Hausdorff measures make it possible to 'catalogue' all these different negligible sets, assigning to each a 'dimension' that may well be a positive integer number (as it is, for example, in the case of a curve, a hyper-plane, or a k-dimensional surface), but that may also take a non-integer value.

The aim of this chapter is to define the notion of Hausdorff measure, in the slightly general setting of a metric space, and discuss its fundamental properties, and see how this family of measures induces a notion of dimension on a metric space. While we shall only begin to scratch the surface of the deep interlink between this type of measures, we shall utilise this link to exhibit a special type of Borel sets whose dimension is generally non-integer and study the fundamental properties of this type of sets. In the general setting of a metric space, we shall also show that the Hausdorff measure $\mathcal{H}^1$ coincides with the classical notion of length for the support of a topological curve.

Moving towards the end of the chapter, we shall focus on the special case of the Euclidean space, $\mathbb{R}^n$, where we shall prove that the Hausdorff measure $\mathcal{H}^n$ coincides with the Borel-Lebesgue measure λ^n. The chapter terminates with the proof of the 'area formula', which shows that the Hausdorff measures $\mathcal{H}^k$ (with k integer) agree with the classical definition of volume for a k-dimensional surface in $\mathbb{R}^n$.

It is worthwhile to remark that this chapter is not expected to furnish a gentle introduction to geometric measure theory, where Hausdorff measures are the most fundamental tool, and where sets with fractional dimension are studied in detail. This chapter has the much humbler aim to show an application of the theory developed thus far, to the case of a family of measures on a metric space, which happen to have a very deep geometrical meaning.

DOI: 10.1201/9781003724766-XII

1. Measure of n-dimensional balls in $\mathbb{R}^n$

The purpose of this section is to derive the general formula for the measure of a ball in $\mathbb{R}^n$. To this end, let us start by recalling that an open ball with centre x and radius r is denoted by $B(x, r)$ and defined by

$$B(x,r) = \{y \in \mathbb{R}^n : \|y - x\| < r\}.$$

The next result shows that the measure of a ball in $\mathbb{R}^n$ is directly proportional to the nth power of the radius of the ball, and that the constant of proportionality is nothing but the measure of the unit ball, for which we shall exhibit a formula.

(1.1) Proposition. *For every x in $\mathbb{R}^n$ and every positive real number r,*

$$\lambda^n\big(B(x,r)\big) = \omega_n r^n,$$

where $\omega_n = \lambda^n(B(0,1))$ is the measure of the unit ball and only depends on the dimension of the ambient space $\mathbb{R}^n$.

Proof. Let us observe that $B(x, r)$ may be seen as the image of the unit ball $B(0,1)$ under the affine transformation T defined by $T(y) = ry + x$, for every y in $\mathbb{R}^n$, whose determinant is clearly $\det(T) = r^n$. Applying the properties IV(2.2)(a) and (d) of the Borel-Lebesgue measure in $\mathbb{R}^n$ yields

$$\lambda^n\big(B(x,r)\big) = \lambda^n\big(T\big(B(0,1)\big)\big) = r^n\lambda^n\big(B(0,1)\big).$$

Setting $\lambda^n(B(0,1)) = \omega_n$, we finally obtain the desired formula

$$\lambda^n\big(B(x,r)\big) = \omega_n r^n. \qquad \square$$

Our ultimate goal for this section is to derive an explicit formula for ω_n, valid for every natural number n. To do this, we shall have to introduce a special function that is going to be needed to write the formula explicitly.

(1.2) Definition. The **Euler's gamma function** is denoted by Γ and is the numeric function on $(0, +\infty)$, defined by

$$\Gamma(x) = \int_0^{+\infty} t^{x-1}e^{-t}\,\mathrm{d}t \qquad \text{for every } x > 0.$$

It is not difficult to show that, for every strictly positive real number x, the integral on the right-hand side of the previous equality is convergent, whence the gamma function is in fact a real function. Further, since the integrand is positive, the gamma function is strictly positive, increasing, and $\Gamma(1) = \int_0^{+\infty} e^{-t}\,\mathrm{d}t = 1$, as it ought to be clear from elementary integration. Furthermore, one can easily prove, using integration by parts, that

$$\Gamma(x+1) = x\Gamma(x).$$

As a consequence of this equality it also follows, by induction, that, for each positive integer number k, the equality $\Gamma(k) = (k-1)!$ holds. Indeed, as we have already mentioned, one has $\Gamma(1) = 1 = 0!$ and, if we assume that the formula holds for a certain positive integer k, that is to say, if we suppose that $\Gamma(k) = (k-1)!$, then we immediately get

$$\Gamma(k+1) = k\Gamma(k) = k(k-1)! = n!$$

We may now prove the following result, which shows the desired explicit formula to calculate the value of ω_n.

(1.3) Proposition. *The measure* ω_n *of the unit ball in* $\mathbb{R}^n$ *is*

$$\omega_n = \frac{\pi^{\frac{n}{2}}}{\Gamma(\frac{n}{2}+1)}.$$

Proof. On the space $\mathbb{R}^{n+1}$ of all pairs (x, r) with $x \in \mathbb{R}^n$ and $r \in \mathbb{R}$, take

$$D = \{(x,r) : x \in \mathbb{R}^n,\ \|x\|^2 < r\}.$$

Integrating the function $(x, r) \mapsto e^{-r}$ over this set gives

$$\begin{aligned}\iint_D e^{-r}\,\mathrm{d}x\,\mathrm{d}r &= \int_{\mathbb{R}^n}\left[\int_{\|x\|^2}^{+\infty} e^{-r}\,\mathrm{d}r\right]\mathrm{d}x\\ &= \int_{\mathbb{R}^n} e^{-\|x\|^2}\,\mathrm{d}x\\ &= \prod_{i=1}^{n}\int_{-\infty}^{+\infty} e^{-x_i^2}\,\mathrm{d}x_i\\ &= \left(\sqrt{\pi}\right)^n = \pi^{\frac{n}{2}}.\end{aligned}$$

On the other hand, reversing the order of integration, and observing that the set $\{x \in \mathbb{R}^n : \|x\|^2 < r\}$ coincides with the open ball $B(0, \sqrt{r})$, by using the formula found in Proposition (1.1), one gets

$$\begin{aligned}\iint_D e^{-r}\,\mathrm{d}x\,\mathrm{d}r &= \int_0^{+\infty}\left[\int_{B(0,\sqrt{r})} e^{-r}\,\mathrm{d}x\right]\mathrm{d}r\\ &= \int_0^{\infty} \lambda^n\big(B(0,\sqrt{r})\big)e^{-r}\,\mathrm{d}r\\ &= \int_0^{+\infty} \omega_n r^{\frac{n}{2}} e^{-r}\,\mathrm{d}r\\ &= \omega_n\Gamma\left(\tfrac{n}{2}+1\right).\end{aligned}$$

Equating these two expression, and rearranging to make ω_n the subject, immediately leads to the desired formula. □

Before moving onto the definition of Hausdorff measures, it is useful to calculate the first few values of ω_n and to ascertain that they give, in the special cases $n = 2$ and $n = 3$, the classical values already known for the area of the unit circle, and the volume of the unit sphere. To this end, let us start by calculating the values $\Gamma(\frac{1}{2})$ and $\Gamma(\frac{3}{2})$. This can easily be done upon recalling the value of the Gaussian integrals

$$\int_{-\infty}^{+\infty} e^{-x^2}\,\mathrm{d}x = \sqrt{\pi}, \qquad \int_{-\infty}^{+\infty} x^2 e^{-x^2}\,\mathrm{d}x = \tfrac{1}{2}\sqrt{\pi}.$$

Now, using the simple change of variable consisting in setting $t = u^2$, whence $\mathrm{d}t = 2u\,\mathrm{d}u$, we get

$$\Gamma(\tfrac{1}{2}) = \int_0^{+\infty} t^{-\frac{1}{2}} e^{-t}\,\mathrm{d}t = \int_0^{+\infty} u^{-1} e^{-u^2}\, 2u\,\mathrm{d}u = 2\int_0^{+\infty} e^{-u^2}\,\mathrm{d}u = \sqrt{\pi}.$$

Similarly,

$$\Gamma(\tfrac{3}{2}) = \int_0^{+\infty} t^{\frac{1}{2}} e^{-t}\,\mathrm{d}t = \int_0^{+\infty} u e^{-u^2}\, 2u\,\mathrm{d}u = 2\int_0^{+\infty} u^2 e^{-u^2}\,\mathrm{d}u = \tfrac{1}{2}\sqrt{\pi}.$$

Utilising these two values, we can easily prove that

$$\omega_0 = 1, \quad \omega_1 = 2, \quad \omega_2 = \pi, \quad \omega_3 = \tfrac{4}{3}\pi, \quad \omega_4 = \tfrac{1}{2}\pi^2, \quad \ldots$$

which shows that the area of the circle is $\lambda^2(B(x,r)) = \pi r^2$ and the volume of the sphere is $\lambda^3(B(x,r)) = \frac{4}{3}\pi r^3$, as expected. Interestingly, when the dimension n is even, the formula in Proposition (1.3) always gives a multiple of an integer power of π. On the other hand, when n is odd, the gamma function of a number of the form $k/2$, with k odd, is always multiple of $\sqrt{\pi}$, thereby cancelling this term with the corresponding value at the numerator, giving again a multiple of an integer power of π.

(1.4) Remark. Let us observe that the formula

$$\omega_n = \frac{\pi^{\frac{n}{2}}}{\Gamma(\frac{n}{2}+1)},$$

which represents the measure of the unit ball in $\mathbb{R}^n$, remains valid when n is substituted with any positive real number s. Indeed, one has

$$\omega_s = \frac{\pi^{\frac{s}{2}}}{\Gamma(\frac{s}{2}+1)}$$

which is still well defined. Clearly, this generalisation, albeit algebraically possible, loses its original meaning of measure of the unit ball, because, when s is not a positive integer, there is no such a thing as an s-dimensional space, at least in the classical sense. However, in the next section, we shall introduce a new definition of dimension that also allows non-integer values.

2. Hausdorff measures and the notion of dimension

This section is devoted to defining the family of the Hausdorff measures, showing its most important properties, and introducing a measure-theoretic notion of 'dimension' on a metric space. Hence, in the sequel, let us assume that (E, d) is an assigned metric space, which shall also be thought of as a measurable space with the Borel σ-field induced by the metric topology. Observe immediately that in this section we are assigning a metric space and not a metrizable space, because the definition of Hausdorff measure heavily depends on the metric, whence the value of measure of a Borel set is bound to change if the metric changes.

With no further ado, let us now give the definition of Hausdorff measure. This shall be done using Method II (see III(9.3)). First off, we shall need to define a pre-measure. Let us consider the set $\mathfrak{X}$ of all subsets X in E with $\operatorname{diam}(X) < \infty$. On this set, let us define the pre-measure

$$\tau_s(X) = 2^{-s}\omega_s \operatorname{diam}(X)^s \qquad \text{for every element } X \text{ of } \mathfrak{X},$$

Where ω_s is the constant defined in Remark (1.4). The rest of the construction is now bound: For every strictly positive real number δ, set

$$\mathcal{H}^s_\delta(A) = \inf\Big\{\textstyle\sum_{k=1}^{\infty} \tau_s(X_k) : X_k \in \mathfrak{X},\ \operatorname{diam}(X_k) \le \delta,\ A \subseteq \bigcup_{k=1}^{\infty} X_k\Big\}$$

for every subset A of E. Then, by virtue of Theorem III(9.3), the set function

$$\mathcal{H}^s(A) = \sup\Big\{\mathcal{H}^s_\delta(A) : \delta > 0\Big\}$$

is a metric outer measure on E.

(2.1) Definition. In the framework described above, for every positive real number s, the restriction of the outer measure $\mathcal{H}^s$ to the Borel σ-field of E shall be called the **s-dimensional Hausdorff measure** on E.

At first glance, this definition may look rather obscure, especially because the pre-measure τ_s is not quite as intuitive as one would hope. However, the term $\operatorname{diam}(X)^s$ looks slightly more intuitive when s is a positive integer. In this case, in fact, it represents the measure of an s-dimensional cube of size $\operatorname{diam}(X)$. Thus, the s-dimensional Hausdorff measure operates by covering a set with a *fine* covering of small sized sets, and approximating the measure with the measure of cuboids whose size coincides with the diameter of the covering. As regard to the constant $2^{-s}\omega_s$, let us observe that, when n is a strictly positive integer, this constant is the measure of the n-dimensional ball with diameter 1. In fact, it turns out that the measure generated using Method II, starting from the pre-measure $\tau(X) = \operatorname{diam}(X)^n$, would assign measure 1 to the n-dimensional ball with diameter 1. Thus, the constant $2^{-s}\omega_s$ may be thought as a normalisation factor between two different units, like multiplying by the factor 1.609 344 to pass from miles to metres.

Having made acquaintance with the definition of Hausdorff measure, it is now time to prove the main properties of this measure. The first of these shows that the Hausdorff measure $\mathcal{H}^0$ is, in fact, a well-known measure. We shall see in §6 and §7 that this is going to happen to $\mathcal{H}^1$ and $\mathcal{H}^n$ as well.

(2.2) Proposition. *$\mathcal{H}^0$ is the counting measure.*

Proof. Let x be a point of E, and δ be a strictly positive real number. It is easy to recognise that $\mathcal{H}^0(\{x\}) = \omega_0 = 1$, whence $\mathcal{H}^0(\{x\}) = 1$. Now, if A is a countable Borel set in E, then $\mathcal{H}^0(A) = \sum_{x\in A} \mathcal{H}^0(\{x\}) = \text{Card}(A)$. On the other hand, if A is infinite, there exists a subset B of A which is countable, whence $\mathcal{H}^0(A) \geq \mathcal{H}^0(B) = \infty$. □

Let us now analyse the behaviour of the Hausdorff measure, $\mathcal{H}^s$, as the parameter s varies over the half-line of all strictly positive real numbers. The next result shows what happens in this case, and, at the same time, will lay the foundation for the definition of Hausdorff dimension.

(2.3) Proposition. *If $0 < \mathcal{H}^s(A) < \infty$, then*

(a) $\mathcal{H}^t(A) = 0$, *for every $t > s$.*

(b) $\mathcal{H}^t(A) = \infty$, *for every $t < s$.*

Proof. Let δ be a strictly positive real number, and let (X_k) be a countable covering of A whose sets all have diameter less than δ. Then,

$$\sum_{k=1}^{\infty} 2^{-s}\omega_s \operatorname{diam}(X_k)^s \leq \left(\tfrac{1}{2}\delta\right)^{s-t}\omega_t^{-1}\omega_s \sum_{k=1}^{\infty} 2^{-t}\omega_t \operatorname{diam}(X_k)^t,$$

that is to say,

$$\mathcal{H}^s_\delta(A) \leq c(t,s)\delta^{s-t}\mathcal{H}^t_\delta(A),$$

and both the two conditions follow immediately, by taking the supremum of the previous inequality with respect to δ. □

The previous proposition shows that, for any Borel measurable subset A of E, not only does the real function $s \mapsto \mathcal{H}^s(A)$ decrease, but the value of this function equals ∞ up to a certain value s_0, and then it drops to 0 for every value which is greater than s_0. This justifies the following definition.

(2.4) Definition. Let A be a Borel measurable set in E. The **Hausdorff dimension** of A is denoted by $\dim(A)$, and it is defined as

$$\dim(A) = \inf\Big\{s \geq 0 : \mathcal{H}^s(A) = 0\Big\}.$$

Let us observe that, for every countable set A in $\mathbb{R}^n$, $\dim(A) = 0$. On the other hand, constructing example of sets with a non-trivial Hausdorff dimension is quite complicated.

Let us complete this section with a very simple, but important remark.

(2.5) Remark. On the real line, $\mathbb{R}$, the 1-dimensional Hausdorff measure $\mathcal{H}^1$ is identical to the Borel-Lebesgue measure λ^1. In order to see this, let us denote by $\mathfrak{X}$ the set of all subsets X of $\mathbb{R}$ with $\operatorname{diam}(X) < \infty$. If A is any Borel subset of $\mathbb{R}$ and δ is any fixed strictly positive real number, from the fact that λ^1 may be constructed, using Method I, starting from the pre-measure $\tau([a,b]) = \operatorname{diam}([a,b]) = b - a$ (see Example III(8.9)), and taking into account that $\omega_1 = 2$, one has

$$\begin{aligned}\lambda^1(A) &= \inf\Big\{\textstyle\sum_{k=1}^{\infty} \operatorname{diam}(X_k) : X_k \in \mathfrak{X},\ A \subseteq \bigcup_{k=1}^{\infty} X_k\Big\}\\ &\leq \inf\Big\{\textstyle\sum_{k=1}^{\infty} \operatorname{diam}(X_k) : X_k \in \mathfrak{X},\ \operatorname{diam}(X_k) \leq \delta,\ A \subseteq \bigcup_{k=1}^{\infty} X_k\Big\}\\ &= \mathcal{H}^1_\delta(A).\end{aligned}$$

To prove the converse, for every integer r, set $\Delta_r = [r\delta, (r+1)\delta]$. Then,

$$\operatorname{diam}(X_k \cap \Delta_r) \leq \delta, \quad \text{and} \quad \textstyle\sum_{r\in\mathbb{Z}} \operatorname{diam}(X_k \cap \Delta_r) \leq \operatorname{diam}(X_k),$$

whence

$$\begin{aligned}\lambda^1(A) &= \inf\Big\{\textstyle\sum_{k=1}^{\infty} \operatorname{diam}(X_k) : X_k \in \mathfrak{X},\ A \subseteq \bigcup_{k=1}^{\infty} X_k\Big\}\\ &\geq \inf\Big\{\textstyle\sum_{k=1}^{\infty} \sum_{r\in\mathbb{Z}} \operatorname{diam}(X_k \cap \Delta_r) : X_k \in \mathfrak{X},\ A \subseteq \bigcup_{k=1}^{\infty} X_k\Big\}\\ &\geq \mathcal{H}^1_\delta(A).\end{aligned}$$

Thus, $\lambda^1 = \mathcal{H}^1_\delta$ for every strictly positive real number δ, and this yields that the measures λ^1 and $\mathcal{H}^1$ are identical.

3. Hausdorff measures and Lipschitz functions

Let us start by the following general result.

(3.1) Proposition. *Let f and g be two continuous maps on a metric space E, and assume that there exists a positive real number c such that*

$$d\big(f(x), f(y)\big) \leq c\,d\big(g(x), g(y)\big) \qquad \text{for every pair } x, y \text{ of elements of } E.$$

Then, for every Borel set A in E, one has

$$\mathcal{H}^s\big(f(A)\big) \leq c^s \mathcal{H}^s\big(g(A)\big).$$

Proof. Let us fix two strictly positive real numbers δ, ε and let (Y_k) be a covering of $g(A)$, constituted of sets with diameter $\operatorname{diam}(Y_k) \leq c^{-1}\delta$, such that $\sum_{k=1}^{\infty}\big(\operatorname{diam}(Y_k)\big)^s \leq \mathcal{H}^s\big(g(A)\big) + c^{-s}\varepsilon$. Then, the sets $X_k = f(g^{-1}(Y_k))$ form a covering for $f(A)$ with $\operatorname{diam}(X_k) \leq c\operatorname{diam}(Y_k) \leq \delta$, so that

$$\mathcal{H}^s_\delta\big(f(A)\big) \leq \textstyle\sum_{k=1}^{\infty}\big(\operatorname{diam}(X_k)\big)^s \leq c^s \mathcal{H}^s\big(g(A)\big) + \varepsilon.$$

The result follows taking the supremum over δ and the infimum over ε. $\square$

The previous proposition unlocks the first and fundamental evidence of the connection that holds between Hausdorff measures and Lipschitz functions. To clarify the subsequent results, let us start by recalling that, if E and F are two metric spaces, and d_E and d_F denote the metrics on E and F, respectively, a map f, defined on E, with values in F, is **Lipschitz continuous** if there exists a positive real number L such that

$$d_F\big(f(x), f(y)\big) \leq L d_E(x, y) \qquad \text{for every pair } x, y \text{ of elements of } E.$$

It is straightforward to verify that a Lipschitz continuous function is, in particular, continuous.

The **Lipschitz constant** is defined as

$$\mathrm{Lip}(f) = \sup\left\{ \frac{d_F\big(f(x), f(y)\big)}{d_E(x, y)} : x, y \in E,\, x \neq y \right\}.$$

Of course, if $L = \mathrm{Lip}(f)$, then $d_F(f(x), f(y)) \leq L d_E(x, y)$ for every $x, y \in E$. So a necessary and sufficient condition for a map f to be Lipschitz continuous is that the Lipschitz constant exists and is finite. Finally, let us also recall that a Lipschitz continuous map is called a **contraction** if its Lipschitz constant is strictly smaller than 1.

The first, immediate, consequence of Proposition (3.1) is the following.

(3.2) Corollary. *Let E be a metric space, let f be a Lipschitz continuous function from E to E, and let us denote by L its Lipschitz constant. Then, for any positive real number s,*

$$\mathcal{H}^s\big(f(A)\big) \leq L^s \mathcal{H}^s(A) \qquad \text{for every Borel subset } A \text{ of } E.$$

In particular, $\dim(f(A)) \leq \dim(A)$.

Proof. It suffices to observe that the hypothesis that the map f is Lipschitz continuous yields the condition expressed in Proposition (3.1) taking as g the identity map on E, and the result follows. □

Let us recall that, if E is a metric space, a bijective map f from E to E, is called a **similitude** if there exists a strictly positive real number r, called the **scale factor** of the similitude, such that

$$d\big(f(x), f(y)\big) = r d(x, y) \qquad \text{for every pair } x, y \text{ of elements of } E.$$

Observe that, in particular, a similitude is Lipschitz continuous. In addition, in the special case when the scale factor r is 1, the map f leaves the distances unchanged. In this special case, the map f is called an **isometry**. In other words, an isometry is a map f, from E to E, such that

$$d\big(f(x), f(y)\big) = d(x, y) \qquad \text{for every pair } x, y \text{ of elements of } E.$$

When $E = \mathbb{R}^n$, a map f is a similitude if and only if there exists a strictly positive real number r such that

$$\|f(x) - f(y)\| = r\|x - y\| \qquad \text{for every pair } x, y \text{ of elements of } \mathbb{R}^n.$$

It is not difficult to recognise that, in this case, a similitude with scale factor r can always be written in the form $f(x) = rO(x) + b$, where O is an orthogonal linear map from $\mathbb{R}^n$ to $\mathbb{R}^n$, that is to say a linear map whose representative matrix has determinant ± 1.

(3.3) Corollary. *If s is a strictly positive real number, A is a Borel subset of E, and f is a similitude, with scale factor r, then $\mathcal{H}^s(f(A)) = r^s\mathcal{H}^s(A)$. In particular, if f is an isometry, then $\mathcal{H}^s(f(A)) = \mathcal{H}^s(A)$.*

Proof. It suffices to apply Corollary (3.2) to both f and f^{-1}. □

Let us now examine in more detail the special case when $E = \mathbb{R}^n$.

(3.4) Remark. A simple application of Corollary (3.2) shows that the Hausdorff measures may be decreased applying a projection over an affine subspace of $\mathbb{R}^n$. Indeed, if H is an affine subspace of $\mathbb{R}^n$ and if f denotes the orthogonal projection onto H, it is not difficult to check that f is a Lipschitz continuous function, and that its Lipschitz constant is 1. Hence,

$$\mathcal{H}^s\big(f(A)\big) \le \mathcal{H}^s(A) \qquad \text{for any Borel measurable subset } A \text{ of } \mathbb{R}^n.$$

Obviously, the same applies for the projection onto a convex subset of $\mathbb{R}^n$.

Moreover, taking this latter statement into account, if H is an n-dimensional hyper-plane in $\mathbb{R}^m$, then there exists an orthogonal projection p, from $\mathbb{R}^n$ to $\mathbb{R}^m$, such that $H = p(\mathbb{R}^n)$, and thus, the restriction of the Hausdorff measure $\mathcal{H}^n$ in $\mathbb{R}^m$ to the hyper-plane H coincides with the image of the Hausdorff measure $\mathcal{H}^n$ in $\mathbb{R}^n$ under the projection p.

The consequence of the previous remark are huge, and they become even huger when paired with the results of §7, which ensures that the Hausdorff measure $\mathcal{H}^n$ in $\mathbb{R}^n$ is nothing but the Borel-Lebesgue measure. Indeed, using this fact and the previous remark, it turns out that the trace of the Hausdorff measure $\mathcal{H}^n$ in $\mathbb{R}^m$ to any n-dimensional hyper-plane coincides with the image of the n-dimensional Borel-Lebesgue measure.

The next result is another consequence of Corollary (3.3) when E is a *metric vector space*, that is to say a vector space which is also a metric space, and such that the vector operations are continuous with respect to the topology defined by the metric, hence, in particular, when E is $\mathbb{R}^n$.

(3.5) Corollary. *Let E be a metric vector space. Then, the Hausdorff measure $\mathcal{H}^s$ is translation-invariant, that is to say, for every vector x of $\mathbb{R}^n$,*

$$\mathcal{H}^s(A + x) = \mathcal{H}^s(A) \qquad \textit{for every Borel set } A.$$

Furthermore, $\mathcal{H}^s$ satisfies the following scaling rule: For every strictly positive real number λ,

$$\mathcal{H}^s(\lambda A) = \lambda^s \mathcal{H}^s(A) \qquad \textit{for every Borel set } A.$$

Proof. Observe that the maps $y \mapsto y + x$ and $y \mapsto \lambda y$ are an isometry and a similitude, respectively, whence the conclusion follows from (3.3). □

Let us observe that, thanks to the previous corollary, in the special case when the underlying space is the Euclidean space $\mathbb{R}^n$, Hausdorff dimension behaves, as one would expect, by assigning dimension n to the whole space and a Hausdorff dimension smaller than or equal to n to any Borel subset of $\mathbb{R}^n$. More precisely, the following proposition holds.

(3.6) Proposition. *In $\mathbb{R}^n$, if $s > n$, then $\mathcal{H}^s = 0$.*

Proof. Let $Q = (0,1)^n$. Since $\lambda^s \mathcal{H}^s(Q) = \mathcal{H}^s(\lambda B) \to \mathcal{H}^s(\mathbb{R}^n)$, as λ tends towards infinity, it suffices to prove that $\mathcal{H}^s(Q) = 0$. To this end, for any strictly positive integer k, select a covering of Q made of k^n open balls with diameter $k^{-1}\sqrt{n}$, and observe that

$$\mathcal{H}^s_{k^{-1}\sqrt{n}}(Q) \leq k^n \cdot 2^{-s}\omega_s \cdot \left(k^{-1}\sqrt{n}\right)^s = n^{\frac{s}{2}} 2^{-s} k^{n-s} \omega_s.$$

The conclusion follows by taking the limit as k goes towards infinity. □

Thus, for every Borel measurable set A of $\mathbb{R}^n$, the Hausdorff dimension $\dim(A)$ is always included in the interval $[0, n]$. In particular, the Hausdorff dimension of the whole space $\mathbb{R}^n$ is $\dim(\mathbb{R}^n) = n$. In §5, we shall aim to show examples of Borel sets whose Hausdorff dimension is not only strictly smaller than n, but in general not even an integer value. Before doing this, however, we shall need to discuss a useful tool. This is the aim of the next section.

4. The Hausdorff metric

The purpose of this short section is to introduce a metric on the set of all compact subsets of a metric space.

(4.1) Definition. Let (E, d) be a metric space, and let us denote by $\mathcal{K}(E)$ the class of all compact subsets of E. If ε is any strictly positive real number, the **ε-neighbourhood** of any subset X of E is the set of all points within distance ε from X, that is to say

$$X_\varepsilon = \bigcup_{x \in X} B(x, \varepsilon).$$

Let us make $\mathcal{K}(E)$ into a metric space by defining a distance between two compact sets X and Y to be the least ε such that the ε-neighbourhood of B contains A and vice versa. More precisely, let us give the following definition.

(4.2) Definition. In the same hypotheses as in Definition (4.1), it is not difficult to recognise that the numeric function δ, defined for every pair X, Y of elements of $\mathcal{K}(E)$ by

$$\delta(X,Y) = \inf\{\varepsilon : X \subseteq Y_\varepsilon \text{ and } Y \subseteq X_\varepsilon\}$$

is a metric on $\mathcal{K}(E)$, and it is called the **Hausdorff metric** on $\mathcal{K}(E)$.

Observe that, if $\delta(X,Y)$ is small, this means that the sets X and Y are close to each other (see Figure XII.1). Let us show that, when E is a complete metric space, the Hausdorff metric makes $\mathcal{K}(E)$ also a complete metric space.

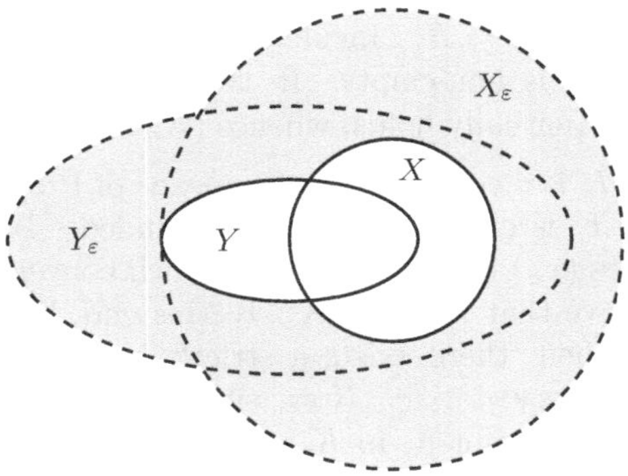

Figure XII.1. Two compact sets X and Y, as well as their ε-neighbourhoods, X_ε and Y_ε. The Hausdorff distance of X and Y is the smallest value of ε for which X lies in the ε-neighbourhoods of Y and vice versa.

(4.3) Lemma. *Let (E,d) be a complete metric space. Then, the set $\mathcal{K}(E)$ with the Hausdorff metric δ is complete.*

Proof. Let us suppose that (K_n) is a Cauchy sequence in $\mathcal{K}(E)$, and let us show that it converges towards a compact set under the Hausdorff metric. To this end, let us denote by K the set of all points x in E for which there exists a sequence (x_k) of points of E, with $x_k \in K_k$ for every index k, and which converges towards x. We wish to prove that K is a non-empty compact set, and that $\delta(K_n, K) \to 0$.

To this end, take a strictly positive real number ε, and observe that, since (K_n) is a Cauchy sequence for the Hausdorff metric, there exists a positive integer c such that, for every pair m, n with $c < n < m$, one has $\delta(K_m, K_n) < \frac{1}{2}\varepsilon$.

By definition, if x is a point of K, there exists a sequence (x_k) in E such that $x_k \in K_k$, for every index k, and $x_k \to x$. This means that there exists a strictly positive integer c' such that, for every integer k, with $c' < k$, one has $d(x, x_k) < \frac{1}{2}\varepsilon$. Therefore, if k is an integer greater than the maximum between c and c', there exists a point y in K_n such that $d(x_k, y) < \frac{1}{2}\varepsilon$, whence $d(x,y) \le d(x,x_k) + d(x_k,y) < \varepsilon$. This shows that $K \subseteq (K_n)_\varepsilon$.

On the other hand, suppose that y is a point of K_n and choose a strictly increasing sequence (k_r) of positive integers, such that

$$k_1 = n, \qquad \delta(K_{k_r}, K_m) < 2^{-r}\varepsilon \quad \text{for every index } m \geq k_r.$$

Let us define a sequence (y_k) as follows: For every strictly positive integer k with $k < n$, choose a point y_k in K_k; set $y_n = y$; if y_{k_r} has been chosen, and $k_r < k < k_{r+1}$, choose a point y_k in K_k such that $d(y_{k_r}, y_k) < 2^{-r}\varepsilon$. Then, the sequence (y_k) is Cauchy by construction, hence it converges towards a point x in E. By definition, the point x lies in K, and

$$d(x, y) = \lim_k d(y_k, y) \leq \textstyle\sum_r d(y_k, y_{k_r}) < \varepsilon,$$

so $y \in K_\varepsilon$. This shows that $K_n \subseteq K_\varepsilon$. Incidentally, the previous construction also shows that the set K is non-empty. In the end, this proves that the inequality $\delta(K, K_n) < \varepsilon$ eventually holds, whence (K_n) converges towards K.

It remains to show that K is a compact set. By means of Proposition A(7.4), it suffices to show that K is closed and totally bounded. To show that K is closed, take a sequence (x_n) of elements of K which converges towards an element x and let us prove that x lies in K. To this end, let ε be a strictly positive real number. Then, there exists a strictly positive integer c such that $d(x_n, x) < \frac{1}{2}\varepsilon$, for every $n > c$. Now, since x_n is by assumption an element of K, there exists a point y_n in K_n such that $d(y_n, x_n) < \frac{1}{2}\varepsilon$. As a result, one has

$$d(y_n, x) \leq d(y_n, x_n) + d(x_n, x) < \varepsilon \qquad \text{for every index } n > c,$$

which shows that (y_n) is such that $y_n \in K_n$ and $y_n \to x$, so $x \in K$.

It now only remains to prove that K is also totally bounded. For this, fix a strictly positive real number ε and observe that, since (K_n) is a sequence of compact sets that converges towards K, there exists a positive integer n such that $K \subseteq (K_n)_{\frac{1}{2}\varepsilon}$. On the other hand, since K_n is compact, there exists a finite set $\{x_1, \ldots, x_p\}$ of points of E such that $K_n \subseteq \bigcup_{k=1}^p B(x_k, \frac{1}{2}\varepsilon)$. Thus,

$$K \subseteq (K_n)_{\frac{1}{2}\varepsilon} \subseteq \textstyle\bigcup_{k=1}^p B(x_k, \varepsilon),$$

and therefore K is totally bounded. □

5. Self-similar sets

This section is devoted to producing some geometrically interesting examples of sets with non-integer Hausdorff dimension. The sets that are going to be considered are called 'self-similar', which means, roughly speaking, that each small part of the set looks like a shrunken copy of the whole set. To start off, in the sequel let us assume that (E, d) is a metric space. Unless otherwise specified, every set shall be assumed to be a Borel subset of E.

(5.1) Definition. Let $\mathcal{S} = \{\sigma_1, \ldots, \sigma_m\}$ be a finite set of similitudes, all with common scale factor $r < 1$. For any Borel set X, let us set

$$\mathcal{S}(X) = \bigcup_{k=1}^{m} \sigma_k(X).$$

If $\mathcal{S}(X) = X$, the set X is said to be **invariant** under $\mathcal{S}$. If, in addition, there exists a positive real number s such that

$$\mathcal{H}^s(X) > 0, \qquad \mathcal{H}^s\big(\sigma_i(X) \cap \sigma_j(X)\big) = 0 \quad \text{for } i \neq j,$$

the set X shall be called **self-similar**.

Observe immediately that the condition expressed by the Hausdorff measure ensures that the self-similar features are not lost by overlaps. The aim of the rest of this section is to show that, for any finite set of similitudes with common scale factor $r < 1$, there exists a unique non-empty compact invariant set. As we shall see in the sequel, if, in addition, the set of similitudes possess a *separating set*, then this unique invariant set turns out to be a self-similar set, and we shall be able to exhibit an open formula, which ultimately depends only on the similitudes, to calculate its Hausdorff dimension.

To this end, let $\mathcal{S} = \{\sigma_1, \ldots, \sigma_m\}$ be a finite family of similitudes with scale factor $r < 1$ and let us define, by induction,

$$\mathcal{S}^0(X) = X, \qquad \mathcal{S}^{k+1}(X) = \mathcal{S}^k\big(\mathcal{S}(X)\big) \quad \text{for every } k \geq 1.$$

If X is an invariant set under $\mathcal{S}$, then, for every index k, one has $\mathcal{S}^k(X) = X$, which means that, for every index k, X is the union of m^k copies of itself, after it has been scaled down by a factor r^k.

The first task consists in proving that every finite family of similitudes have a unique non-empty compact invariant set. This can be done by utilising the Hausdorff metric, which was introduced in the previous section, but we shall also need a simple but elegant fixed-point theorem named after the Polish mathematician Stefan Banach (1892–1945) and the Neapolitan mathematician Renato Caccioppoli (1904–1959).

(5.2) Theorem (Banach-Caccioppoli). *Let (E, d) be a complete metric space, and f a contraction. Then, there exists a unique fixed-point of f, that is to say, a unique point x^* in E such that $f(x^*) = x^*$. Moreover, for any point x in E, the sequence (x_n) defined inductively by*

$$x_0 = x, \qquad x_{n+1} = f(x_n) \qquad \text{for } n \geq 1, \tag{5.3}$$

converges towards x^*.

Proof. Let us denote by q the Lipschitz constant of f which is, by assumption, a positive real number strictly smaller than 1. Let us start by observing that, if a fixed-point exists, it must be unique. Indeed, if x^* and x^{**} were two fixed-points of f, then $x^* = f(x^*)$ and $x^{**} = f(x^{**})$, then

$$d(x^*, x^{**}) = d\big(f(x^*), f(x^{**})\big) \leq q d(x^*, x^{**}),$$

which yields $d(x^*, x^{**}) = 0$, whence $x^{**} = x^*$.

In addition, if a sequence (x_n) is defined by (5.3), and $x_n \to x^\bullet$, then, since f is Lipschitz continuous, hence continuous, by taking the limit as n approaches ∞, the equation $x_{n+1} = f(x_n)$ becomes $x^\bullet = f(x^\bullet)$.

Thus, it only remains to prove that such a sequence (x_n) always converges. To this end, let x be any point of E, and consider the sequence (x_n) defined by (5.3). By induction, it is not difficult to recognise that

$$d(x_{n+1}, x_n) \leq q^n d(x_1, x_0)$$

for every strictly positive integer n, which in turn, for any pair n, m of strictly positive real numbers, with $m > n$, yields

$$\begin{aligned} d(x_n, x_m) &\leq \textstyle\sum_{k=0}^{m-n-1} d(x_{n+k}, x_{n+k+1}) \\ &\leq \textstyle\sum_{k=0}^{m-n-1} q^{n+k} d(x_1, x_0) \\ &\leq q^n d(x_1, x_0) \textstyle\sum_{k=0}^{\infty} q^k \\ &\leq q^n (1-q)^{-1} d(x_1, x_0), \end{aligned}$$

which shows immediately that (x_n) is a Cauchy sequence and, since (E, d) is a complete metric space, it possesses a limit x^* in E. □

We are finally ready to prove that every finite family of similitudes with scale factor strictly smaller than 1 admits a unique compact invariant set.

(5.4) Theorem. *Let $\mathcal{S} = \{\sigma_1, \dots, \sigma_m\}$ be a finite family of similitudes with scale factor $r < 1$. Then, there exists a unique non-empty compact set K which is invariant under $\mathcal{S}$.*

Furthermore, if X is any non-empty compact subset of E, the sequence of the iterates $(\mathcal{S}^k(X))$ converges towards K in the Hausdorff metric. In particular, if $\mathcal{S}(X) \subseteq X$, then the sequence $(\mathcal{S}^k(X))$ is monotone, and

$$K = \textstyle\bigcap_{k=0}^{\infty} \mathcal{S}^k(X).$$

Proof. Let X and Y be two elements of $\mathcal{K}(E)$ and observe that by definition of the Hausdorff metric, if the ε-neighbourhood $(\sigma_i(X))_\varepsilon$ contains $\sigma_i(Y)$ for every index i, then $\big(\bigcup_{i=1}^m \sigma_i(X)\big)$ contains $\bigcup_{i=1}^m \sigma_i(Y)$, and vice versa. Hence,

$$\delta\big(\mathcal{S}(X), \mathcal{S}(Y)\big) \leq \max_{i=1,\dots,m} \delta\big(\sigma_i(X), \sigma_i(Y)\big)$$

and therefore

$$\delta\big(\mathcal{S}(X), \mathcal{S}(Y)\big) \le r\delta(X, Y).$$

Thus, the map $\mathcal{S}$, from $\mathcal{K}(E)$ to $\mathcal{K}(E)$ is a contraction, and since the metric space $(\mathcal{K}(E), \delta)$ is complete, Banach-Caccioppoli's fixed-point theorem guarantees that there exists a unique fixed-point, K, which is the limit of the sequence $(\mathcal{S}^k(X))$, for any choice of X in $\mathcal{K}(E)$. In particular, if $\mathcal{S}(X) \subseteq E$, then $(\mathcal{S}^k(X))$ is a decreasing sequence of non-empty compact sets containing K, with intersection $\bigcap_{k=0}^{\infty} \mathcal{S}^k(X)$ which must be equal to K. □

It is now finally time to see some examples of invariant sets. The following three are amongst the most famous and fundamental. Observe that all these examples are invariant under the families of similitudes used to generate them. Let us also notice that each of these sets is made of successive copies of themselves, which are either disjoint or they meet in a single point, whence they are all self-similar sets.

(5.5) Example (Cantor's set). Given a real number β, with $0 < \beta < \frac{1}{2}$, let us consider the set of similitudes $\mathcal{S} = \{\sigma_1, \sigma_2\}$, where

$$\sigma_1(x) = \beta x, \qquad \sigma_2(x) = \beta x + 1 - \beta \qquad \text{for every } x \text{ in } [0, 1].$$

Let us set

$$C_\beta = \bigcap_{k=0}^{\infty} \mathcal{S}^k\big([0, 1]\big).$$

Then C_β is, by construction an invariant set, which is customarily referred to as **Cantor's set**, in honour of the German-Russian mathematician, logician and philosopher Georg Cantor (1845–1918), and a picture of the first nine iterations of this set $C_{1/3}$ are shown in Figure XII.2. We shall see that this is, in fact, a self-similar set, and we shall calculate its Hausdorff dimension.

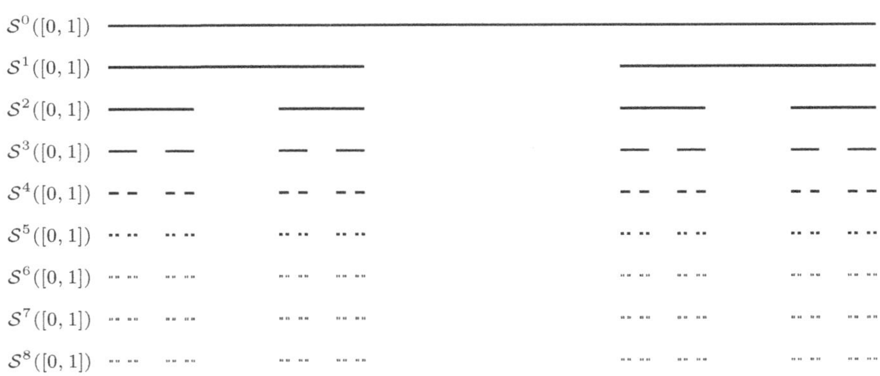

Figure XII.2. The first nine iterations of the Cantor's set with $\beta = 1/3$.

(5.6) Example (Sierpiński's triangle). Let T be the equilateral triangle in $\mathbb{R}^2$ with vertices in $(0,0)$, $(1,0)$, and $(\frac{1}{2}, \frac{1}{2}\sqrt{3})$, and let us consider the set of similitudes $\mathcal{S} = \{\sigma_1, \sigma_2, \sigma_3\}$ defined for every x in $\mathbb{R}^2$ by

$$\sigma_j(x) = \tfrac{1}{2}x + b_j, \qquad \text{where } b_1 = (0,0),\ b_2 = (\tfrac{1}{2}, 0),\ b_3 = (\tfrac{1}{4}, \tfrac{1}{4}\sqrt{3}).$$

The set

$$G = \textstyle\bigcap_{k=0}^{\infty} \mathcal{S}^k(T)$$

is called **Sierpiński's triangle**, in honour of the Polish mathematician Wacław Sierpiński (1882–1969), and a picture of the first five iterations of this set is shown in Figure XII.3.

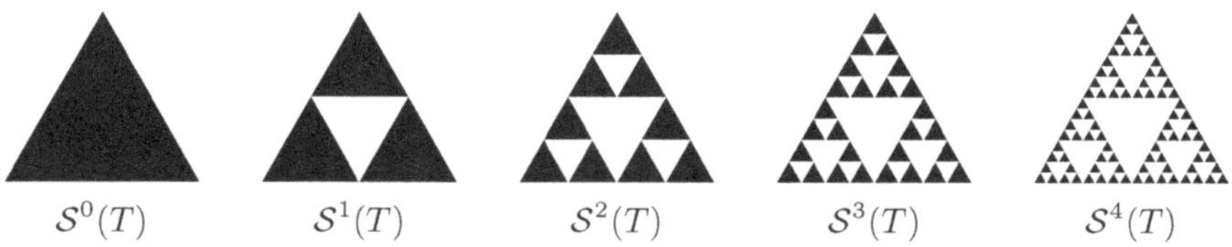

Figure XII.3. The first five iterations of the Sierpiński's triangle.

(5.7) Example (Koch's curve). On the Euclidean plane $\mathbb{R}^2$, take the segment line L with extremes $(0,0)$ and $(1,0)$. At each iteration, the next is constructed replacing the middle third of each line segment by the two legs of the equilateral triangle based on that middle third (there are two such triangles, thus one needs to make a definite choice). To achieve this, let us consider the set of similitudes $\mathcal{S} = \{\sigma_1, \sigma_2, \sigma_3, \sigma_4\}$ defined by

$$\sigma_j(x) = \tfrac{1}{3}O_j(x) + b_j$$
$$b_1 = (0,0), \qquad b_2 = (\tfrac{1}{3}, 0), \qquad b_3 = (\tfrac{1}{2}, \tfrac{1}{6}\sqrt{3}), \qquad b_4 = (\tfrac{2}{3}, 0),$$
$$O_1 = O_4 = I, \qquad O_2 = R(\tfrac{1}{3}\pi), \qquad O_3 = R(-\tfrac{1}{3}\pi)$$

where $R(\theta)$ denotes the rotation counter-clockwise through the angle θ.

The set

$$K = \lim_k \mathcal{S}^k(L),$$

where the limit is taken with respect to the Hausdorff metric, is called the **Koch's curve**, named after the Swedish mathematician Helge von Koch (1870–1924), and the first five iterations are shown in Figure XII.4.

Our aim is to show under which conditions an invariant set is self-similar, and to find a formula to calculate its dimension. To this end, let us introduce some further notation for the actions of $\mathcal{S} = \{\sigma_1, \ldots, \sigma_m\}$ on a point, a set, and measures. Precisely, let x be a point, X a Borel set, μ a Borel measure, $I = (i_1, \ldots, i_k)$ a sequence of k indices in $\{1, \ldots, m\}$, and let us

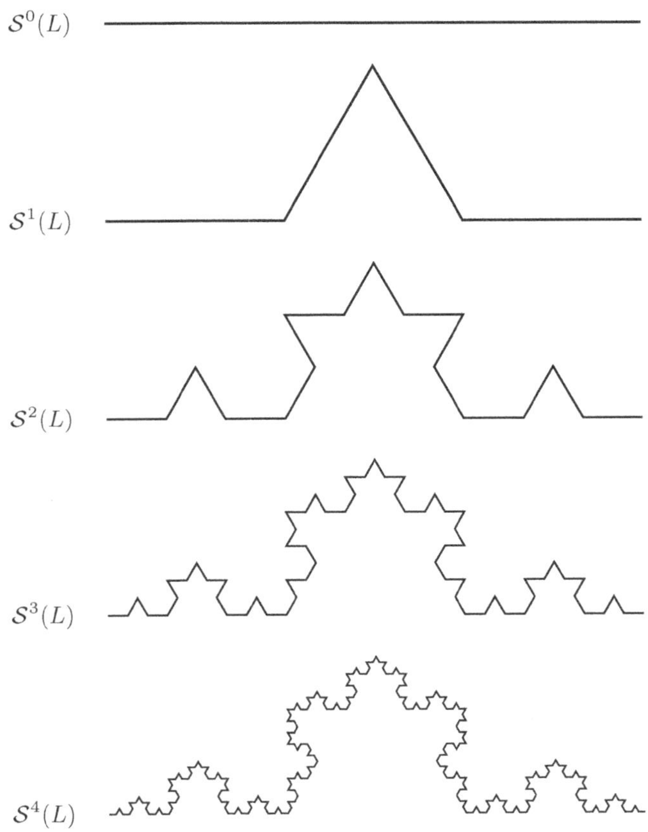

Figure XII.4. The first five iterations of the Koch's curve.

denote by $\Delta(m,k)$ the collection of all these sequences of indices. Let us also set

$$\sigma_I(x) = (\sigma_{i_1} \circ \cdots \circ \sigma_{i_k})(x), \qquad \sigma_I(X) = (\sigma_{i_1} \circ \cdots \circ \sigma_{i_k})(X),$$
$$\sigma_I(\mu) = (\sigma_{i_1} \circ \cdots \circ \sigma_{i_k})(\mu).$$

Finally, if $I = (i_1, \ldots, i_p) \in \Delta(m,p)$ and $J = (j_1, \ldots, j_q) \in \Delta(m,q)$, let us set

$$IJ = (i_1, \ldots, i_p, j_1, \ldots, j_q),$$

which is an element of $\Delta(m, p+q)$.

An important property of an invariant set under a finite family of similitudes, is that it is always possible to find a Borel probability measure concentrated on this set which is invariant under each of these similitudes, in the sense that is shown in the following theorem.

(5.8) Theorem. *Let $\mathcal{S} = \{\sigma_1, \ldots, \sigma_m\}$ be a finite family of similitudes with scale factor $r < 1$, and let K be a non-empty compact set that is invariant under $\mathcal{S}$. Then, there exists a Borel probability measure μ on $\mathbb{R}^n$, concentrated on K, such that, for every strictly positive integer k, one has*

$$\mu = m^{-k} \textstyle\sum_{I\in\Delta(m,k)} \sigma_I(\mu). \tag{5.9}$$

Proof. It is sufficient to construct the measure μ on K, for it can then be extended to $\mathbb{R}^n$ by setting $\mu(K^c) = 0$. To this end, let us pick a point x in K and let us consider the Dirac measure ε_x on x. For every strictly positive integer k, let us define

$$\mu^k = m^{-k} \textstyle\sum_{I\in\Delta(m,k)} \sigma_I(\varepsilon_x).$$

Observe that μ^k is thus a Borel probability measure on K. Furthermore, for every continuous function f defined on K, one has

$$\textstyle\int f \,\mathrm{d}\mu^k = m^{-k} \sum_{I\in\Delta(m,k)} f\big(\sigma_I(x)\big).$$

Let us prove that the sequence (μ^k) is weakly convergent as $k \to \infty$. To do this, let f be a continuous function defined on K, and let us start by proving that $(\int f \,\mathrm{d}\mu^k)$ is a Cauchy sequence. If ε is a strictly positive real number, then, by definition, there exists a strictly positive integer number c such that $|f(x) - f(y)| < \varepsilon$ whenever x and y are in K and $d(x,y) < r^c \operatorname{diam}(K)$.

Suppose that $p > q \geq c$, take I in $\Delta(m,p)$, J in $\Delta(m, q-p)$, and observe that, since $\sigma_{IJ}(x) = \sigma_I(\sigma_J(x)) \in \sigma_I(K)$ and $\operatorname{diam}(\sigma_I(K)) = r^c \operatorname{diam}(K)$, then

$$\Big| f\big(\sigma_I(x)\big) - f\big(\sigma_{IJ}(x)\big) \Big| < \varepsilon.$$

Taking the average over the indices J gives

$$\Big| f\big(\sigma_I(x)\big) - m^{-(q-p)} \textstyle\sum_{J\in\Delta(m,q-p)} f\big(\sigma_{IJ}(x)\big) \Big| < \varepsilon.$$

Taking now the average over the indices I yields $\left| \int f \,\mathrm{d}\mu^p - \int f \,\mathrm{d}\mu^q \right| < \varepsilon$.

Thus the sequence $(\int f \,\mathrm{d}\mu^k)$ is a Cauchy sequence, whence it converges for every continuous function f on K, thereby defining a positive linear functional on $\mathcal{C}(K)$. By virtue of Riesz-Markov's representation theorem, this positive linear functional is in fact a Daniell's functional, and it is represented by a Borel measure. Let us denote by μ such a measure.

Clearly $\mu(K) = \int I_K \,\mathrm{d}\mu = \lim_k \int I_K \,\mathrm{d}\mu^k = 1$, so μ is a probability measure on K. If $I = (i_1, \ldots, i_k)$ is a sequence of k indices in $\{1, \ldots, m\}$, one has

$$\sigma_I(x) \in \sigma_I(K), \quad \operatorname{diam}(\sigma_I(K)) = r^k \operatorname{diam}(K), \quad K = \textstyle\bigcup_{J\in\Delta(m,k)} \sigma_J(K).$$

So, the collection of all the $\sigma_I(x)$, as I varies over $\Delta(m,k)$, is dense in K, which yields that the support of μ coincides with K. Finally, by definition of the sequence (μ^k), one also has

$$\mu^{k+r} = m^{-k} \sum_{I\in\Delta(m,k)} \sigma_I(\mu^r)$$

and taking the limit as r approaches ∞, both μ^{k+r} and μ^r converge weakly towards μ, which yields (5.9), as desired. □

Invariant sets are interesting, but even more so are those for which it is possible to calculate the Hausdorff dimension explicitly, i.e. self-similar sets. the rest of this section is devoted to show a condition that makes an invariant set self-similar.

(5.10) Definition. Let $\mathcal{S} = \{\sigma_1, \ldots, \sigma_m\}$ be a finite family of similitudes with common scale factor $r < 1$. We shall say that the family $\mathcal{S}$ possesses a **separating set** if there exists a non-empty bounded open set U such that:

$$\text{(5.11)} \qquad \mathcal{S}(U) \subseteq U, \qquad \sigma_i(U) \cap \sigma_j(U) = \emptyset \qquad \text{for } i \neq j.$$

The existence of a separating set is more delicate than it might look like at first glance: In fact, the first of conditions (5.11) will fail if U is too small, while the second will fail if U is too big. All the examples introduced above admit a separating set.

(5.12) Example. It is not difficult to recognise that, for the Cantor set C_β a separating set is $U = (0,1)$; for the Sierpiński's triangle, a separating set is the topological interior of the region T; and for Koch's curve, a separating set is the interior of the triangle with vertices $(0,0)$, $(0,1)$, and $(\frac{1}{2}, \frac{1}{6}\sqrt{3})$, that is to say, the interior of the convex hull of the initial broken line $\mathcal{S}(L)$.

(5.13) Lemma. *Let c, d, ϱ be three positive numbers. Suppose that $(U_i)_{i\in I}$ is a collection of disjoint open sets such that U_i contains a ball of radius $c\varrho$ and is contained in a ball of radius $d\varrho$. Then, no ball of radius ϱ intersect more than $(1+2d)^n c^{-n}$ of the sets $\overline{U}_i$.*

Proof. If B is a ball of radius ϱ, and $B \cap \overline{U}_i \neq \emptyset$, then $\overline{U}_i$ is contained on the ball concentric with B with radius $(1+2d)\varrho$. Hence, if N of the $\overline{U}_i$'s intersect B, there are N disjoint balls of radius $c\varrho$ contained in a ball of radius $(1+2d)\varrho$. Adding up their n-dimensional Lebesgue measures, it is easy to check that $N(c\varrho)^n \leq [(1+2d)\varrho]^n$, so $N \leq (1+2d)^n c^{-n}$. □

With the aid of this packing lemma, we can now prove the following important result that shows that an invariant set that possesses a separating set is always self-similar, and gives a formula, which only depends on the scale factor r and the number m of similitudes, for the Hausdorff dimension of such a set.

(5.14) Theorem. *Suppose that $\mathcal{S} = \{\sigma_1, \ldots, \sigma_m\}$ is a family of similitudes with scale factor $r < 1$ that admits a separating set U, let K be the unique non-empty compact set that is invariant under $\mathcal{S}$, and let $s = \log_{1/r} m$. Then,*

(a) $0 < \mathcal{H}^s(K) < \infty$; *in particular, K has Hausdorff dimension s.*

(b) $\mathcal{H}^s\big(\sigma_i(K) \cap \sigma_j(K)\big) = 0$ *for $i \neq j$.*

Proof. For every positive integer k, observe that

$$K = \mathcal{S}^k(K) = \bigcup_{I\in\Delta(m,k)} \sigma_I(K),$$

where $\operatorname{diam}(\sigma_I(K)) = r^k \operatorname{diam}(K)$, so, setting $\delta_k < r^k \operatorname{diam}(K)$, one has

$$\mathcal{H}^s_{\delta_k}(K) \leq \sum_{I\in\Delta(m,k)} \operatorname{diam}(\sigma_I(K))^s = m^k r^{ks} \operatorname{diam}(K)^s = \operatorname{diam}(K)^s.$$

Since δ_k tends towards zero, as $k \to \infty$, it follows that

$$\mathcal{H}^s(K) \leq \operatorname{diam}(K)^s < \infty.$$

In order to show that $\mathcal{H}^s(K) > 0$, let us choose three strictly positive real numbers c, d, δ, so that the separating set U, which is open by assumption, contains a ball with radius cr^{-1}, and it is contained into a ball of radius d, and let us set $N = (1 + 2d)^n c^{-n}$. We are going to prove $\mathcal{H}^s_\delta(K) \geq N^{-1}$ by showing that, if $(U_j)_{j\geq 1}$ is any covering of K by sets with diameter at most δ, then $\sum_{j=1}^{\infty} \operatorname{diam}(U_j)^s \geq N^{-1}$. In fact, since any set with diameter δ is contained in a (closed) ball of radius δ, it suffices to show that if $K \subseteq \bigcup_{j=1}^{\infty} B_j$ holds, where B_j is a ball of radius $\delta_j \leq \delta$, then $\sum_{j=1}^{\infty} \delta_j^s \geq N^{-1}$.

To this end, let μ be the probability measure concentrated on K given by Theorem (5.8). Let us prove that, if B is any ball with radius $\delta \leq 1$, then $\mu(B) \leq N\delta^s$, for the desired conclusion follows: In fact,

$$1 = \mu(K) \leq \sum_{j=1}^{\infty} \mu(B_j) \leq N \sum_{j=1}^{\infty} \delta_j^s.$$

To prove this claim, let k be the only positive integer such that $r^k < \delta \leq r^{k-1}$. Thus, by virtue of equation (5.9),

$$\mu(B) = m^{-k} \sum_{I\in\Delta(m,k)} \sigma_I(\mu)(B).$$

Since $K \subseteq \overline{U}$, one has $\operatorname{supp}(\sigma_I(\mu)) = \sigma_I(K) \subseteq \sigma_I(\overline{U})$, so $\sigma_I(\mu)(B) = 0$ unless B intersects $\sigma_I(\overline{U})$. On the other hand, (5.11) shows that the sets $\sigma_I(U)$ are all disjoint, and each of them contains a ball of radius $cr^{k-1} \geq c\delta$ and is contained in a ball of radius $dr^k < d\delta$. By virtue of Lemma (5.13), B can intersect at most N of the $\sigma_I(\overline{U})$'s. Thus,

$$\mu(B) \leq Nm^{-k} = Nr^{ks} \leq N\delta^s,$$

as claimed.

Finally, since every σ_j in $\mathcal{S}$ decreases distances by a factor r, one has

$$\mathcal{H}^s\big(\sigma_j(K)\big) = r^s\mathcal{H}^s(K) = m^{-1}\mathcal{H}^s(K)$$

hence $\mathcal{H}^s(K) = \sum_{j=1}^m \mathcal{H}^s(\sigma_j(K))$. On the other hand, one also has $K = \bigcup_{j=1}^m \sigma_j(K)$, and this can only happen if $\mathcal{H}^s(\sigma_i(K)\cap\sigma_j(K)) = 0$ for $i \neq j$. □

As a result of the previous theorem it is now possible to calculate the dimension of any self-similar set which possesses a separating set.

(5.15) Example. The Hausdorff dimension of the Cantor's set is $\log_{1/\beta} 2$; the Hausdorff dimension of the Sierpiński's triangle is $\log_2 3$; and the Hausdorff dimension of the Koch's curve is $\log_3 4$.

6. $\mathcal{H}^1$ and the classical notion of length

Let (E, d) be a metric space. A **curve** is any continuous map γ, which is defined on an interval of the real line, and takes values in E. For every interval $[a, b]$ included in the set of definition of the curve γ, the **support** of γ is the Borel set $\gamma([a, b])$. The **length** of γ over $[a, b]$ is classically defined as

$$L_\gamma(a, b) = \sup\Big\{\textstyle\sum_{i=1}^m d\big(\gamma(t_i), \gamma(t_{i-1})\big) : a = t_0 < t_1 < \cdots < t_m = b\Big\},$$

where the supremum is taken over all finite subdivisions $(t_0, \ldots, t_m)$ of the interval $[a, b]$. An approximation of $L_\gamma(a, b)$ is shown in Figure XII.5.

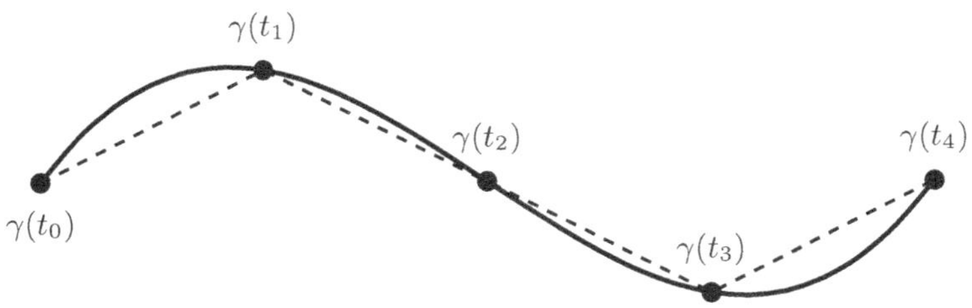

Figure XII.5. The topological curve is approximated with a subdivision with five points. The length of the curve is approximated by the sum of the distances (which are represented in the diagram by the dashed straight lines) of each consecutive pair of points in the subdivision.

The next theorem shows that the support of a curve always has Hausdorff dimension 1, and if the curve is injective its Hausdorff measure coincides with the length of the curve.

(6.1) Theorem. *Let γ be a curve in (E, d), and let C be the support of γ on the interval $[a, b]$, included in the set of definition of γ. Then,*

$$\mathcal{H}^1(C) \leq L_\gamma(a, b).$$

In addition, if γ is injective, then the previous inequality is in fact an equality.

Proof. In the hypotheses of the theorem, the set C is compact, hence, if ε is a strictly positive real number, from the open covering $\{B(\gamma(t), \varepsilon) : t \in [a, b]\}$ it is possible to extract a finite sub-covering,

$$\Big\{B\big(\gamma(t_0), \varepsilon\big), \cdots, B\big(\gamma(t_m), \varepsilon\big)\Big\}.$$

With no loss of generality, we may assume that $a = t_0 < t_1 < \cdots < t_m = b$. Let us set $C_i = \gamma([t_{i-1}, t_i])$, and observe that $C = \bigcup_{i=1}^m C_i$. Furthermore, by refining the subdivision $(t_0, \ldots, t_m)$, if necessary, we may assume that, for every strictly positive index i, one has $\operatorname{diam}(C_i) \le d(\gamma(t_i), \gamma(t_{i-1}))$. Thus, if δ is a strictly positive real number, by refining even further, if necessary, the subdivision $a = t_0 < t_1 < \cdots < t_m = b$ so that $\operatorname{diam}(C_i) < \delta$, one has

$$\mathcal{H}^1_\delta(C) \le \textstyle\sum_{i=1}^n \operatorname{diam}(C_i) \le \sum_{i=1}^n d(\gamma(t_i), \gamma(t_{i-1})) \le L_\gamma(a, b).$$

Taking the supremum with respect to δ, this gives the desired inequality.

Let us now assume that γ is injective, and let us prove that, then, the converse inequality also holds. To this end, let us start by proving that

$$\mathcal{H}^1(C) \ge d\big(\gamma(a), \gamma(b)\big).$$

To see this, consider the real function p defined on C by $p(u) = d(\gamma(a), u)$, and observe that, if u and v are two points in C, one has

$$\big|p(u) - p(v)\big| = \big|d\big(\gamma(a), u\big) - d\big(\gamma(a), v\big)\big| \le d(u, v)$$

which shows that p is Lipschitz continuous, and Corollary (3.2) yields

$$\mathcal{H}^1\big(p(C)\big) \le \mathcal{H}^1(C).$$

Furthermore, $p(C)$ must contain the interval $[0, d(\gamma(a), \gamma(b))]$ and therefore,

$$\mathcal{H}^1(C) \ge \mathcal{H}^1\big(p(C)\big) \ge \mathcal{H}^1\big(\big[0, d(\gamma(a), \gamma(b))\big]\big) = d\big(\gamma(a), \gamma(b)\big).$$

Let us now consider a subdivision $a = t_0 < t_1 < \cdots < t_m = b$, of the interval $[a, b]$, and, for every strictly positive index i, set $C_i = \gamma([t_{i-1}, t_i])$. Then, one clearly has $C = \bigcup_{i=1}^m C_i$, and, thanks to the fact that γ is injective, this gives

$$\mathcal{H}^1\big(\textstyle\bigcup_{i=1}^m (C_{i-1} \cap C_i)\big) = \mathcal{H}^1\big(\{\gamma(t_1), \ldots, \gamma(t_{m-1})\}\big) = 0.$$

Thus,

$$\mathcal{H}^1(C) = \textstyle\sum_{i=1}^m \mathcal{H}^1(C_i) \ge \sum_{i=1}^m d\big(\gamma(t_i), \gamma(t_{i+1})\big).$$

Taking the supremum over all possible finite subdivisions of the interval $[a, b]$ gives $\mathcal{H}^1(C) \ge L_\gamma(a, b)$, and this completes the proof. □

7. $\mathcal{H}^n$ is the Lebesgue measure in $\mathbb{R}^n$

The aim of this section is to prove that, in $\mathbb{R}^n$, the measure $\mathcal{H}^n$ coincides with the Borel-Lebesgue measure λ^n. This is not a simple fact, and not an obvious fact either, since the Borel-Lebesgue measure is defined as the product of n copies of the 1-dimensional Borel-Lebesgue measure, while $\mathcal{H}^n$ is computed in terms of fine coverings. For the very nature of the Hausdorff measures, the first step in this direction consists in proving an important result, due to G. Vitali, regarding coverings, which incidentally is a fundamental tool in Geometric Measure Theory, and it is used to create a connection between local and global properties of a measure.

Keeping in mind our goal to prove the equality between the n-dimensional Hausdorff measure and Lebesgue measure, we shall derive Vitali's covering theorem from the so-called '$5r$-covering theorem', which is the following.

(7.1) Theorem (The $5r$-covering theorem). *Let E be a separable metric space, and let $\mathcal{B}$ be a family of open balls in E such that*

$$\sup\{\operatorname{diam}(B) : B \in \mathcal{B}\}$$

is finite. Then, there is a finite or countable subset $\mathcal{B}^*$ of pairwise disjoint elements of $\mathcal{B}$ such that

$$\bigcup_{B\in\mathcal{B}} B \subseteq \bigcup_{C\in\mathcal{B}^*} 5C.$$

In addition, if B belongs to $\mathcal{B}$, there exists C in $\mathcal{B}^*$ such that $B \subseteq 5C$.

Proof. For every positive integer n, denote by $\mathcal{B}_n$ the set of all balls in $\mathcal{B}$ whose diameter belongs to the interval $(2^{-n-1}m, 2^{-n}m]$, where

$$m = \sup\{\operatorname{diam}(B) : B \in \mathcal{B}\},$$

which is finite by hypothesis. By induction, let $\mathcal{B}_0^*$ be the maximal subset of $\mathcal{B}$ whose elements are pairwise disjoint, which exists by Zorn's lemma. Now, assuming that $\mathcal{B}_0^*, \ldots, \mathcal{B}_k^*$ have been selected, and let $\mathcal{B}_{k+1}^*$ be the maximal subset of $\mathcal{B}_{k+1}$ constituted of pairwise disjoint sets which are also disjoint from every element of $\mathcal{B}_0^* \cup \cdots \cup \mathcal{B}_k^*$, which again exists by virtue of Zorn's lemma. Finally, let us set

$$\mathcal{B}^* = \bigcup_{k=0}^{\infty} \mathcal{B}_k^*.$$

Let us observe that $\mathcal{B}^*$ is a countable set of pairwise disjoint elements of $\mathcal{B}$. Indeed, $\mathcal{B}$ is a collection of pairwise disjoint subsets of E, hence it is countable, because E is separable. Furthermore, every ball B in $\mathcal{B}$ intersects a ball C in $\mathcal{B}^*$ such that $B \subseteq 5C$. To see this, let B be an element of $\mathcal{B}$, and observe that there must be a positive integer k such that B belongs to $\mathcal{B}_k$. Now, if B lies in $\mathcal{B}_k^*$, there is nothing to prove, for it suffices to take C to be the set B itself, so let us assume that B does not belong to $\mathcal{B}_k^*$, which implies that $k > 0$, and, by construction, that B intersects a ball C that belongs to the union $\mathcal{B}_0^* \cup \cdots \cup \mathcal{B}_{k-1}^*$. As a result, such a ball C must have diameter $2^{-k-1}m$. Since the diameter of B is, by assumption, less than or equal to $2^{-k}m$, the triangle inequality yields that $B \subseteq 5C$, as required. From this, it also follows immediately that

$$\bigcup_{B \in \mathcal{B}} B \subseteq \bigcup_{C \in \mathcal{B}^*} 5C,$$

and the theorem is fully proved. □

(7.2) Remark. Let us observe that the constant 5 which appears in the previous theorem is not optimal. By taking, in lieu of the factor 2^{-n}, the factor c^{-n}, with $c > 1$, in the definition of the partition $\mathcal{B}_n$, the final value would reduce from 5 to $1 + 2c$. As a result, any constant larger than $1 + 2c$ gives the correct statement for the previous theorem, except 3.

The reason why we have introduced the previous theorem is because it constitutes a fundamental tool to prove the following result originally found by G. Vitali (1908) in the special case when the dimension is $n = 1$, and extended in the present form by H. Lebesgue (1910).

(7.3) Theorem (Vitali's covering theorem). *Let E be a Borel subset of $\mathbb{R}^n$ with finite Lebesgue measure, and δ a strictly positive real number. Then, there exists a countable family $\{C_i : i \in I\}$ of pairwise disjoint closed balls contained in E with diameter less than δ such that*

$$\lambda^n\big(E \setminus \bigcup_{i \in I} C_i\big) = 0.$$

Proof. Let us denote by $\mathcal{B}$ the set of all balls with centre in a point of E and diameter smaller than or equal to δ. With no loss of generality, we may assume that $\delta = 1$. Then, by virtue of Theorem (7.1), there exists a countable subset $\mathcal{B}^*$ of $\mathcal{B}$ of pairwise disjoint balls, such that, for every ball B in $\mathcal{B}$, there exists a ball C in $\mathcal{B}^*$ such that $B \subseteq 5C$.

Let us take a strictly positive real number r, and let us denote by Z the set of all points of E that do not belong to any ball from $\mathcal{B}^*$ but that belong to the ball $B(0, r)$ with centre at the origin and radius r. It suffices to show that Z is negligible with respect to the Lebesgue measure λ^n for every given r.

To this end, let us denote by $\mathcal{C}(r)$ the subset of $\mathcal{B}^*$ constituted by all those balls that meet $B(0, r)$, and notice that $\mathcal{C}(r)$ must be countable, because such is $\mathcal{B}^*$. Let $(C_i)_{i \geq 1}$ be an enumeration of the elements of $\mathcal{C}(r)$, and let us consider a point z in Z. For every strictly positive integer p, observe that, by construction, z does not belong to the closed set $K = C_1 \cup \cdots \cup C_p$ but, at the same time, from the fact that $\mathcal{B}$ is a covering of E and $z \in E$, we can find a ball B in $\mathcal{B}$, which contains z, belongs to $\mathcal{C}(r)$, and it is disjoint from K.

By the properties of $\mathcal{B}^*$, the ball B intersects some ball C_i in $\mathcal{B}^*$, and it is contained in $5C_i$. But, since K and B are disjoint, it must be $i > p$. As a result, z lies in $5C_i$ for some $i > p$, and therefore

$$Z \subseteq \bigcup_{i=p}^{\infty} 5C_i.$$

This gives, for every strictly positive integer p, the inequality

$$\lambda^n(Z) \leq \sum_{i=p}^{\infty} \lambda^n(5C_i) = 5^n \sum_{i=p}^{\infty} \lambda^n(C_i). \tag{7.4}$$

Since the balls of $\mathcal{C}(r)$ are contained in $B(0, r+2)$, and these balls are pairwise disjoint, this gives

$$\sum_{i=1}^{\infty} \lambda^n(C_i) < \infty.$$

Thus, the term on the right-hand side of (7.4) converges towards 0 as p goes to infinity, which shows that Z is negligible with respect to the Lebesgue measure λ^n, as desired. □

Sadly, Vitali's covering theorem is not sufficient to prove the desired equality between the Lebesgue measure λ^n and the Hausdorff measure $\mathcal{H}^n$ in $\mathbb{R}^n$, for there is another important ingredient, known as the 'isodiametric inequality', that needs to be addressed first. In order to prove this inequality, we shall also a technical tool, called 'symmetrisation'. To introduce the latter, let us start by assuming that $n > 1$, so that the space $\mathbb{R}^n$ may be decomposed into the product $\mathbb{R}^{n-1} \times \mathbb{R}$, and, correspondingly every point x in $\mathbb{R}^n$ may be written in the form

$$x = (\hat{x}, x_n), \qquad \text{where } \hat{x} \in \mathbb{R}^{n-1} \text{ and } x_n \in \mathbb{R}.$$

Furthermore, for any subset E of $\mathbb{R}^n$, let us denote by $E(\hat{z})$ the section of E relative to the point $\hat{z}$ in $\mathbb{R}^{n-1}$, that is to say the set defined by

$$E(\hat{z}) = \big\{t \in \mathbb{R} : (\hat{z}, t) \in E\big\}.$$

With this notation, we may now give the following definition.

(7.5) Definition. Let E be a subset of $\mathbb{R}^n$. The **symmetrisation** of E, is denoted by $S_n(E)$, and is defined as

$$S_n(E) = \Big\{x = (\hat{x}, x_n) \in \mathbb{R}^n : |x_n| \leq \tfrac{1}{2}\lambda^1\big(E(\hat{x})\big)\Big\}.$$

In the same hypotheses as the previous definition, let us observe immediately that the set $S_n(E)$ is symmetric with respect to the hyper-plane with equation $x_n = 0$ (see Figure XII.6).

(7.6) Remark. By virtue of Fubini's theorem, if E is Borel measurable in $\mathbb{R}^n$, then $S_n(E)$ is Borel measurable in $\mathbb{R}^{n-1}$, the map $\hat{z} \mapsto \lambda^1(S_n(E(\hat{z}))$ is Borel measurable, and one has

$$\lambda^n(E) = \textstyle\int \lambda^1\big(E(\hat{z})\big)\, \mathrm{d}\hat{z} = \int \lambda^1\big(S_n\big(E(\hat{z})\big)\big)\, \mathrm{d}\hat{z} = \lambda^n\big(S_n(E)\big),$$

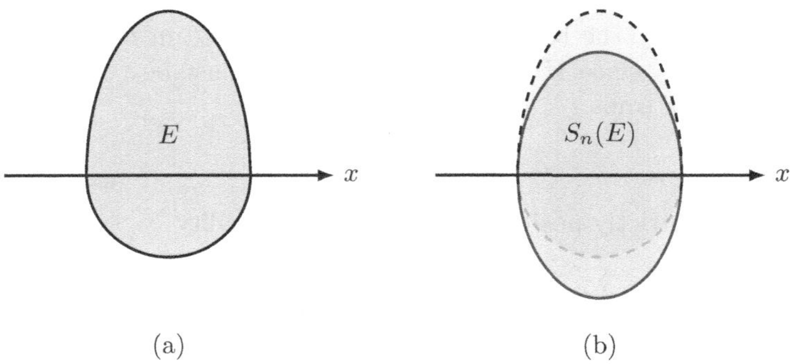

Figure XII.6. Diagram (a) shows a region E in $\mathbb{R}^2$ which is not symmetrical with respect to the x-axis. Diagram (b) shows the symmetrisation, $S_n(E)$, of E.

which means that the Lebesgue measure is invariant under the operation of symmetrisation. On the other hand, the diameter decreases under symmetrisation, that is to say,

$$\operatorname{diam}\big(S_n(E)\big) \le \operatorname{diam}(E).$$

In order to demonstrate this, let $x = (\hat{x}, x_n)$ be a point in $S_n(E)$, and let us consider two points $u(x) = (\hat{u}(x), u_n(x))$ and $v(x) = (\hat{v}(x), v_n(x))$ in the closure of E such that

$$\begin{aligned} &\hat{u}(x) = \hat{v}(x) = \hat{x}, \\ &u_n(x) \le z_n \le v_n(x) \qquad \text{for every } z = (\hat{z}, z_n) \in E \text{ with } \hat{z} = \hat{x}. \end{aligned}$$

Given now x and y in $S_n(E)$, observe that, by construction of $S_n(E)$ and by the second of the two relations above, it follows that

$$|x_n - y_n| \le \max\Big\{|v_n(x) - u_n(y)|, |u_n(x) - v_n(y)|\Big\}.$$

Further, by the first of the two relations above, and since $\|z\| = \|\hat{z}\|^2 + |z_n|^2$,

$$\|x - y\| \le \max\Big\{\big\|v(x) - u(y)\big\|, \big\|u(x) - v(y)\big\|\Big\} \le \operatorname{diam}(\overline{E}) = \operatorname{diam}(E).$$

Since the points x and y are arbitrary in $S_n(E)$, taking the supremum on the left-hand side of the previous relation gives the desired inequality.

(7.7) Theorem (Isodiametric inequality). *Among all Borel measurable sets of fixed diameter, the balls have the maximum volume. In other words,*

$$\lambda^n(E) \le 2^{-n}\omega_n \operatorname{diam}(E)^n \qquad \textit{for every Borel subset } E \textit{ of } \mathbb{R}^n.$$

Proof. Let us start by observing that, substituting E with $\overline{E}$ does not affect either side of the isodiametric inequality. Hence, with no loss of generality, we may assume that E is a closed subset of $\mathbb{R}^n$. Let S_i denote the symmetrisation with respect to the ith coordinate axis. In other words, let us set

$$S_i = L_{in} \circ S_n \circ L_{in},$$

where L_{in} is the diffeomorphism of $\mathbb{R}^n$ that permutes the ith and the nth coordinates. By induction, let $E_0 = E$, and, for every index $i = 1, \ldots, n$, let $E_i = S_i(E_{i-1})$. Since $\lambda^n(E_n) = \lambda^n(E)$, and since $\operatorname{diam}(E_n) \leq \operatorname{diam}(E)$, it suffices to prove the isodiametric inequality for E_n. By construction, however, the Borel measurable set E_n is symmetric under reflection with respect to each coordinate hyper-plane, so that $x \in E_n$ if and only if $-x \in E_n$. Thus, E_n is included in the ball B of centre at the origin and radius $r = \frac{1}{2}\operatorname{diam}(E_n)$. Since $\lambda^n(B) = \omega_n r^n$, the inequality follows. □

At last, we now have all the ingredients needed to prove the announced equality between the n-dimensional Hausdorff measure and the Lebesgue measure λ^n in $\mathbb{R}^n$. Precisely, the following result holds.

(7.8) Theorem. *If E is a Borel subset of $\mathbb{R}^n$, and if δ is a strictly positive real number, then*

$$\lambda^n(E) = \mathcal{H}^n(E) = \mathcal{H}^n_\delta(E).$$

Proof. Let E be a Borel subset of $\mathbb{R}^n$, and let us start by observing that, for any strictly positive real number δ, one has

$$\mathcal{H}^n_\delta(E) \leq 2^{-n}\omega_n(\sqrt{n})^n\lambda^n(E). \tag{7.9}$$

In fact, if (X_k) is countable covering of E made of closed cuboids of $\mathbb{R}^n$, and if $r(X_k)$ denotes the side length of the cuboid X_k, then $\sqrt{n}r(X_k) = \operatorname{diam}(X_k)$. If $\operatorname{diam}(X_k) \leq \delta$ for every index k, one thus has

$$\mathcal{H}^n_\delta(E) \leq \omega_n \textstyle\sum_{k=1}^\infty 2^{-n}\operatorname{diam}(X_k)^n = 2^{-n}\omega_n(\sqrt{n})^n \sum_{k=1}^\infty r(X_k)^n,$$

and the conclusion follows taking the infimum over all possible coverings (X_k).

Let us now prove that $\lambda^n(E) \geq \mathcal{H}^n(E)$ holds. With no loss of generality, we may assume that $\lambda^n(E)$ is finite. In this case, given a pair ε, δ of strictly positive real numbers, let us consider an open set U which contains E and such that $\lambda^n(U) \leq \lambda^n(E) + \varepsilon$, and consider a countable family $\{C_i : i \in I\}$ of pairwise disjoint closed balls contained in E with diameter less than δ, and such that $\lambda^n(U \setminus F) = 0$, where $F = \bigcup_{i\in I} C_i$, which exists by virtue of Vitali's covering theorem. Then,

$$\begin{aligned}\lambda^n(E) + \varepsilon \geq \lambda^n(U) &= \lambda^n\big(\textstyle\bigcup_{i\in I} C_i\big)\\ &= \textstyle\sum_{i\in I}\lambda^n(C_i)\\ &= 2^{-n}\omega_n \textstyle\sum_{i\in I}\operatorname{diam}(C_i)^n \geq \mathcal{H}^n_\delta(F).\end{aligned}$$

Thanks to (7.9), this means that $\mathcal{H}^n_\delta(U \setminus F) = 0$, whence $\mathcal{H}^n(E \setminus F) = 0$ and therefore $\mathcal{H}^n_\delta(E \setminus F) = 0$ as well. Furthermore,

$$\mathcal{H}^n_\delta(E) \le \mathcal{H}^n_\delta(E \cap F) + \mathcal{H}^n_\delta(E \setminus F) \le \mathcal{H}^n_\delta(F) \le \lambda^n(E) + \varepsilon,$$

and the conclusion follows from the fact that δ and ε are chosen arbitrarily.

Finally, let δ be a strictly positive real number, and let (X_k) be a countable covering of E with sets with diameter smaller than δ. By virtue of the isodiametric inequality, one has

$$\begin{aligned}\lambda^n(E) &\le \lambda^n\big(\textstyle\bigcup_{k=1}^\infty X_k\big)\\ &\le \textstyle\sum_{k=1}^\infty \lambda^n(X_k)\\ &\le 2^{-n}\omega_n \textstyle\sum_{k=1}^\infty \operatorname{diam}(X_k)^n\end{aligned}$$

whence

$$\lambda^n(E) \le \mathcal{H}^n_\delta(E) \le \mathcal{H}^n(E),$$

which completes the proof. □

8. The area formula

Let n be a strictly positive integer, and let us observe that, on the Euclidean space $\mathbb{R}^n$, the family of Hausdorff measures $\mathcal{H}^s$, with $s \in [0, n]$, allows to classify and measure all Borel measurable sets which are negligible with respect to λ^n. In this section, we shall focus on a specific class of this type of sets, which have integer dimension $0 \le k \le n$, and we shall prove that, when they are 'sufficiently regular', it is possible to calculate their Hausdorff measure with a simple formula.

Precisely, let us start by giving the following definition, which identifies what we mean by a 'sufficiently regular' set.

(8.1) Definition. Let k be an integer with $1 \le k \le n$. A **k-dimensional surface** or, simply, a **k-surface** is a set M of $\mathbb{R}^n$ with the following property: For every x in M, there exists a neighbourhood V of x in $\mathbb{R}^n$, an open set U in $\mathbb{R}^k$, and an injective continuously differentiable map f from U to V, called a **parametrisation** of $M \cap V$, such that $f(U) = M \cap V$ and such that the Jacobian $Df(x)$ is injective for each x in U.

Let us observe immediately that, in the same hypotheses as the previous definition, a k-surface M can be covered by a countable collection of open sets for which the intersection with M have a parametrisation, so for our purposes, it suffices to assume that $M = M \cap V$ has a global parametrisation.

Let us begin by proving the area formula in the linear case. Precisely, let L be a linear map from $\mathbb{R}^k$ to $\mathbb{R}^n$, and let L^* from $\mathbb{R}^n$ to $\mathbb{R}^k$ denote its transpose. Then, then L^*L is a positive semidefinite linear operator on $\mathbb{R}^n$, and its determinant is therefore positive, whence we may set

$$JL = \sqrt{\det(L^*L)}$$

(8.2) Proposition. *Let n be a strictly positive integer and k an integer with $1 \leq k \leq n$, and let L be a linear map from $\mathbb{R}^k$ to $\mathbb{R}^n$. Then,*

$$\mathcal{H}^k\big(L(A)\big) = JL \cdot \lambda^k(A) \qquad \textit{for every Borel subset } A \textit{ of } \mathbb{R}^k.$$

Proof. Assume that $k = n$. Then, $\det(L^*L) = (\det(L))^2$, so $JL = |\det(L)|$, and since the Hausdorff measure $\mathcal{H}^n$ coincides with the Lebesgue measure λ^n, the assertion reduces to Proposition XI(2.2)(d). If $k < n$, the space $\mathbb{R}^n$ can be decomposed into the product $\mathbb{R}^k \times \mathbb{R}^{n-k}$, and let us denote by P and Q the projections of $\mathbb{R}^n$ onto the first and the second component, so that, for every point x of $\mathbb{R}^n$, one has the decomposition $x = (Px, Qx)$.

Let us denote by R the rotation that maps the range of L into the subspace of $\mathbb{R}^k \times \mathbb{R}^{n-k}$ with equation $Qx = 0$. Thus, the composite map RL, maps any vector x of $\mathbb{R}^n$, into a vector of $\mathbb{R}^n$ which belongs to the range of P, which means whose last $n - k$ components are zero; whence $PRL = RL$.

Let us denote by M the linear map, from $\mathbb{R}^n$ to $\mathbb{R}^k$, that maps a vector x of $\mathbb{R}^n$ into the vector of $\mathbb{R}^k$ obtained from x by taking only the first k components. It is easy to check that the transpose M^* of M is the linear map, from $\mathbb{R}^k$ to $\mathbb{R}^n$, that to each vector y of $\mathbb{R}^k$ associates the vector of $\mathbb{R}^n$ obtained by adjoining $n - k$ zeros to the k components of y. Moreover, $M^*M = P$. Thus, considering the linear map $T = MRL$,

$$\begin{aligned} T^*T &= L^*R^*M^*MRL \\ &= L^*R^*PRL \\ &= L^*R^*RL \\ &= L^*L. \end{aligned}$$

As a result, one has $JT = JL$. Thus, since R^* and M^* are isometries, and since $R^*M^*T = R^*M^*MRL = R^*PRL = R^*RL = L$, it follows by virtue of Corollary (3.3) that

$$\mathcal{H}^k\big(L(A)\big) = \mathcal{H}^k\Big(R^*\big(M^*(T(A))\big)\Big) = \mathcal{H}^k\big(M^*(T(A))\big) = \mathcal{H}^k\big(T(A)\big).$$

On the other hand $\mathcal{H}^k$ coincides with λ^k in $\mathbb{R}^k$, and the conclusion follows. □

This partial result will be one of the two ingredients to prove the general formula. The second ingredient is the following.

(8.3) Lemma. *Let M be a k-surface on $\mathbb{R}^n$, let U be an open set in $\mathbb{R}^k$, and let f, from U to $\mathbb{R}^n$, be a parametrisation of M. Then, for every real number t strictly greater than 1, there exists a sequence (B_j) of pairwise disjoint Borel subsets of U such that $U = \bigcup_j B_j$, and a sequence (L_j) of linear maps from $\mathbb{R}^k$ to $\mathbb{R}^n$ such that, for every x in B_j and z in $\mathbb{R}^k$*

$$(8.4) \qquad t^{-1}\|L_j(z)\| \le \|Df(x)z\| \le t\|L_j(z)\|,$$

and, for every pair x, y of elements of B_j,

$$(8.5) \qquad t^{-1}\|L_j(x) - L_j(y)\| \le \|f(x) - f(y)\| \le t\|L_j(x) - L_j(y)\|.$$

Proof. Let us fix a strictly positive real number ε and a real number s strictly greater than 1 such that

$$t^{-1} + \varepsilon < s^{-1} < 1 < s < t - \varepsilon,$$

and let $\mathcal{L}$ be a countable dense subset of the set of all linear maps from $\mathbb{R}^k$ to $\mathbb{R}^n$ (for instance, $\mathcal{L}$ may be taken as the set of all linear maps whose representative matrix has all rational coefficients). For L in $\mathcal{L}$ and m positive integer, let us denote by $E(L, m)$ the set of all elements x of U such that

$$s^{-1}\|L(z)\| \le \|Df(x)z\| \le s\|L(z)\|$$
$$t^{-1}\|L(x) - L(y)\| \le \|f(x) - f(y)\| \le t\|L(x) - L(y)\|$$

hold, for every $z \in \mathbb{R}^k$, and every $y \in U$ with $\|y - x\| < m^{-1}$.

The definition of $E(L, m)$ is unaffected if y and z are restricted to lie in countable dense subsets of $\mathbb{R}^n$ and $\mathbb{R}^k$, respectively; hence $E(L, m)$ is defined by countably many inequalities involving continuous functions, and therefore it is a Borel set. Furthermore, with no loss of generality, every set $E(L, m)$ may be assumed to have diameter less than m^{-1}.

Let us prove that the family

$$(8.6) \qquad \{E(L, m) : L \in \mathcal{L},\ m \in \mathbb{N}^*\}$$

is a covering of U.

To this end, let x be an element of U, and let us set

$$\delta_0 = \inf\{\|Df(x)z\| : \|z\| = 1\}.$$

Since $Df(x)$ is injective, δ_0 is strictly positive. Let us choose a strictly positive real number δ so that $\delta \le (s-1)\delta_0$ and $\delta \le (1-s^{-1})\delta_0$, and pick an element L of $\mathcal{L}$ such that $\|L - Df(x)\| < \delta$. Then,

$$\begin{aligned}
\|L(z)\| &\le \|Df(x)z\| + \|L(z) - Df(x)z\| \\
&\le \|Df(x)z\| + \delta\|z\| \\
&\le \|Df(x)z\| + (s-1)\delta_0\|z\| \\
&\le \|Df(x)z\| + (s-1)\|Df(x)z\| = s\|Df(x)z\|,
\end{aligned}$$

and similarly, $\|L(z)\| \geq s^{-1}\|Df(x)z\|$. This establishes the first inequality and also shows that L is injective, so $\eta = \inf\{\|L(z)\| : \|z\| = 1\}$ is strictly positive. Since f is differentiable at x, there is a positive integer m such that

$$\|f(y) - f(x) - Df(x)(y - x)\| \leq \varepsilon\eta\|y - x\| \leq \varepsilon\|L(y - x)\|$$

for $\|x - y\| < m^{-1}$. On the other hand

$$\begin{aligned}\|f(y) - f(x)\| &\leq \|f(y) - f(x) + Df(x)(y - x)\| + \|Df(x)(y - x)\| \\ &\leq \varepsilon\|L(y) - L(x)\| + s\|L(y) - L(x)\| \\ &\leq t\|L(y) - L(x)\|,\end{aligned}$$

and similarly $\|f(y) - f(x)\| \geq t^{-1}\|L(y) - L(x)\|$, which proves also the second inequality. In other words, the point x belongs to the set $E(L, m)$.

Thus, the family (8.6) is a countable covering of U constituted of Borel sets with diameter smaller than m^{-1}. Hence, denoting by $(E_j)_{j\geq 1}$ an enumeration of this countable covering, it is now possible to define $B_1 = E_1$ and, inductively, assuming that the sets $B_1, \ldots, B_{j-1}$ have been defined,

$$B_j = E_j \setminus \textstyle\bigcup_{k=1}^{j-1} B_k.$$

The conclusion of the theorem follows by taking the sequence $(L_j)_{j\geq 1}$ of linear maps, such that, for every index j, L_j is the linear map in $\mathcal{L}$ associated to E_j whence to B_j. □

We are finally ready to state and prove the main result of this section, which is a masterpiece of the theory of Hausdorff measures: Indeed, this very powerful formula links together concepts that, in principle, are very different, one being the Hausdorff measure, whence the measure of sets calculate using fine coverings, and the image measure calculated using the parametrisation of a k-surface to transfer the Lebesgue measure from the space $\mathbb{R}^k$ to the k-surface.

(8.7) Theorem (The area formula). *Let M be a k-dimensional surface, and let f, from U to $\mathbb{R}^n$ be a parametrisation of M. If A is a Borel subset of U, then $f(A)$ is a Borel subset of $\mathbb{R}^n$ and*

$$\mathcal{H}^k\big(f(A)\big) = \int_A Jf(x)\,\mathrm{d}x, \tag{8.8}$$

where

$$Jf(x) = \sqrt{\det\big(Df(x)^* Df(x)\big)}.$$

Furthermore, if g is a Borel measurable function on $f(U)$, then

$$\int g(y)\,\mathcal{H}^k(\mathrm{d}y) = \int g\big(f(x)\big)Jf(x)\,\mathrm{d}x, \tag{8.9}$$

in the sense that, if either of the integrals in the previous equality exists, then both exist and the equality holds.

Proof. Let us start by observing that, if (8.8) holds, then (8.9) also holds: In fact, (8.8) coincides with (8.9) when g is the indicator function of a Borel subset of M, and by linearity this ensures that (8.9) keeps holding for every simple function g on M; thanks to Beppo Levi's monotone convergence theorem, the result also extends to all positive measurable functions g on M, and, finally, linearity ensures that (8.9) also holds for every semi-integrable function g on M. Thus, it suffices to prove (8.8).

To this end, observe that U is an open subset of $\mathbb{R}^k$, hence it is σ-compact. It follows that if A is closed in U, then A is also σ-compact, and therefore $f(A)$ is σ-compact as well, for f is continuous. The collection of all subset A of U such that $f(A)$ is a Borel set is therefore a σ-field that contains all closed sets, whence it is the Borel σ-field.

Given a real number t strictly greater than 1, let $(B_j)_{j\geq 1}$ and $(L_j)_{j\geq 1}$ be as in Lemma (8.3), and, for every strictly positive integer j, let $A_j = A \cap B_j$. From Proposition (3.1) and from (8.4), it follows that

$$t^{-k}\mathcal{H}^k\big(L_j(X)\big) \leq \mathcal{H}^k\big(Df(x)X\big) \leq t^k\mathcal{H}^k\big(L_j(X)\big)$$

for every x in A_j and every Borel subset X of $\mathbb{R}^k$. Hence, by virtue of Proposition (8.2), and by dividing throughout by $\lambda^k(X)$,

$$t^{-k}JL_j \leq Jf(x) \leq t^k JL_j$$

for every point x of A_j. Therefore, taking the integral with respect to the Lebesgue measure, it follows that

$$t^{-k}JL_j \cdot \lambda^k(A_j) \leq \int_{A_j} Jf(x)\,\mathrm{d}x \leq t^k JL_j \cdot \lambda^k(A_j).$$

On the other hand, from Proposition (3.1) and from (8.5), it follows that

$$t^{-k}\mathcal{H}^k\big(L_j(A_j)\big) \leq \mathcal{H}^k\big(f(A_j)\big) \leq t^k\mathcal{H}^k\big(L_j(A_j)\big),$$

and again, by virtue of Proposition (8.2), it follows that

$$\mathcal{H}^k\big(L_j(A_j)\big) = JL_j \cdot \lambda^k(A_j).$$

Thus, putting all together, we finally get

$$\begin{aligned} t^{-2k}\mathcal{H}^k\big(f(A_j)\big) &\leq t^{-k}\mathcal{H}^k\big(L_j(A_j)\big) = t^{-k}JL_j \cdot \lambda^k(A_j) \\ &\leq \int_{A_j} Jf(x)\,\mathrm{d}x \leq t^k JL_j \cdot \lambda^k(A_j) \\ &= t^k\mathcal{H}^k\big(L_j(A_j)\big) \leq t^{2k}\mathcal{H}^k\big(f(A)\big). \end{aligned}$$

Since (A_j) is a measurable partition of A, adding up over j, finally gives

$$t^{-2k}\mathcal{H}^k\big(f(A)\big) \leq \int_A Jf(x)\,\mathrm{d}x \leq t^{2k}\mathcal{H}^k\big(f(A)\big),$$

and the proof is completed by taking the limit as t tends towards 1. □

Let us complete this section by showing a few examples of application of the area formula, which show how the classical formulae of length, area, and volume, are in fact special cases of Hausdorff measures.

(8.10) Example (Length of a smooth curve). We have already seen in §6 that the Hausdorff measure $\mathcal{H}^1$ coincides with the classical definition of length for a topological curve. When the curve is smooth, the area formula offers an even more explicit expression, which again corresponds to the traditional formula for the length of a curve. Precisely, a smooth curve in $\mathbb{R}^n$ can be written in the form

$$\gamma(t) = \big(\gamma_1(t), \ldots, \gamma_n(t)\big),$$

where the function $\gamma_1, \ldots, \gamma_n$ are differentiable on a suitable interval $[a, b]$. It is not difficult to recognise that the Jacobian matrix of this function is the vector $D\gamma = \dot{\gamma} = (\dot{\gamma}_1, \ldots, \dot{\gamma}_n)$ where the 'dot' denotes the first derivative of the function. The corresponding Jacobian is therefore

$$J\gamma(t) = \sqrt{\det\big(D\gamma(t)^* D\gamma(t)\big)} = \|\dot{\gamma}(t)\|.$$

Thus, the length of the support $C = \gamma([a, b])$ is, by the area formula

$$\text{length of } C = \mathcal{H}^1(C) = \int_a^b \|\dot{\gamma}(t)\| \, \mathrm{d}t$$

(8.11) Example (Surface area of a graph). Let us now assume that g is a real continuously differentiable function defined on an open set Ω in $\mathbb{R}^n$, and define the function f by setting

$$f(x) = \big(x, g(x)\big) \qquad \text{for every } x \text{ in } \Omega.$$

Setting $\nabla g = (g_{x_1}, \ldots, g_{x_n})$, the Jacobian matrix of f can be written as

$$Df = \begin{pmatrix} 1 & \cdots & 0 \\ \vdots & \ddots & \vdots \\ 0 & \cdots & 1 \\ g_{x_1} & \cdots & g_{x_n} \end{pmatrix}.$$

The corresponding Jacobian is therefore

$$Jf = \sqrt{1 + \|\nabla g\|^2}.$$

Setting $G = f(\Omega)$, observe that G is the graph of the function g defined on Ω. Then, the surface area of G is given by

$$\text{surface area of } G = \mathcal{H}^n(G) = \int_\Omega \sqrt{1 + \|\nabla g(x)\|^2} \, \mathrm{d}x.$$

(8.12) Example (Hyper-volume of a k-surface). Let M be a k surface in $\mathbb{R}^n$, let U be an open set in $\mathbb{R}^k$, f a parametrisation of M defined on U. Let A be a Borel set of $f(U)$, and set $B = f^{-1}(A)$. Let us define

$$g_{ij} = \frac{\partial f}{\partial x_i} \cdot \frac{\partial f}{\partial x_j} \qquad \text{for } i, j = 1, \ldots, n.$$

Then $Df^* Df = (g_{ij})$ and therefore $Jf = \sqrt{g}$, where $g = \det(g_{ij})$. Thus, applying the area formula,

$$\text{volume of } A \text{ in } M = \mathcal{H}^k(A) = \int_B \sqrt{g(x)}\, \mathrm{d}x.$$

Appendix A

Filters and Ultrafilters

1. Convergence according to Moore-Smith

A well-known type of convergence, which naturally extends the notion of convergence of a sequence, and which is broad enough to cover the usual needs of Mathematical Analysis, is the convergence of a *directed function* (or *net*). Let us recall the fundamental definitions concerning this type of convergence, which may also be referred to as *convergence according to Moore-Smith.*

(1.1) Definition. On a set E, a **pre-order** relation is a reflexive and transitive relation. Such a relation, usually denoted by $\leq$, is said to be **filtering** if, for every pair x, y of elements of E, there exists an element z of E such that $x \leq z$ and $y \leq z$. A set endowed with such a relation is therefore called a **pre-ordered filtering set**, or simply a **directed set**.

(1.2) Definition. A mapping f from a directed set E into a set F is called a **directed function** with values in F, or also a **directed family** of elements of F. Moreover, for a subset A of F, it is said that the directed function f is **eventually** in A if there exists an element x_0 of E such that, for every element x of E, the implication

$$x_0 \leq x \Rightarrow f(x) \in A$$

holds. If, in addition, the set F is equipped with a topology, given a point y of F, the directed function f is said to **converge** (according to Moore-Smith) towards the point y (or that y is a **limit** for f) if, for each neighbourhood V of y, it happens that f is eventually in V.

The concept of convergence of a directed function can be seen as a special case of the concept of convergence with respect to a filter, which was first introduced by the French mathematician Henri Cartan (1904–2008) and stunningly developed by N. Bourbaki in their beautiful *General Topology.*

DOI: 10.1201/9781003724766-A

2. Filters and filter bases

In the sequel, unless otherwise specified, E denotes a non-empty set.

(2.1) Definition. A **filter base** on E is a non-empty collection $\mathcal{B}$ of subsets of E, which does not contain the empty set, and such that the intersection of any pair (and therefore also of any finite family) of elements of $\mathcal{B}$ contains an element of $\mathcal{B}$. In other words, $\mathcal{B}$ is a filter base if, for every $A, B \in \mathcal{B}$, there exists $C \in \mathcal{B}$ such that $C \subseteq A \cap B$.

(2.2) Examples.

(a) If E is a topological space, a fundamental system of neighbourhoods of a fixed point x is a filter base on E.

(b) On a measure space $(E, \mathcal{E}, \mu)$, for every strictly positive real number ε, let us denote by $\mathcal{B}_\varepsilon$ the collection of all measurable sets whose complement has a measure strictly smaller than ε, that is to say,

$$\mathcal{B}_\varepsilon = \{A \in \mathcal{E} : \mu(A^c) < \varepsilon\}.$$

Then, the set $\mathcal{B} = \bigcup_{\varepsilon>0} \mathcal{B}_\varepsilon$ is a filter base on E.

(2.3) Definition. A **filter** on E is a non-empty set $\mathcal{F}$ of subsets of E, with the following properties:

(F1) $\emptyset \notin \mathcal{F}$

(F2) $A \in \mathcal{F},\ B \in \mathcal{F} \Rightarrow A \cap B \in \mathcal{F}$,

(F3) $A \in \mathcal{F},\ A \subseteq B \subseteq E \Rightarrow B \in \mathcal{F}$.

In the hypotheses of the previous definition, recall that condition (F2) can be rephrased by saying that a filter is *stable* under the operation of finite intersection, while condition (F3) expresses the property that every subset of E containing an element of $\mathcal{F}$ belongs to $\mathcal{F}$. This may also be described by saying that a filter is **anti-hereditary**. Albeit obvious, it is worth to just remark the fact that a filter is, in particular, a filter base.

(2.4) Examples.

(a) If E is a topological space, the set of all neighbourhoods of a fixed point x, which shall hereinafter be denoted by $\mathcal{U}_E(x)$, or simply $\mathcal{U}(x_0)$, when there is no risk of ambiguity, is a filter.

(b) If E is an infinite set, the set consisting of the complements of the finite subsets of E is a filter. In particular, when E is the set $\mathbb{N}$ of all positive integers, this filter shall be called the **Fréchet filter**.

Given a filter base $\mathcal{B}$, the term 'base' naturally induces one to ask oneself whether there exists a filter that is 'generated' by the given filter base. To answer (affirmatively) to this question, let us first prove the following simple lemma, which shows that the intersection of a family of filters is a filter itself.

(2.5) Lemma. *The intersection of an arbitrary non-empty family of filters on E is a filter on E.*

Proof. Let $\mathfrak{D}$ be a non-empty family of filters on E, and let us denote by $\mathcal{F}$ the intersection of all the sets in the family. Clearly, if $\emptyset$ does not belong to any filter in the family $\mathfrak{D}$, it does not belong to their intersection either. If A and B are two elements of $\mathcal{F}$, for A and B are, in particular, elements of every filter $\mathcal{G}$ in $\mathfrak{D}$, by virtue of (F2), the set $A \cap B$ also belongs to every such $\mathcal{G}$, whence to $\mathcal{F}$. Similarly, if B is a subset of E and A is an element of $\mathcal{F}$ such that $A \subseteq B$, then from the fact that A belongs to every filter $\mathcal{G}$ in $\mathfrak{D}$ and from (F3), it follows that $B \in \mathcal{G}$, whence B is an element of $\mathcal{F}$, and this proves that $\mathcal{F}$ is indeed a filter on E. □

The previous lemma justifies the following definition.

(2.6) Definition. Given a filter base $\mathcal{B}$ on E, the intersection of all filters on E which contains $\mathcal{B}$ is a filter on E which contains $\mathcal{B}$, and it is called the filter **generated** by $\mathcal{B}$. It is easy to check that this coincides with the set consisting of all subsets of E which contain some element of $\mathcal{B}$.

(2.7) Examples.

(a) If $\mathcal{F}$ is a filter, the filter generated by $\mathcal{F}$ is $\mathcal{F}$ itself.

(b) If E is a topological space and $\mathcal{B}$ is a fundamental system of neighbourhoods of a fixed point x, then the filter generated by $\mathcal{B}$ coincides with the filter of all neighbourhoods of x.

(c) Let E be a directed set with respect to a relation $\leq$. For each element x_0 in E, the set $E(x_0) = \{x \in E : x_0 \leq x\}$ shall be called the **section** of E, relative to the element x_0. The set $\mathfrak{S}_E$ of all sections on E is a filter base, for if x_1 and x_2 are two elements of E, then, by hypothesis there is an element x_0 in E such that $x_1 \leq x_0$ and $x_2 \leq x_0$, and therefore

$$E(x_0) \subseteq E(x_1) \cap E(x_2).$$

The filter generated by $\mathfrak{S}_E$ is called the **section filter** of E.

3. Ultrafilters

In the class $\mathcal{P}(E)$ of all subsets of E, the relation of inclusion is a pre-order, which induces a pre-order on the class of all filters, and by extension on the class of all filter bases. This observation leads to the following definition.

(3.1) Definition. Given two filter bases $\mathcal{B}$ and $\mathcal{B}'$ on the same set E, $\mathcal{B}'$ is said to be **finer** than $\mathcal{B}$, or that $\mathcal{B}$ is **coarser** than $\mathcal{B}'$, if the filter generated by $\mathcal{B}$ is contained in the filter generated by $\mathcal{B}'$. If these two filters are distinct, then $\mathcal{B}'$ is said to be **strictly finer** than $\mathcal{B}$, or $\mathcal{B}$ is **strictly coarser** than $\mathcal{B}'$.

As a consequence of the previous definition, the class of all filters on E turns out to be a pre-ordered set. Maximal elements of this class are particularly important and deserve a specific name.

(3.2) Definition. An **ultrafilter** on E is a filter $\mathcal{U}$ on E that is maximal in the set of all filters on E. In other words, $\mathcal{U}$ is an ultrafilter on E if no other filter on E is strictly finer than $\mathcal{U}$. An **ultrafilter base** is a filter base whose generated filter is an ultrafilter.

The following proposition shows that the pre-ordered set of all filters on E always have maximal elements.

(3.3) Proposition. *The union of an arbitrary family of filters on E is a filter on E. Consequently, every filter on E is contained in an ultrafilter.*

Proof. The first assertion is easy to check out. In order to prove the second, let $\mathcal{F}$ be a filter and let $\mathfrak{G}$ be the family of all filters on E which contain $\mathcal{F}$. This is obviously a pre-ordered set with the order relation defined in (3.1). If $\mathfrak{C}$ is a chain in $\mathfrak{G}$, that is to say a totally ordered subset of $\mathfrak{G}$, then, its union contains every element of $\mathfrak{C}$ and belongs to $\mathfrak{G}$, hence the existence of an ultrafilter which contains $\mathcal{F}$ follows from Zorn's lemma, observing such a filter is by definition a maximal element of the class $\mathfrak{G}$. □

(3.4) Proposition. *Let $\mathcal{F}$ be an ultrafilter on a non-empty set E. Then, for every pair A, B of subsets of E, the following implication holds:*

$$A \cup B \in \mathcal{F} \Rightarrow A \in \mathcal{F} \text{ or } B \in \mathcal{F}.$$

Proof. By contradiction, suppose that A, B are two subsets of E such that

$$A \cup B \in \mathcal{F}, \qquad A \notin \mathcal{F}, \qquad B \notin \mathcal{F}.$$

Set

$$\mathcal{G} = \{X \in \mathcal{P}(E) : A \cup X \in \mathcal{F}\}.$$

The hypothesis $A \notin \mathcal{F}$ implies $\emptyset \notin \mathcal{G}$. Moreover, $\mathcal{G}$ contains $\mathcal{F}$: indeed, since $\mathcal{F}$ is anti-hereditary, if X is an element of $\mathcal{F}$, so is $A \cup X$. Finally, as is easily verified, $\mathcal{G}$ is itself hereditary and stable under finite intersections. Thus, $\mathcal{G}$ is a filter, finer than $\mathcal{F}$. Thanks to the hypothesis $A \cup B \in \mathcal{F}$, we have $B \in \mathcal{G}$. Since, however, $B \notin \mathcal{F}$, we see that $\mathcal{G}$ is strictly finer than $\mathcal{F}$. Since this contrasts with the hypothesis that $\mathcal{F}$ is an ultrafilter, the proposition is proved. □

The next proposition offers a handy characterisation of ultrafilters.

(3.5) Proposition. *Let $\mathcal{B}$ be a filter base on E. If, for every subset A of E, one has either $A \in \mathcal{B}$ or $A^c \in \mathcal{B}$, then $\mathcal{B}$ is an ultrafilter.*

Proof. Let $\mathcal{U}$ be an ultrafilter containing $\mathcal{B}$. Then, $\mathcal{U}$ coincides with $\mathcal{B}$. In fact, if $A \in \mathcal{U}$, then $A^c \notin \mathcal{U}$, otherwise it would be $\emptyset = A \cap A^c \in \mathcal{U}$, hence $A^c \notin \mathcal{B}$, and therefore $A \in \mathcal{B}$. □

4. Limits and cluster points of a filter

One of the main features of a filter on a topological space is that one can use them to define, in a very elegant manner, the notion of convergence. To this end, let us now assume that E is a topological space.

(4.1) Definition. Let $\mathcal{B}$ be a filter base on E. A point x is called a **limit** point (or simply a **limit**) of $\mathcal{B}$, if $\mathcal{B}$ is finer than the filter of all neighbourhoods of x. In this case, we shall also say that $\mathcal{B}$ **converges** towards x. In other words, a filter base $\mathcal{B}$ converges towards a point x if the filter $\mathcal{U}(x)$ of all neighbourhoods of x is included in the filter generated by $\mathcal{B}$.

From the definitions, the following proposition is immediate.

(4.2) Proposition. *A filter base $\mathcal{B}$ on E converges towards x if and only if every set of a fundamental system of neighbourhoods of x contains a set of $\mathcal{B}$.*

The transitivity of inclusion, in the class of all filters on E, yields that, if a filter is finer then the filter of neighbourhoods of a point, then any filter which is finer than the given one is also finer than the filter of neighbourhoods of the same point. Thus, if a filter converges towards a point x, any filter which is finer than the given one also converges towards the same point. This proves the following proposition.

(4.3) Proposition. *If a filter base $\mathcal{B}$ on E converges towards a point x, every filter (base) finer than $\mathcal{B}$ also converges towards x. In particular, for a filter base $\mathcal{B}$ to converge towards a point x it is necessary and sufficient that every ultrafilter which is finer than $\mathcal{B}$ converges towards x.*

The next definition is a relaxation of the notion of limit point.

(4.4) Definition. A point x in E is an **cluster point** of a filter base $\mathcal{B}$ if it lies in the closure of every set in $\mathcal{B}$.

From the definition of cluster point and filter base, keeping into account that a fundamental system of neighbourhoods of a point is a filter base for the filter of neighbourhoods of that point, the following proposition immediately holds.

(4.5) Proposition. *A point x is a cluster point of a filter base $\mathcal{B}$ on E if and only if every set of a fundamental system of neighbourhoods of x meets every set of $\mathcal{B}$.*

The next proposition shows that the statement 'x is a cluster point of the filter $\mathcal{F}$' is equivalent to the statement 'there is a filter which is finer than $\mathcal{F}$ which converges towards x'. Precisely, the following holds.

(4.6) Proposition. *A point x is a cluster point of a filter base $\mathcal{B}$ on E if and only if there exists another filter, finer than $\mathcal{B}$, which converges towards x.*

Proof. Assume that x is a cluster point of the filter base $\mathcal{B}$. By definition, this means that every neighbourhood of x meets every set belonging to $\mathcal{B}$. Consider the family $\mathcal{B}'$ of sets obtained by intersecting each neighbourhood of x with each set of $\mathcal{B}$. By definition, none of these intersections is empty. Moreover, the family is stable under finite intersections: If two such sets are taken, their intersection is again of the same form, since the intersection of two neighbourhoods of x is a neighbourhood of x, and the intersection of two sets of $\mathcal{B}$ belongs again to the filter base. Hence, $\mathcal{B}'$ is a filter base. The filter $\mathcal{F}$ generated by $\mathcal{B}'$ is finer than $\mathcal{B}$, because each set of $\mathcal{B}$ already belongs to $\mathcal{B}'$. Furthermore, $\mathcal{F}$ converges towards x, for every neighbourhood of x contains an element of $\mathcal{B}'$, hence an element of $\mathcal{F}$.

Conversely, suppose that there exists a filter $\mathcal{F}$, finer than $\mathcal{B}$, which converges towards x. The convergence means that every neighbourhood of x contains some element of $\mathcal{F}$. On the other hand, the filter $\mathcal{F}$ being finer than $\mathcal{B}$ means that for every set of $\mathcal{B}$, one can find a set of $\mathcal{F}$ the latter of which is included in the former. If one now takes any neighbourhood U of x and any set B in $\mathcal{B}$, there exist two sets F', F'' in $\mathcal{F}$ such that $F \subseteq U$ and $F' \subseteq B$. Since $\mathcal{F}$ is a filter, $F' \cap F'' \neq \emptyset$. Consequently, $B \cap U \neq \emptyset$, and this is precisely the condition for x to be a cluster point of $\mathcal{B}$. □

(4.7) Corollary. *An ultrafilter $\mathcal{U}$ converges towards a point x if and only if x is a cluster point of $\mathcal{U}$.*

Proof. Assume that the ultrafilter $\mathcal{U}$ converges towards x. By definition, this means that every neighbourhood of x belongs to $\mathcal{U}$. In particular, every neighbourhood of x meets every element of $\mathcal{U}$, so x is a cluster point of $\mathcal{U}$.

Conversely, assume that x is a cluster point of $\mathcal{U}$. Then every neighbourhood of x meets every element of $\mathcal{U}$. For a fixed neighbourhood U of x, consider its complement U^c. Since $\mathcal{U}$ is an ultrafilter, either U belongs to $\mathcal{U}$ or its complement does. But if the complement belonged to $\mathcal{U}$, then U would fail to meet that set, contradicting the assumption that x is a cluster point of $\mathcal{U}$. Hence U itself belongs to $\mathcal{U}$. As this holds for every neighbourhood of x, the ultrafilter $\mathcal{U}$ converges towards x. □

(4.8) Proposition. *Assume that E is a Hausdorff space. If a filter $\mathcal{F}$ converges towards a point x, then it is the only cluster point of $\mathcal{F}$. In particular, no filter on E has more than one limit point.*

Proof. By contradiction, suppose that the filter $\mathcal{F}$ has two distinct cluster points x and y, with $x \neq y$. Since E is Hausdorff, there exists a closed neighbourhood U of x such that $y \notin U$, and, by Proposition (4.5), there exists a set M in $\mathcal{F}$ such that $M \subseteq U$, thus $M \cap U^c = \emptyset$. Since U^c is an open neighbourhood of y, this proves that y cannot lie in the closure of M, whence y cannot be a cluster point for $\mathcal{F}$. □

5. Convergence of a map with respect to a filter

Let us now assume that E and F are two topological spaces, $\mathcal{B}$ is a filter base on E, and let f be a map from E to F. It is not difficult to check that the set

$$f(\mathcal{B}) = \{f(B) : B \in \mathcal{B}\}$$

is a filter base on F. This is called the **image** of $\mathcal{B}$ under f.

(5.1) Definition. A point y of F is a **limit point** (or simply a **limit**) of f with respect to the filter base $\mathcal{B}$ if it is a limit point of the filter base $f(\mathcal{B})$. If y is a limit of f with respect to $\mathcal{B}$, we shall also say that f **converges** towards y with respect to $\mathcal{B}$. Similarly, a point y in F is a **cluster point** of f with respect to the filter base $\mathcal{B}$ if it is a cluster point of $f(\mathcal{B})$.

(5.2) Proposition.

(a) *A point y in F is a limit of f with respect to a filter $\mathcal{F}$ on E if and only if, for each neighbourhood V of y, there exists a set M in $\mathcal{F}$ such that $f(M) \subseteq V$ — i.e. $f^{-1}(V) \in \mathcal{F}$, for every neighbourhood V of y.*

(b) *A point y in F is a cluster point of f with respect to a filter $\mathcal{F}$ if and only if for each neighbourhood V of y and each set M in $\mathcal{F}$ there exists a point $x \in M$ such that $f(x) \in V$.*

Proof. (a) Suppose that y is a limit of f with respect to the filter $\mathcal{F}$. This means that for every neighbourhood V of y the set $f^{-1}(V)$ belongs to $\mathcal{F}$. Equivalently, for each such neighbourhood there exists a set M in $\mathcal{F}$ with the property that $f(M)$ is contained in V. Conversely, if this condition holds, then every neighbourhood V of y satisfies $f^{-1}(V) \in \mathcal{F}$, which is precisely the definition of y being a limit of f with respect to $\mathcal{F}$.

(b) Suppose now that y is a cluster point of f with respect to the filter $\mathcal{F}$. By definition, this means that, for every neighbourhood V of y, and every set M in $\mathcal{F}$, the image $f(M)$ meets V. Equivalently, for every neighbourhood V of y, and every set M of $\mathcal{F}$, there exists an element x of M such that $f(x)$ belongs to V. Conversely, if this condition is satisfied, then every neighbourhood of y meets the image of every set of $\mathcal{F}$, which is precisely the definition of y being a cluster point of f with respect to $\mathcal{F}$. □

(5.3) Examples.

(a) A **sequence** (x_n) on a topological space E is a map $n \mapsto x_n$ from $\mathbb{N}$ to E. If (x_n) converges towards x with respect to the Fréchet filter, we shall also say that x is the **limit of the sequence** (x_n) **as** n **tends to infinity**, and providing that E is a Hausdorff space (so that the limit is unique), we shall use the notation $\lim_n x_n = x$. It is not difficult to check that this, in fact, is equivalent to the classical definition of convergence.

(b) If f is a directed function defined on a directed set E, with values on a topological space F, a point y in F is a limit of f with respect to the section filter of E if and only if f converges towards y according to Moore-Smith, that is to say in the sense of Definition (1.2).

Let us continue to assume that E and F are two topological spaces. Then, given a point x_0 of E, one may always consider on E the filter $\mathcal{U}_E(x_0)$ of all neighbourhoods of x_0. If a point y_0 in F is a limit of a map f, from E to F, with respect to this filter, we shall also say that f converges towards the point y_0 **at the point** x_0 and, assumed that F is a Hausdorff space, the limit of f at x_0 shall be denoted by

$$\lim_{x \to x_0} f(x).$$

Let us recall that, in topology, a map f from a topological space E to a topological space F is 'continuous' at a point x_0 in E, if the inverse image of every open neighbourhood of $y_0 = f(x_0)$ is an open neighbourhood of x_0 in E. Keeping in mind that the sets of neighbourhoods $\mathcal{U}_E(x_0)$ and $\mathcal{U}_F(y_0)$ are filters, this gives the following immediate result.

(5.4) Proposition. *A map f from a topological space E to a topological space F is continuous at a point x_0 in E if and only if $f(x_0)$ is the limit of f at x_0, that is to say, if and only if*

$$\lim_{x \to x_0} f(x) = f(x_0).$$

(5.5) Corollary. *Let E and F be two topological spaces, and f a map from E to F which is continuous at the point x_0. Then, for every filter base $\mathcal{B}$ on E which converges towards x_0, the filter base $f(\mathcal{B})$ converges towards $f(x_0)$.*

Conversely, if, for every ultrafilter $\mathcal{U}$ on E which converges towards x_0, the ultrafilter base $f(\mathcal{U})$ converges towards $f(x_0)$, then f is continuous at x_0.

Proof. Suppose first that f is continuous at x_0. Thanks to the previous proposition, this means that

$$\lim_{x \to x_0} f(x) = f(x_0),$$

that is to say, the image $f(\mathcal{U}_E(x_0))$ of the filter $\mathcal{U}_E(x_0)$ under f converges towards $f(x_0)$. In other words, the filter base $f(\mathcal{U}_E(x_0))$ is finer than $\mathcal{U}_F(f(x_0))$. Thus, if $\mathcal{B}$ is finer than $\mathcal{U}_E(x_0)$, then $f(\mathcal{B})$ shall be finer than $f(\mathcal{U}_E(x_0))$, hence finer than $\mathcal{U}_F(f(x_0))$, which means that $f(\mathcal{B})$ converges towards $f(x_0)$.

Let us now assume that f is not continuous at x_0. Then, there exists an open neighbourhood V of $f(x_0)$ such that, for every open neighbourhood U of x_0, the set $U \cap f^{-1}(V^c)$ is not empty. This means that the class obtained by adjoining the set $f^{-1}(V^c)$ to the filter of all neighbourhoods of x_0 is a filter base. By Proposition (3.3), there exists an ultrafilter $\mathcal{U}$ which contains this filter base. By construction, this ultrafilter converges towards x_0, but its image meets V^c, hence does not converge towards $f(x_0)$. □

(5.6) Remark. Let us recall that, with the phrase 'a family $(x_t)_{t \in T}$ of elements of E' we mean a map $t \mapsto x_t$, defined on the set of indices, T, with values in the set E. If T is equipped with a filter, the family $(x_t)_{t \in T}$ may be called a **filtered family** of elements of E. If the image of the filter on T under this map converges towards an element x of E, we shall also say that the filtered family $(x_t)_{t \in T}$ **converges** towards x, and if E is a Hausdorff space, whereby the limit is unique, one also writes

$$x = \lim_{t \in T} x_t$$

to say that the limit of the filtered family $(x_t)_{t \in T}$ is x.

6. The words 'eventually' and 'frequently'

(6.1) Definition. Let f be a map from E to F, let $\mathcal{B}$ be a filter base E, and let A be a subset of F.

(a) We shall say that f is **eventually** in A with respect to $\mathcal{B}$ if there exists an element B of $\mathcal{B}$ such that $f(x) \in A$ for every $x \in B$.

(b) We shall say that f is **frequently** in A with respect to $\mathcal{B}$ if, for every element B of $\mathcal{B}$, there exists $x \in B$ such that $f(x) \in A$.

(6.2) Proposition. *In the same hypotheses as in Definition (6.1), assume that f is eventually in A. Then f is also frequently in A.*

Proof. Assume that f is eventually in A with respect to the filter base $\mathcal{B}$. By definition, this means that there exists some $B_0 \in \mathcal{B}$ such that $f(x) \in A$ for every $x \in B_0$. Now consider an arbitrary $B \in \mathcal{B}$. Since $\mathcal{B}$ is a filter base, the intersection $B \cap B_0$ is non-empty. Choose $x \in B \cap B_0$. Then $x \in B$ and $f(x) \in A$. Thus every $B \in \mathcal{B}$ contains some point whose image under f lies in A, that is to say that f is frequently in A. □

(6.3) Proposition. *In the same hypotheses as in Definition (6.1), assume that f is eventually in A. Assume further that f is frequently in a certain subset B of F. Then f is frequently in $A \cap B$.*

Proof. Assume that f is eventually in A with respect to the filter base $\mathcal{B}$, so there exists $B_0 \in \mathcal{B}$ such that $f(x) \in A$ for every $x \in B_0$. Assume further that f is frequently in B, meaning that for each $M \in \mathcal{B}$ there is a point $x \in M$ such that $f(x) \in B$. Now let M be an arbitrary element of $\mathcal{B}$. Since $\mathcal{B}$ is a filter base, the intersection $M \cap B_0$ is non-empty and belongs to the filter generated by $\mathcal{B}$. By the assumption, there exists $x \in M \cap B_0$ with $f(x) \in B$. But for all $x \in B_0$, one has $f(x) \in A$. Thus for this choice of x one has $f(x) \in A \cap B$. This shows that f is frequently in $A \cap B$. □

7. A characterisation of compactness through filters

If E is a non-empty set, a **covering** of E is a collection of subsets of E, whose union is E. In general topology, it is a customary to define a compact as a topological space in which from every open covering it is always possible to extract a finite open covering. Even though this definition is well-established, it does have the flaw of being rather technical and may fail to show the most important features that a compact set hold. Using filters and ultrafilters, it is possible to give an equivalent definition of compact set which does not have the same flaws.

(7.1) Proposition. *Let E be a topological space. The following conditions are equivalent:*

(a) *Every filter on E has at least one cluster point.*

(b) *Every ultrafilter on E is convergent.*

(c) *Every family of closed subsets of E whose intersection is empty contains a finite subfamily whose intersection is empty.*

(d) *Every open covering of E contains a finite open covering of E.*

Proof. (a) $\Rightarrow$ (b): If an ultrafilter has a cluster point, then it converges to this point (Corollary (4.7)).

(b) $\Rightarrow$ (a): If $\mathcal{F}$ is a filter on E, then there is an ultrafilter finer than $\mathcal{F}$. Since this ultrafilter converges to a point x, x is a cluster point for $\mathcal{F}$ by virtue of Proposition (4.6).

(a) $\Rightarrow$ (c): Let $\mathcal{G}$ be a family of closed subsets of E with empty intersection. If every finite subfamily of $\mathcal{G}$ has a non-empty intersection, then $\mathcal{G}$ generates a filter which, by hypothesis, has a cluster point. Since every element of $\mathcal{G}$ is closed, this point belongs to all the sets of $\mathcal{G}$. This contradicts the fact the intersection of all elements of $\mathcal{G}$ is empty.

(c) $\Rightarrow$ (a): If (c) is false, then there is a filter $\mathcal{F}$ on E which has no cluster point; hence the closure of the sets of $\mathcal{F}$ form a family of closed subsets of E which contradicts (c).

(c) $\Leftrightarrow$ (d): This follows by taking the complement. □

(7.2) Definition. A topological space E is **compact** if one (hence each) of the equivalent conditions stated in the previous proposition holds.

Compact subsets of a metric space can be characterised by a slightly simpler condition, which may be useful in proofs. To this end, let us start by recalling the following definition.

(7.3) Definition. Let E be a metric space. We shall say that a subset K of E is **totally bounded** if, for every strictly positive real number ε, there exists a finite family of open balls of radius ε whose union covers K.

(7.4) Proposition. *Let (E, d) be a metric space, and let K be a subset of E. Then, the following conditions are equivalent:*

(a) *K is compact.*

(b) *K is closed and totally bounded.*

Proof. Let us start by proving (a) $\Rightarrow$ (b). To this end, assume that K is compact. Then every filter on K has a cluster point in K, which implies that K is closed. In addition, if K were not totally bounded, there would exist a strictly positive real number ε such that no finite family of open balls of radius ε covers K. The family $\mathcal{B}$ of complements of all finite unions of such balls generates a filter base on K. In particular, if x is any point of K, the set $K \setminus B(x, \varepsilon)$ belongs to $\mathcal{B}$ and is disjoint from $B(x, \frac{1}{2}\varepsilon)$. Hence, x cannot be a cluster point for $\mathcal{B}$, and this proves that K must be totally bounded.

Let us now prove (b) $\Rightarrow$ (a). Suppose that K is closed and totally bounded, and let $\mathcal{F}$ be a filter on K. For every strictly positive integer n, the fact that K is totally bounded ensures that there exists a finite family $\mathcal{B}_n$ of open balls of radius $1/n$ which covers K. Since $\mathcal{B}_n$ covers E, there must exist an element M of $\mathcal{F}$ and an element B_n of $\mathcal{B}_n$ whose intersection is not empty; hence B_n belongs to $\mathcal{F}$. The countable family of these balls, as n varies, forms a filter base whose cluster point lies in K by closure. Hence $\mathcal{F}$ has a cluster point in K, and K is compact. □

8. Comparison with Moore-Smith's theory

In Example (2.7)(c) we have shown that a directed set may be seen as a set equipped with a filter, and in Example (5.3)(b) we have seen that the convergence according to Moore-Smith is a special case of convergence with respect to a filter, and in particular with respect to the section filter. Conversely, the following proposition shows that the theory of convergence with respect to a filter can be reduced to the theory of convergence according to Moore-Smith.

(8.1) Proposition. *Given a filter base $\mathcal{B}$ on E and a mapping f from E into a set F, it is possible to construct a set E', equipped with a filtering pre-order relation, and a mapping f' from E' into F, such that, for every subset A of F, the following conditions are equivalent:*

(a) *f is eventually in A with respect to $\mathcal{B}$.*

(b) *f' is eventually in A with respect to the pre-order relation of E', that is to say, with respect to the section filter of E'.*

Proof. It suffices to set $E' = \{(B, x) : B \in \mathcal{B}, x \in B\}$, denote by f' the mapping from E' into F defined by $f'(B, x) = f(x)$, and equip E' with the filtering pre-order relation defined as follows:

$$(B_0, x_0) \leq (B, x) \Leftrightarrow B \subset B_0.$$

□

Bibliography

[1] P. Acquistapace. "Appunti di Analisi Convessa". Dispense del corso di Istituzioni di Analisi Matematica 2. Pisa, 2017.

[2] P. Acquistapace. "Appunti di Analisi Funzionale". Dispense del corso di Istituzioni di Analisi Matematica 1. Pisa, 2022.

[3] L. Ambrosio, E. Brué, and D. Semola. *Lectures on Optimal Transport.* Unitext 169. Springer, 2024.

[4] L. Ambrosio, G. Da Prato, and A. Mennucci. *Introduction to Measure Theory and Integration.* Edizioni della Normale, 2011.

[5] G. de Barra. *Measure Theory and Integration.* Wiley, 1981.

[6] H. Bauer. *Measure and Integration Theory.* De Gruyter, 2001.

[7] P. Billingsley. *Convergence of Probability Measures.* Wiley, 1968.

[8] P. Billingsley. *Probability and Measure.* Wiley, 1995.

[9] V.I. Bogachev. *Measure Theory.* V. I. Springer, 2007.

[10] V.I. Bogachev. *Measure Theory.* V. II. Springer, 2007.

[11] N. Bourbaki. *General Topology, Chapters 1-4.* Springer, 1995.

[12] N. Bourbaki. *General Topology, Chapters 5-10.* Springer, 1998.

[13] N. Bourbaki. *Integration, Chapters 1-6.* Springer, 2004.

[14] N. Bourbaki. *Topological Vector Spaces, Chapters 1-5.* Springer, 2003.

[15] H. Brézis. *Analyse fonctionelle. Théorie et application.* Dunod, 1983.

[16] D.L. Cohn. *Measure Theory.* Birkhäuser, 2013.

[17] J.P. Daniell. "A general form of integral". In: *Annals of Mathematics.* Second Series v. 19. n. 4 (1918), pp. 279–294.

[18] J.P. Daniell. "Properties of the General Integral". In: *Annals of Mathematics.* Second Series v. 21. n. 3 (1920), pp. 203–220.

[19] C. Dellacherie and P.-A. Meyer. *Probabilités et potentiel, Ch. I-IV.* Hermann, 1975.

[20] J.L. Doob. *Measure Theory.* Graduate Texts in Mathematics. Springer, 1994.

[21] R.M. Dudley. *Real Analysis and Probability.* V. 10. Cambridge Studies in Advanced Mathematics 74. Cambridge University Press, 2002.

[22] J. Dugundji. *Topology.* Allyn and Beacon, 1966.

[23] N. Dunford and J.T. Schwartz. *Linear Operators.* Wiley, 1958.

[24] G. Edgar. *Measure, Topology, and Fractal Geometry.* Undergraduate Texts in Mathematics. Springer, 2000.

[25] C. Evans and R.F. Gariepy. *Measure Theory and Fine Properties of Functions.* CRC Press, 1992.

[26] K. Falconer. *Fractal Geometry. Mathematical Foundations and Applications.* Wiley, 2014.

[27] H. Federer. *Geometric Measure Theory.* Classics in Mathematics. Springer, 1969.

[28] W. Fleming. *Functions of Several Variables.* Springer-Verlag, 1977.

[29] G.B. Folland. *Real Analysis. Modern Techniques and Their Applications.* John Wiley, 1999.

[30] D.H. Fremlin. *Measure Theory.* V. I-V. Torres Fremlin, 2008.

[31] D.H. Fremlin. *Topological Riesz Spaces and Measure Theory.* Cambridge University Press, 1974.

[32] D.J.H. Garling. *Analysis on Polish Spaces and an Introduction to Optimal Transportation.* LMS Student Texts № 89. Cambridge University Press, 2018.

[33] R. Godement. *Analyse mathématique.* V. IV. Springer, 2003.

[34] P.R. Halmos. *Lectures on Booelan Algebras.* Springer, 1974.

[35] P.R. Halmos. *Measure Theory.* Van Nostrand, 1950.

[36] G.A. Hunt. *Martingales et processus de Markov.* Dunod, 1966.

[37] J.E. Hutchinson. "Fractals and self similarity". In: *Indiana University Mathematics Journal* v. 30. n. 5 (1981), pp. 713–747.

[38] C.T. Ionescu Tulcea. "Mesures dans les espaces produits". In: *Atti Accad. Naz. Lincei Rend* v. 7 (1949), pp. 208–211.

[39] M. Jiřina. "On regular conditional probabilities". In: *Czechoslovak Mathematical Journal* v. 09. n. 3 (1959), pp. 445–451.

[40] O. Kallenberg. *Foundation of Modern Probability.* Springer, 2001.

[41] O. Kallenberg. *Random Measures, Theory and Applications.* Springer, 2017.

[42] J.L. Kelley. *General Topology.* Graduate Texts in Mathematics. Springer, 1955.

[43] S. Lang. *Real and Functional Analysis.* Springer, 1993.

[44] G. Letta. *Appunti di Analisi.* Editrice Tecnico-Scientifica, 1973.

[45] G. Letta. *Argomenti scelti di teoria della misura.* Quaderni dell'Unione Matematica Italiana № 54. Pitagora Editrice, 2013.

[46] G. Letta. *Teoria elementare dell'integrazione.* Boringhieri, 1976.

[47] W.A.J. Luxemburg and A.C. Zaanen. *Riesz Spaces.* Mathematical studies v. 1. North-Holland, 1971.

[48] F. Maggi. *Sets of Finite Perimeter and Geometric Variational Problems. An Introduction to Geometric Measure Theory.* Cambridge Studies in Advanced Mathematics 135. Cambridge University Press, 2012.

[49] P. Mattila. *Geometry of Sets and Measures in Euclidean Spaces. Fractals and rectifiability.* Cambridge Studies in Advanced Mathematics 44. Cambdirge University Press, 1995.

[50] E.J. McShane and T.A. Botts. *Real Analysis.* Van Nostrand, 1959.

[51] J. Neveu. *Bases mathématiques du calcul des probabilités.* Masson et Cie, 1964.

[52] K.R. Parthasarathy. *Probability Measures on Metric Spaces.* Academic Press, 1967.

[53] N. Pintacuda. *Probabilità.* Decibel. Zanichelli, 1994.

[54] D. Revuz. *Markov Chains.* North-Holland, 1984.

[55] D. Revuz. *Mesure et intégration.* Hermann, 1997.

[56] C.A. Rogers. *Hausdorff Measures.* Cambridge University Press, 1970.

[57] H.L. Royden. *Real Analysis.* Prentice Hall, 1988.

[58] W. Rudin. *Principles of Mathematical Analysis.* McGraw-Hill, 1964.

[59] W. Rudin. *Real and Complex Analysis.* McGraw-Hill, 1974.

[60] J.S. Taylor. *An Introduction to Measure and Integration.* Cambridge University Press, 1973.

[61] F. Treves. *Topological Vector Spaces, Distributions and Kernels.* Dover, 1967.

[62] C. Villani. *Optimal Transport: Old and New.* Grundlehren der mathematischen Wissenshaften 338. Springer, 2009.

[63] C. Villani. *Topics in Optimal Transportation.* American Mathematical Society, 2003.

[64] A.C. Zaanen. *Integration.* North-Holland, 1967.

Index